NEUROMETHODS

Series Editor
Wolfgang Walz
University of Saskatchewan,
Saskatoon, Canada

For further volumes:
http://www.springer.com/series/7657

Optogenetics: A Roadmap

Edited by

Albrecht Stroh

Focus Program Translational Neurosciences and Institute for Microscopic Anatomy and Neurobiology, University Medical Center of the Johannes Gutenberg-University Mainz, Mainz, Germany

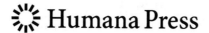 Humana Press

Editor
Albrecht Stroh
Focus Program Translational Neurosciences and Institute
for Microscopic Anatomy and Neurobiology
University Medical Center of the Johannes
Gutenberg-University Mainz
Mainz, Germany

ISSN 0893-2336   ISSN 1940-6045   (electronic)
Neuromethods
ISBN 978-1-4939-8479-4   ISBN 978-1-4939-7417-7   (eBook)
DOI 10.1007/978-1-4939-7417-7

Printed on acid-free paper

This Humana Press imprint is published by Springer Nature
The registered company is Springer Science+Business Media, LLC
The registered company address is: 233 Spring Street, New York, NY 10013, U.S.A.

Preface to the Series

Experimental life sciences have two basic foundations: concepts and tools. The *Neuromethods* series focuses on the tools and techniques unique to the investigation of the nervous system and excitable cells. It will not, however, shortchange the concept side of things as care has been taken to integrate these tools within the context of the concepts and questions under investigation. In this way, the series is unique in that it not only collects protocols but also includes theoretical background information and critiques which led to the methods and their development. Thus it gives the reader a better understanding of the origin of the techniques and their potential future development. The *Neuromethods* publishing program strikes a balance between recent and exciting developments like those concerning new animal models of disease, imaging, in vivo methods, and more established techniques, including, for example, immunocytochemistry and electrophysiological technologies. New trainees in neurosciences still need a sound footing in these older methods in order to apply a critical approach to their results.

Under the guidance of its founders, Alan Boulton and Glen Baker, the *Neuromethods* series has been a success since its first volume published through Humana Press in 1985. The series continues to flourish through many changes over the years. It is now published under the umbrella of Springer Protocols. While methods involving brain research have changed a lot since the series started, the publishing environment and technology have changed even more radically. Neuromethods has the distinct layout and style of the Springer Protocols program, designed specifically for readability and ease of reference in a laboratory setting.

The careful application of methods is potentially the most important step in the process of scientific inquiry. In the past, new methodologies led the way in developing new disciplines in the biological and medical sciences. For example, Physiology emerged out of Anatomy in the nineteenth century by harnessing new methods based on the newly discovered phenomenon of electricity. Nowadays, the relationships between disciplines and methods are more complex. Methods are now widely shared between disciplines and research areas. New developments in electronic publishing make it possible for scientists that encounter new methods to quickly find sources of information electronically. The design of individual volumes and chapters in this series takes this new access technology into account. Springer Protocols makes it possible to download single protocols separately. In addition, Springer makes its print-on-demand technology available globally. A print copy can therefore be acquired quickly and for a competitive price anywhere in the world.

Saskatoon, Canada   *Wolfgang Walz*

Preface

Optogenetics—as the name suggests—coalesces multiple disciplines in Biomedicine, Optics, and Biotechnology, and now, almost 12 years after the publication of the first seminal paper, it may be safe to state that optogenetics truly revolutionized neuroscience.

For the first time, a causal interrogation of neuronal circuitry with millisecond precision, cell-type specificity, and minimal invasiveness became a reality. As a result, optogenetic techniques have pervaded almost all disciplines and fields of neuroscience over the years, from applications in brain slices to in vivo application in rodents, zebrafish, *C. elegans*, and nonhuman primates, and with clinical applications now just on the horizon. What is more, optogenetics allowed for the combination with already well-established readouts, such as single-cell electrophysiology, electric population recordings, functional magnetic resonance imaging, and, more recently, optical methods such as 2-photon calcium imaging.

The research questions involving optogenetics are equally diverse, including the modulation of stem cell differentiation in mouse brain, of *C. elegans* behavior, and the causal dissection of sensory processing in nonhuman primates.

While many scholarly articles and books exist on each of these methods and animal models, from viral vectors to electrophysiology and behavior, there are unique requirements for each step of the optogenetic approach. In fMRI, for instance, imaging coils need a lead-through for an optic fiber. Likewise, viral vectors need to accommodate the sequences of promoter, opsin, and fluorophore. Some research questions may not require any sophisticated cellular targeting tools, but rather the implementation of a complicated readout interfering with optical interrogation, while others may center on a novel promoter in combination with fairly standard opsins and analytical tools.

And maybe most importantly, a project involving optogenetics has to entail tailored control experiments addressing both a potential interaction of light with the readout method and the potential effect of light itself on neuronal physiology.

Still, even after more than a decade, a successful implementation of optogenetics requires expertise in multiple fields, and each step of the way poses individual challenges and takes considerable time and efforts.

This book should guide both the optogenetics newbie and the expert through all the steps required for the implementation of optogenetics in neuroscience. It should empower the reader to identify the critical aspects of each methodological step and to decide on the necessary level of complexity to address the respective research question.

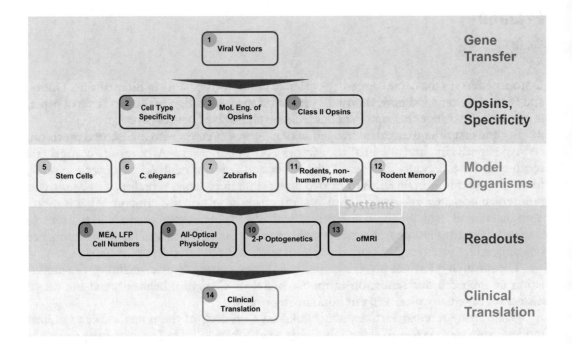

This book is structured along the optogenetics work flow, starting with viral vectors, followed by targeting strategies, choice of opsins, animal models, and readouts, and closing with applications in systems neuroscience and an outlook on clinical applications. Due to the unique format of this series, the reader will be able to gain in-depth knowledge of each procedure, yet each chapter only deals with those aspects of the method that are relevant from an optogenetics perspective. For example, the chapter on viral vectors deals exclusively with those viruses that can be used for gene delivery of opsins, and in the chapter on optical readouts, the complex problem of the crosstalk of opsins and optical reporters will be discussed in detail. The chapters comprise a review-like introduction of the current state of the art in the field, followed by a methodological step-by-step manual. Furthermore, a Notes section describes typical roadblocks and offers hands-on advice, and each chapter concludes with an outlook. Clearly, this book cannot cover all procedures and approaches in detail, but it should serve as a guide covering the most relevant aspects. I am very thankful to the series editor Wolfgang Walz and the entire team at Springer for supporting this concept. Lastly, I wish to express my deepest gratitude to all authors, many of whom I had the pleasure of working with since the early days of optogenetics.

May this book be of help and of inspiration particularly to the new generation of neuroscientists.

Mainz, Germany   *Albrecht Stroh*

Contents

Contributors

ANTOINE ADAMANTIDIS • *Department of Neurology, Inselspital University Hospital, University of Bern, Bern, Switzerland; Department of Clinical Research, Inselspital University Hospital, University of Bern, Bern, Switzerland*

RAAG D. AIRAN • *Department of Radiology, Neuroimaging and Neurointervention, Stanford University, Stanford, CA, USA*

ALLYSON L. ALEXANDER • *Department of Neurosurgery, Stanford University, Stanford, CA, USA; Department of Neurosurgery, University of Colorado School of Medicine, Aurora, CO, USA*

ISABELLE ARNOUX • *Focus Program Translational Neurosciences and Institute for Microscopic Anatomy and Neurobiology, University Medical Center of the Johannes Gutenberg University Mainz, Mainz, Germany*

ANDRÉ BERNDT • *Department of Bioengineering, University of Washington, Seattle, WA, USA*

JOHANNES BOLTZE • *Department of Translational Medicine and Cell Technology, Fraunhofer Research Institution for Marine Biotechnology and Cell Technology, Lübeck, Germany; Institute for Medical and Marine Biotechnology, University of Lübeck, Lübeck, Germany*

KARL EMANUEL BUSCH • *Centre for Discovery Brain Sciences, Edinburgh Medical School: Biomedical Sciences, University of Edinburgh, Edinburgh, Scotland*

I-WEN CHEN • *Wavefront-Engineering Microscopy group, Neurophotonics Laboratory, CNRS UMR8250, Paris Descartes University, Paris Cedex 06, France*

ILKA DIESTER • *Faculty of Biology, Albert Ludwig University of Freiburg, Freiburg, Germany; Bernstein Center for Computational Neuroscience Freiburg, Albert Ludwig University of Freiburg, Freiburg, Germany; BrainLinks-BrainTools, Albert Ludwig University of Freiburg, Freiburg, Germany*

JAN DOERING • *Focus Program Translational Neurosciences and Institute for Microscopic Anatomy and Neurobiology, University Medical Center of the Johannes Gutenberg University Mainz, Mainz, Germany*

ERIK ELLWARDT • *Focus Program Translational Neurosciences (FTN) and Immunology (FZI), Rhine Main Neuroscience Network (rmn²), Department of Neurology, University Medical Center of the Johannes Gutenberg University of Mainz, Mainz, Germany*

VALENTINA EMILIANI • *Wavefront-Engineering Microscopy group, Neurophotonics Laboratory, CNRS UMR8250, Paris Descartes University, Paris Cedex 06, France*

KATHARINA ELISABETH FISCHER • *Centre for Discovery Brain Sciences, Edinburgh Medical School: Biomedical Sciences, University of Edinburgh, Edinburgh, Scotland*

TING FU • *Focus Program Translational Neurosciences and Institute for Microscopic Anatomy and Neurobiology, University Medical Center of the Johannes Gutenberg University Mainz, Mainz, Germany*

MARTA GAJOWA • *Wavefront-Engineering Microscopy group, Neurophotonics Laboratory, CNRS UMR8250, Paris Descartes University, Paris Cedex 06, France*

INBAL GOSHEN • *Edmond and Lily Safra Center for Brain Sciences (ELSC), The Hebrew University, Jerusalem, Israel*

GOLAN KARVAT • *Faculty of Biology, Albert Ludwig University of Freiburg, Freiburg, Germany; Bernstein Center for Computational Neuroscience Freiburg, Albert Ludwig University of Freiburg, Freiburg, Germany*

HANNAH K. KIM • *Department of Neurosurgery, Stanford University, Stanford, CA, USA*

HEIKO J. LUHMANN • *Institute of Physiology, University Medical Center of the Johannes Gutenberg University Mainz, Mainz, Germany*

RODRIGO J. DE MARCO • *German Resilience Center, University Medical Center, Johannes Gutenberg University Mainz, Mainz, Germany; Developmental Genetics of the Nervous System, Max Planck Institute for Medical Research, Heidelberg, Germany*

EIRINI PAPAGIAKOUMOU • *Wavefront-Engineering Microscopy group, Neurophotonics Laboratory, CNRS UMR8250, Paris Descartes University, Paris Cedex 06, France; Institut national de la santé et de la recherche médicale–Inserm, Paris, France*

ALEXIS PICOT • *Wavefront-Engineering Microscopy group, Neurophotonics Laboratory, CNRS UMR8250, Paris Descartes University, Paris Cedex 06, France*

PIERRE-HUGUES PROUVOT • *Focus Program Translational Neurosciences and Institute for Microscopic Anatomy and Neurobiology, University Medical Center of the Johannes Gutenberg University Mainz, Mainz, Germany*

LIMOR REGEV • *Edmond and Lily Safra Center for Brain Sciences (ELSC), The Hebrew University, Jerusalem, Israel*

EMILIANO RONZITTI • *Wavefront-Engineering Microscopy group, Neurophotonics Laboratory, CNRS UMR8250, Paris Descartes University, Paris Cedex 06, France*

EDUARDO ROSALES JUBAL • *Focus Program Translational Neurosciences and Institute for Microscopic Anatomy and Neurobiology, University Medical Center of the Johannes Gutenberg University Mainz, Mainz, Germany; Institute of Biomedical Sciences, Universidad Autónoma de Chile, Santiago, Chile*

SOOJIN RYU • *German Resilience Center, University Medical Center, Johannes Gutenberg University Mainz, Mainz, Germany*

MIRIAM SCHWALM • *Focus Program Translational Neurosciences and Institute for Microscopic Anatomy and Neurobiology, University Medical Center of the Johannes Gutenberg University Mainz, Mainz, Germany; GRADE Brain, Goethe Graduate Academy, Goethe University Frankfurt, Frankfurt am Main, Germany*

IVAN SOLTESZ • *Department of Neurosurgery, Stanford University, Stanford, CA, USA*

ALBRECHT STROH • *Focus Program Translational Neurosciences and Institute for Microscopic Anatomy and Neurobiology, University Medical Center of the Johannes Gutenberg University Mainz, Mainz, Germany*

KIMBERLY R. THOMPSON • *Drug Discovery, Circuit Therapeutics, Menlo Park, CA, USA*

CHRIS TOWNE • *Gene Therapy, Circuit Therapeutics, Menlo Park, CA, USA*

NATHALIE ALEXANDRA VLADIS • *Centre for Discovery Brain Sciences, Edinburgh Medical School: Biomedical Sciences, University of Edinburgh, Edinburgh, Scotland*

JENQ-WEI YANG • *Institute of Physiology, University Medical Center of the Johannes Gutenberg University Mainz, Mainz, Germany*

OFER YIZHAR • *Department of Neurobiology, Weizmann Institute of Science, Rehovot, Israel*

Chapter 1

A Hitchhiker's Guide to the Selection of Viral Vectors for Optogenetic Studies

Kimberly R. Thompson and Chris Towne

Abstract

The very first article to describe optogenetics in neural systems used viruses as delivery vectors (Boyden et al., Nat Neurosci 8(9):1263–1268, 2005). Since then, viral-mediated gene delivery has become the method of choice for opsin expression in the field. There are many classes of viruses, each with unique attributes that can be taken advantage of to serve specific experimental needs. For example, precise cellular targeting can be achieved by exploiting the propensity of different vectors to transduce specific cell types. Distinct anatomical inputs or outputs to defined regions can be identified and manipulated by choosing vectors for opsin expression with retrograde or anterograde trafficking abilities. Some vectors also have the capability to spread between synaptically connected neurons, and this holds great potential for the determination of structure–function relationships across complex networks. Here we review the major viral vector types used in optogenetic studies and offer a detailed protocol for the production of adeno-associated virus, which has become the most popular vector for optogenetic applications. This chapter is intended to provide an understanding of basic principles in vectorology and to serve as a user's guide to aid in the selection of appropriate vector. The engineering of recombinant viruses promises to expand the level of experimental precision and control, and may one day even lead to effective optogenetic therapies.

Key words Viral-mediated gene delivery, Viral vector, Optogenetics, AAV, Lentivirus, Canine adenovirus, Herpes simplex virus, Rabies virus, Anterograde, Retrograde

1 Introduction

A central goal in neuroscience is the definition of precise cell populations and network connections that constitute functional circuits controlling complex behaviors and emotional states. The unprecedented level of genetic, spatial, and temporal resolution provided by optogenetic tools has enabled the exploration of detailed experimental hypotheses in awake, behaving animals (see Chaps. 11 and 12). Several methods can be used to introduce opsins into live cells including DNA transfection, electroporation, and transgenic expression systems. However, expression through viral-mediated gene delivery provides a superior ability to flexibly target any brain region in a variety of species of wild-type or

Albrecht Stroh (ed.), *Optogenetics: A Roadmap*, Neuromethods, vol. 133,
DOI 10.1007/978-1-4939-7417-7_1, © Springer Science+Business Media LLC 2018

transgenic models. The resulting tidal wave of investigation across a number of neurobiological disciplines is rapidly accelerating our understanding of neural systems.

Viruses make ideal vehicles for transgene delivery because they naturally penetrate into host cells and efficiently exploit the native transcription machinery to drive the expression of viral genes. Engineered vectors can be generated by substituting the viral genome with sequences that encode for opsin expression under the control of a specific gene promoter. There are numerous viral expression systems, each with unique properties to suit different experimental applications. Here we review the viruses most commonly used for optogenetic studies that include lentivirus, adeno-associated virus (AAV), canine adenovirus, herpes simplex virus, and rabies virus (Table 1). We highlight the advantages and disadvantages of each and provide a discussion of the practical factors and caveats that should be considered when making a choice of vector. AAVs currently represent the most widely applied means of delivering opsins into intact systems. These can be obtained through virus production facilities such as the vector cores at Stanford University, University of North Carolina, Chapel Hill, and University of Pennsylvania. However, they can also be easily produced in biosafety level 1-certified tissue culture facilities within 1–2 weeks (biosafety level 2 for protocols involving helper virus). We therefore describe a detailed step-by-step protocol for the production of AAV vectors and provide additional guidance on quality control measures and viral titering. Future directions of the field will be discussed in conclusion.

2   Viral Vector Expression Systems

Viral expression systems have greatly facilitated optogenetic studies in rodent, zebrafish, and primate models [1, 2]. They are fast, versatile, and can be implemented without the need for transgenic lines. Viral vectors drive robust levels of opsin expression, which is critical for efficient optical activation. Furthermore, they allow for intersectional targeting strategies to achieve both genetic and anatomical specificity. Each viral vector has distinct advantages and disadvantages that should be considered when planning an experiment.

2.1 Lentiviral Vectors

Lentiviruses (LV) have enveloped, single-stranded RNA genomes and belong to the retroviral family [3, 4]. After transduction, they stably integrate into the host chromosome to achieve permanent expression that is passaged to the progeny of dividing precursors. LVs are therefore suitable for use in mitotic as well as post-mitotic cell types, and they have been used successfully to transduce stem cells [5] (see Chap. 5). In post-mitotic cells, LV vectors integrate at

Table 1
Viral vectors commonly utilized for optogenetic studies

	Lentivirus	Adeno-associated virus	Canine adenovirus	Herpes virus	rabies virus
Structure size	100 nm	22 nm	~90 nm	~200 nm	75 × 180 nm
Viral family	Retrovirus	Parvovirus	Adenovirus	Herpesvirus	Rhabdovirus
Description	Two positive-strand, single-strand RNA. Genome integrating. For use in mitotic and post-mitotic cell populations. Drives stable long-term expression. Easy to produce	Single-strand DNA. Limited packaging capacity prevents use with large promoters. Different serotypes with unique tropism and transduction efficiency	Double-strand DNA. Replication-incompetent retrograde tracer that allows the identification of input neurons to defined regions	Double-strand DNA. Drives strong, rapid expression with broad tropism. Can be used to identify input neurons. Traffics most strongly in retrograde direction	Negative-strand, single-strand RNA. Retrograde transsynaptic tracer. RVdG variants restrict infection to monosynaptically connected cells
Allows retrograde labeling	No (for VSVg pseudotyped)	Some serotypes (e.g., AAV6, 9)	Yes	Yes	Yes (strongly)
Host genome integration	Yes	No	No	No	No
Max transgene capacity	9 kb	4.5 kb	30–36 kb	100–150 kb	1–3 kb
Cross synapses	No	No	No	No	Yes
Bio-safety level	BSL2	BSL1	BSL2	BSL2	BSL2

random, whereas integration preferentially occurs into active gene sites in dividing cells [6]. Little or no inflammatory or immune responses are found with LVs compared to earlier gene delivery vectors such as adenovirus that contained remnant viral genes [7–9].

The transduction preference for specific cell types, known as viral tropism, is determined for LV vectors by the particular glycoproteins that are expressed on the particle surface. These determine which membrane receptor the virus can bind to and thereby gain access into a host cell. Tropism can be modified by pseudotyping, which is the expression of glycoproteins from other viruses that function to alter or enhance the native tropism [10]. The most common pseudotyping method for LVs is the vesicular stomatitis virus glycoprotein (VSVg) that has wide tropism thereby facilitating infection into both neurons and glia [11]. The genome capacity for LV vectors is 8–10 kb for maximal packaging efficiency, which allows for the expression of larger transgene constructs when compared to AAV vectors that have smaller packaging capacity [12]. However, the larger viral particle size of 100 nm limits the diffusion of LVs in vivo and as such, limits the spread of the vector compared to the smaller AAVs [13, 14]. Lentivirus can be easily produced using standard tissue culture techniques [2] and so has become widely adopted for optogenetic studies in rodents [15, 16] and in primates [17, 18].

2.2 Adeno-Associated Virus

Adeno-associated virus (AAV) belongs to the parvovirus family owing to its single-stranded DNA genome that does not contain an envelope membrane [19]. After receptor-mediated endocytosis at the somatic compartment and transfer into the nucleus, the transgene-encoded proteins are synthesized, transported anterogradely out to the axon terminals and integrated into the membrane to allow optogenetic modulation of axonal projections. Retrograde transport of the AAV particle from the nerve ending to the cell body is also possible for various serotypes of AAV [19–21]. Recombinant AAV genomes do not readily integrate into the genome and rather exist as episomal DNA concatemers (multiple DNA strands aligned head to tail) within the nucleus [22]. Because of this, the AAV genomes become diluted upon cell division, rendering AAVs less desirable for use in mitotic populations [23] (see Chap. 5). In nondividing cells such as neurons, AAVs have become the vector of choice as they drive long-term stable expression with low immunogenicity [24]. Additionally, they have proven to be safe in clinical trials in humans targeting the brain [25] and other tissues [24]. The primary shortcoming of AAV vectors is their <4.7 kb packaging capacity which is limited compared to other vectors. Although most optogenetic proteins fit easily within this genome, the limited space restricts use of very large promoter fragments that confer expression specificity [12].

Regardless, a high level of neural specificity can be achieved with the reduced human Synapsin (hSyn) or CaM kinase II alpha (CaMKIIα) promoters that are frequently used in optogenetic studies. Interestingly, self-complementary AAVs (scAAVs) have recently been developed that overcome the rate-limiting step of DNA synthesis required by the single-stranded DNA genome for transgene expression. These scAAVs have been engineered to contain already double-stranded DNA, which significantly shortens viral incubation time but at the expense of a further limited packaging capacity by about one half (2.4 kb) [26, 27].

There are numerous naturally occurring AAV serotypes with distinct capsid proteins that confer specific properties such as infectivity for a specific cell type, preference to enter via axon terminals, or tissue diffusion [28]. Different serotypes have distinct sequences required for replication and packaging of the viral genome that are known as inverted terminal repeats (ITRs). By pseudotyping with various packaging systems, recombinant AAV hybrids can be formed. For example, AAV serotype 2 (AAV2) is the best-characterized natural serotype and it has been widely used to generate recombinant forms such as AAV2/5 (commonly referred to simply as AAV5; nomenclature adopted throughout this chapter) that has the ITRs from AAV2 and the capsid proteins from AAV5 [28]. Molecular engineering is constantly deriving new AAV hybrids that express novel tropisms with improved potency and cell type-selectivity [29].

AAV serotypes 1, 2, 5, 6, 8, and 9 have been successfully used to express opsins in rodent models while AAV 1, 5, and 8 have been utilized in primates (see Table 2 for references). A pronounced difference has been observed between AAV2 and AAV5 in the degree of viral spread through tissue, mainly due to differences in relative distribution of potential binding partners throughout the tissue [16, 20]. AAV5 diffuses broadly from the injection site and provides the best coverage for labeling large structures like the hippocampus or striatum. In comparison, injection of AAV2 results in relatively restricted expression patterns, which can be an advantage for targeting smaller structures like the habenula, or when it is desired, to restrict modulation to a subregion, such as the prelimbic region of the cortex. As a general guide, the spread of the virus in the brain increases from AAV2 < 1 & 6 < 5 < 8 < 9 but this must be confirmed empirically for each region and cell type. Of interest, AAV9, when injected peripherally, has been found to cross the blood–brain barrier to infect both neurons and glia cells of the brain. This attribute is currently being explored for its potential utility as a vector for gene therapy [49–51]. One recently developed recombinant serotype is AAV-DJ, which is a hybrid capsid derived from eight different serotypes that displays very high infectivity across a broad range of cell types [52]. AAV-DJ is gaining popularity in optogenetic approaches since researchers are finding that viral

Table 2
AAV serotypes used in optogenetic studies

AAV serotype	Species	Structure	Optogenetic references
1,2,5,8,9, DJ	Rodent	CNS: Brain	[16, 30–34]
2	Rodent	CNS: Retina	[35–37]
6,8	Rodent	PNS: Sensory	[38, 39]
6	Rodent	PNS: Cochlea	[40]
6	Rodent	PNS: Motor	[41]
1,9	Rodent	Cardiac Muscle	[42–44]
9	Rodent	Skeletal muscle	[45]
1,5,8	Primate	CNS: Brain	[17, 46–48]

incubation times can be significantly shortened due to the robust potency of the vector [31–33].

2.3 Canine Adenovirus

Adenovirus, belonging to the adenoviral family, contains double-stranded DNA with no surrounding membrane. Its usefulness as a viral vector in gene therapy has been limited due to toxicity issues and to the presence of preexisting neutralizing antibodies in the majority of the population [53]. For this reason, canine adenovirus [47] was developed based on the CAV-2 vector strain that has low levels of seropositivity in human populations [54]. CAV-2 has great potential for optogenetic studies because it preferentially infects axon terminals and travels retrogradely to the neuron cell body, allowing the precise identification of input neurons to a defined region [54–57]. CAV vectors have a scalable production process and can be produced at clinically relevant titers. They are replication defective and do not integrate into the host genome [19]. They have the additional advantage of supporting a large packaging capacity of up to around 30 kb. In optogenetic studies, CAV-2 has been used to drive opsin expression directly [58] or indirectly through the expression of Cre-recombinase [59].

2.4 Herpes Simplex Viral Vectors

Herpes simplex virus (HSV) is an enveloped, doubled-stranded DNA virus from the herpesvirus family. It has broad tropism and is known to drive strong and rapid expression within 1–2 weeks in a range of species including rodents and zebrafish [60–62]. Some issues with stability of expression due to inflammatory reactions have been reported with HSV approaches [60, 63]. However, a major advantage of HSV is that it has an extremely large (150 kb) packaging capacity that can accommodate long sequences of foreign DNA such as full-size promoter sequences for tight cell type-specific expression. Replication incompetent HSV vectors do not cross synapses but they have the ability to transduce axon terminals and travel retrogradely to the neuron cell body. This provides a means for targeting neurons based on connectivity. However, HSV is also known to traffic to some extent in an anterograde direction so care must be taken when designing experiments. Several optogenetic studies have taken advantage of HSV-1 for cell body [64, 65] or projection-specific manipulations [66].

2.5 Rabies Virus   Rabies virus (RV) is an enveloped negative-strand RNA virus from the rhabdovirus family that selectively infects neurons but not glia [67]. RV vectors have become powerful tools for circuit mapping because they infect axon terminals and then spread exclusively in a retrograde direction between synaptically connected neurons [68, 69]. This allows for the transsynaptic labeling of neuronal circuits across a theoretically unlimited number of synapses. For some peripheral sensory neurons, however, the strict retrograde directionality is not maintained [70, 71]. RV has been used as an effective tool to optogenetically decipher input-specific functional roles for neurons projecting to midbrain dopamine neurons [72].

A major advancement in the field was achieved with the development of a method to restrict viral spread to only the cells that are monosynaptically connected to the infected neurons. This glycoprotein (G)-deleted rabies virus (RVdG) approach has become extremely useful for circuit mapping because it both enables the spread of RV to be monosynaptically restricted and allows the primary infection to be restricted to genetically defined cell populations [73, 74]. Indeed, RVdG has facilitated many elegant neuronal tracing studies that precisely map inputs to defined cell types [75–78]. RVdG vectors can accommodate genome sizes of around 1–3 kb and variants have also been developed that successfully encode ChR2 [79, 80]. A similar approach was recently developed known as "tracing the relationship between input and output" or TRIO (also cTRIO for cell type-specific tracing), that combines with CAV-driven retrograde targeting to allow for an increasingly sophisticated definition of multiple nodes of connectivity [81, 82]. Such unbiased anatomical mapping has enormous potential if it can be combined with optogenetics to establish causal relationships between circuit activity and behavior. However, one caveat for optogenetic use is that RVdV infection invariably kills neurons after 2 weeks of expression. Multiple observations suggest that cells are viable for at least 1 week after infection so it is advisable to conduct studies within this time frame [73, 74, 80]. Recently, novel engineered forms of RVdG have been published that have improved cell heath and tracing efficiency [83, 84].

3   Considerations for Selection of Expression System

The choice of viral vector is complex and depends on the particular details of the experimental question at hand. Table 3 outlines the major factors influencing choice of vector and provides guidance to aid in the decision-making. The most widely applied vectors in optogenetics are LV and AAV. One of the first considerations for choice of vector is whether the target cell type is mitotic or post-mitotic. As discussed above, LVs stably integrate into the host genome and are therefore suitable for both types, whereas AAVs

Table 3
Major experimental factors impacting the choice of viral vector

Factor	Decision guidance
Species	LV and AAV are commonly used in rodents and primates but do not work well for zebrafish. RV and HSV effective for rodent, primate, and zebrafish models
Cell type mitotic or post-mitotic	LV integrates into the genome so suitable for both dividing and non-diving cells. AAV, CAV, HSV, and RV all remain episomal and so are suitable for post-mitotic cell populations only
Vector genome size	The maximal transgene packaging capacity increases from RV < AAV < LV < CAV < HSV. AAV is sufficient to drive an opsin tagged with a fluorescent protein under the control of a common gene promoter but does not accommodate large cell type-specific promoters. CAV and HSV allow for larger cell type-specific promoters and for multiple promoters or transgenes to be driven from the same vector
Duration of expression	LV, AAV, and CAV drive stable expression over long durations. Toxicity issues have been associated with HSV and RV, which limit their utility for long-term study. Notably, the window for optimal cell health is under 2 weeks following transduction with RV
Expression level	AAVs can be produced as high titer stocks and infect cells with multiple vectors so expression is generally high. CAV drives similar high-level expression. Expression with LV and HSV tends to be more modest
Viral spread	The small size of AAV facilitates diffusional spread over wide areas of tissue. LVs have a large particle size and show restricted diffusion in vivo, which makes it effective for targeting small regions or subfields of a structure
Antero/retrograde labeling	In general LV and AAV infect neurons at the soma and the transgene products travel in an anterograde direction towards the axonal projections. CAV and RV are taken up at the axon terminals and transported retrogradely to identify the input neurons to a defined region. HSV is transported in both directions, although most strongly in the retrograde direction.
Viral production difficulty	LV is easily produced in tissue culture facilities. AAV production is a bit more complicated but can be produced using standard tissue culture techniques and an ultracentrifuge. CAV, HSV, and RV vectors are not as straightforward to produce.

remain episomal and are therefore unable to support stable, long-term expression in dividing cells. A main challenge in viral targeting is that the genome size contained in a viral capsid is limited. LV can package larger gene fragments of up to 9 kb [12] compared with AAV systems where 4.7 kb is the upper limit [85, 86]. Additionally, AAVs spread more effectively in tissue than LVs because they are smaller and able to diffuse further away from the injection site. AAV is therefore recommended when larger area coverage is desired, whereas LV is preferred when greater spatial specificity is needed, for example when it is desired to target subfields of a structure, such as the CA1 region of the hippocampus [87]. Although AAV and LV can infect a wide range of species including rodents and primates, they are not suitable for expression in zebrafish, for which HSV and RV are more effective [62, 88].

When it comes to choice of vector for retrograde labeling studies, several factors should be considered. HSV provides the largest packaging capacity of up to around 150 kb, which is sufficiently large to accommodate a full-sized cell type-specific promoter, transgene and fluorescent reporter. This can be especially useful if multiple transgenes are driven from the same promoter or if multiple promoters are desired. HSV, however, typically drives lower expression levels and is associated with toxicity [63]. Although CAV has the capacity to encode a smaller transgene of up to around 30 kb, this is sufficient for most opsin inserts and the level of expression and long-term cell health are comparatively higher. Among transsynaptic vectors, RV provides the most faithful representation of anatomical projections because the labeling is very robust. It is also the vector of choice for specificity since it travels exclusively in the retrograde direction.

4   Potential Pitfalls of Viral Gene Delivery

Viral-mediated gene delivery is associated with certain limitations that must be taken into account when designing, executing, and interpreting optogenetic experiments. In general, viral expression reduces with distance from the injection site due to multiplicity of infection, resulting in different levels of opsin expression across the cells so expression should not be expected to be uniform throughout the entirety of a targeted structure. There may also be significant adverse effects on cell health due to high levels of non-endogenous protein expression so caution must be taken when interpreting effects (see Chap. 8 for examples of overexpression). This is especially true when working with RV vectors that are known to have long-term toxicity issues. Since it is likely that cell function is compromised well before signs of cell death, it is critical to design appropriate control conditions. Comparing behavioral "light off" values across different cohorts is central to evaluating deleterious viral effects. Likewise, histological analysis after the experiment should be performed to confirm lack of cell toxicity. Indeed, even the relatively nontoxic AAV vectors may cause cellular abnormalities when expressed over long time periods at high titer, especially when used in combination with a strong gene promoter [89]. Users should also be aware that contaminants might be present in viral preparations, such as impurities and endotoxin, which could influence immune responses and cell health, highlighting the importance of quality control for sensitive applications (discussed below).

5   A Simple AAV Production Protocol for Optogenetic Studies

AAV has become the vector of choice for optogenetic studies in rodents due to its high neural tropism and relatively low inflammatory response compared to other vectors. The main limitation of AAV, its relatively small packaging capacity, is not a major impediment since the size of the microbial opsins readily fits within the recombinant viral genome. For the above reasons we therefore present an optogenetic vector production protocol for AAV.

This section describes a simple and robust method to produce high-quality AAV of any serotype using off-the-shelf materials in a short time frame. AAV is generated through an adenovirus free triple transfection method as described previously [90, 91] that is devoid of helper viruses and thus permits the production to be performed in a biosafety level 1 laboratory. HEK 293 cells are transfected via calcium phosphate with (1) an opsin plasmid that contains the optogenetic gene sequence flanked by inverted terminal repeats (ITR), (2) a packaging plasmid that contains the Rep (replication) and Cap (capsid) genes required to recognize the ITRs and package their flanked sequences into the virion, and (3) a helper plasmid that supplies the remaining adenovirus proteins required for AAV construction. Forty-eight hours after transfection, the AAV particles are generated and remain within the nucleus. The cells are lysed and the homogenates are added to a chemical gradient where ultracentrifugation separates the AAV from cell debris. Without access to an ultracentrifuge, heparin columns can be used instead to purify AAV serotypes that bind heparin (AAV2 and AAV6) [92]. However, chemical gradients (iodixanol or cesium chloride) are preferable as they offer higher purity and can be used to purify all serotypes [93]. We describe the iodixanol protocol in this chapter, as this method is more rapid than cesium chloride (2 h vs. 3 days) while producing virus of equal titer and quality [94]. After purification, the purified AAV undergoes titration and quality control prior to stereotaxic injection into the rodent brain. The complete workflow is summarized in Fig. 1.

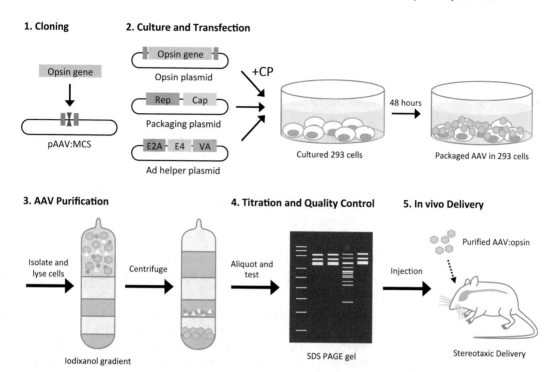

Fig. 1 Workflow of AAV production for optogenetic studies. The opsin of interest is cloned into the AAV backbone between the inverted terminal repeats (ITRs) to create the opsin plasmid. Recombinant AAV is produced by triple transfection of the opsin plasmid along with the AAV structural plasmid and Adenovirus helper plasmid into 293 cells. Forty-eight hours later, cell homogenates are loaded onto an iodixanol gradient and centrifuged to separate the AAV from the cellular debris. The AAV then undergoes titration and quality control prior to stereotaxic injection into the rodent brain

5.1 Cloning, Cell Culture and Transfection

1. AAV expression plasmids (pAAV:MCS, multiple cloning site), packaging plasmids and adenovirus helper plasmids are available for purchase from multiple vendors. Note that the packaging plasmid encodes for the serotype and should be specified at time of purchase. The opsin expression plasmid can be obtained from the Deisseroth laboratory at Stanford by following instructions on the Stanford University Optogenetics Resource Center (http://web.stanford.edu/group/dlab/optogenetics/). This site is maintained regularly and lists all the available opsins (>100) published by the laboratory. Alternatively, novel opsin genes can be cloned directly into the pAAV:MCS expression plasmid using conventional molecular biology techniques.

2. Maintain 293 cells with growth media in a humidified incubator with 5% CO_2. Passage cells 1:3 twice per week towards maintaining a stable high-quality cell line.

3. Split confluent 293 cells on the afternoon before transfection at 1:4 into 15 cm plates. 12 × 15 cm plates are used for a typical

production run capable of producing enough virus for a substantial number of optogenetic studies in rodents.

4. The next day the cells should be approximately 60% confluent. Two hours before transfection, change the growth medium to fresh media minus serum.

5. Add 180 μg of the opsin plasmid and a 1:1:1 molar ratio of the packaging and helper plasmids to 2.78 mL of 2.5 M $CaICl_2$. Bring the volume to 27.75 mL with H_2O. Prepare three individual 50 mL conical tubes containing 9.25 mL of 2× BES buffered saline (BBS) (Sigma-Aldrich, St Louis, MO, 14280). Add 9.25 mL of the DNA/$CaCl_2$ solution dropwise (under gentle vortexing) to each of the 9.25 mL 2× BES buffered saline containing tubes.

6. Incubate the mix for approximately 3 min at room temperature. The solution should turn opaque but no solids should be visible.

7. Remove the 293 cells from the incubator. Add 4.63 mL of the transfection mix gently to the cells of each 15 cm plate by distributing the drops uniformly across the surface. Mix the solution in a cross "T" manner and return cells to the incubator.

5.2   AAV Isolation

1. Four days following transfection, scrape the cells from the bottom of the plates and centrifuge for 20 min at 4700 rpm ($3433 \times g$) in a standard benchtop centrifuge.

2. Aspirate the media supernatant and resuspend the cells in a total volume of 7 mL of 1× gradient buffer (GB). GB will be used in the iodixanol gradient and a 10× solution is prepared by combining 10 mL 1 M Tris (pH 7.6), 30 mL 5 M NaCl, 10 mL 1 M $MgCl_2$ and 50 mL ddH_2O. This solution is then sterilized using a 0.22 μm vacuum filter.

3. Pool the cells into a single 50 mL conical tube. Cells are lysed through multiple freeze/thaw cycles. Submerge the tube into a container of liquid nitrogen for 10 min. Loosen the cap in a biosafety hood to equilibrate air pressure and then re-tighten the cap. Transfer the tube to 37 °C water bath until the cells are thawed.

4. In a biosafety hood, triturate the cell suspension through a 23G needle attached to a 20 mL syringe to aid cell lysis while avoiding the introduction of air bubbles.

5. Repeat the freeze/thaw cycles an additional three times and then transfer the cells to a 37 °C water bath.

6. Add 50 U/mL benzonase and incubate at 37 °C for 1 h. Swirl the tube every 15 min to help break down the DNA to reduce viscosity of the solution.

7. Centrifuge the suspension at 3700 rpm (2128 × *g*) for 3 min at 18 °C. Decant the supernatant into a fresh 50 mL conical tube and move immediately onto the next step or freeze at −20 °C to proceed on another day.

5.3 AAV Purification Through Iodixanol Gradient

1. Prepare iodixanol solutions of four different concentrations (15, 25, 40 and 58%). Iodixanol can be purchased from Sigma under the synonym OptiPrep™ Density Gradient Medium (Sigma-Aldrich, D1556). The 15, 25, 40 and 58% solutions are prepared by adding 40, 46.67, 64 or 96.67 mL iodixanol to 15, 11.2, 9.6 and 3.33 mL of 10× GB, 32, 0, 0 and 0 mL of 5 M NaCl, 72, 56, 22.4 and 0 mL ddH$_2$O, and 0, 280, 0 and 280 μL of 5 mg/mL Phenol Red, respectively. The Phenol Red aids in visualization of the layers within the tube.

2. The iodixanol gradient is built in a quick-seal centrifuge tube (Beckman Coulter, Brea, CA, 344326) and started by adding the 15% solution to the bottom of the tube using 20 mL syringe with a long 18G needle (Fig. 2a).

3. The subsequent layers are added below the previous layer by placing the syringe needle each time to the bottom of the tube

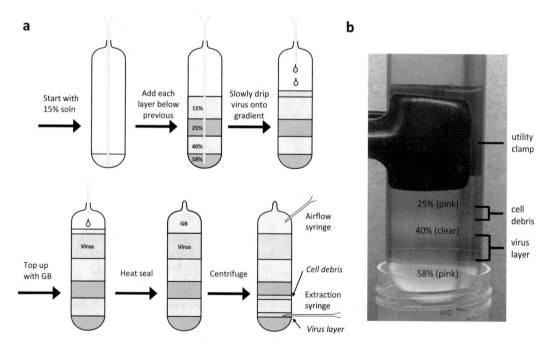

Fig. 2 The use of iodixanol gradients in AAV production. (**a**) Schematic of various steps in the generation of an iodixanol gradient and its use to separate AAV from cellular debris. (**b**) A picture of an iodixanol gradient post centrifugation prior to extraction by syringe. A faint grey band can be observed between the 40% and 58% layers where the virus is located

and injecting slowly. Avoid the generation of bubbles that could result in mixing of the layers.

4. After the gradient has been built, the transfection homogenate containing the AAV from the previous section is carefully added to the top of the gradient in a drop wise fashion. This step can be achieved using a small 20G needle.

5. Add 1× GB to fill the remaining empty volume of the tube and seal the tube using a heating device.

6. Centrifuge the gradient at 48,000 rpm in a Beckman ultracentrifuge Type 70Ti rotor ($169,538 \times g$) for 130 min at 18 °C. Conversions are available online and should be used to ensure a constant K factor.

7. Gently remove the tube and place it in a utility clamp attached to a stand in the back of the biosafety hood (Fig. 2b). Care should be taken to avoid agitating the tube that would result in disturbing the layers and the separated virus.

8. Pierce a hole near the top of the tube using an 18G needle to let airflow in. As the initial poke can sometime result in spurting of the solution, first place a piece of laboratory tissue paper around the tip of the needle to absorb the leakage. Likewise, a 50 mL conical tube can be placed under the tube to catch further run off.

9. To harvest the AAV, insert an 18G needle attached to a 5 mL syringe 1–2 mm below the interface between the 40 and 58% layers with the bevel of the needle facing up. The virus is visible as a subtle grey tinge to the gradient. Extract carefully 3–5 mL of solution, first with the beveled needle opening facing upwards (approx. 2.5 mL), then with the bevel facing downwards to take the remained of the vector solution (approx. 1.5 mL). Note that most of the cell debris collects at the interface between the 25 and 40% layers. Transfer the solution to a 15 mL conical tube and go immediately to the next step.

5.4 AAV Concentration and Storage

1. Add 5 mL of 1× GB to the purified AAV prep to aid in diluting out the iodixanol. Mix with a pipette.

2. Divide the mixture into two equal portions and load equally onto two Amicon Ultra-4 Centrifugal Filter Unit with Ultracel-100 membrane (Cole-Parmer, Vernon Hills, IL, #EW-29969-78). Centrifuge at 4700 rpm ($3433 \times g$) for 30 min at 18 °C using a standard benchtop centrifuge.

3. Aspirate flow through and add 3× of the remaining volume of 1× GB buffer to the viral concentrate. Mix gently with a pipette and centrifuge at 4700 rpm ($3433 \times g$) for an additional 20 min at 18 °C.

4. Approximately 50–150 μL of AAV concentrate should remain after centrifugation. Aliquot the mixture into low adhesion cryovials and store at 4 °C for short-term storage (<2 weeks) or freeze at −80 °C for long-term storage. You have successfully generated an optogenetic AAV construct.

6   AAV Titration and Quality Control

A wide array of protocols exists to measure the properties of produced recombinant AAV vectors. Titration methods report the dose of the vector by a number of different means. Likewise, quality control procedures give information regarding purity that may have the potential to confound experiments, i.e., injection of a "dirty" virus that causes inflammation and subsequent behavioral deficits.

6.1   Titer by Quantitative PCR

Quantitative PCR is the most common method used to quantify AAV stocks. The majority of optogenetic research articles report AAV in these terms, i.e., vector genomes per milliliter (vg/mL). The method uses primers (DNA probes) against the packaged DNA vector genome that facilitate their amplification, under the polymerase chain reaction (PCR), to levels that can be measured. Briefly, the contaminating nonpackaged DNA (that would otherwise artificially inflate the signal) is first degraded by using the endonuclease DNAse. The capsid (protein shell of the vector) is then digested with sodium hydroxide to release the packaged DNA. Primers are added to the mixture and, along with a standard curve of known DNA amounts, quantitative PCR determines the absolute amount of vector genomes per volume. It is important to note that this titration method is fast and reliable, however, does not discriminate between incompletely assembled AAV particles and AAV particles capable of carrying out a successful infection, i.e., having the correct topology required to bind its cognate receptor, enter the cell, traffic through the cytoplasm and deliver the genome to the nucleus. Indeed, nonfunctional AAV vectors will still be measured as long as the DNA sequences that match the probes are contained within the vector.

6.2   Titer by In Vitro Infection

Unlike quantitative PCR, the in vitro infection method can be used as a surrogate test for AAV infectivity in vivo. Titration can be achieved in various ways that typically involve infecting 293 cells (or other cells) and recording the number of cell infection (transduction) events per volume, i.e., transducing units per milliliter (tu/mL) as described previously [95]. Transduction is normally measured by quantifying the number of cells transduced, either on a plate using microscopy, or through FACS analysis. This method is ideal when it is known that both the AAV serotype and the promoter can function to infect and drive efficient expression in a particular cell type. However, situations may arise where an AAV does not efficiently infect the cell type of interest (despite being potent in vivo) or has a promoter that does not express, e.g., the commonly used human synapsin promoter (hSyn) that does not express in 293 cells. In the latter example, use of a neural cell line or primary neuronal culture would facilitate titration. Despite the clear benefits of in vitro titration, care should be taken to use this method only when comparing AAV vectors of the same serotype and promoter. Comparing in vitro titers of AAV vectors with different serotypes (that may infect the cell culture and brain at different rates) or different promoters (that may express in cell culture and brain at different levels) would not yield valid information.

6.3 SDS-PAGE for Purity Analysis

In addition to viral titer, AAV purity is an important property that can affect transduction efficiency [96] and safety profile [97] of an AAV preparation. The simplest method to determine vector purity is to perform sodium dodecyl sulfate–polyacrylamide gel electrophoresis (SDS–PAGE) followed by interrogation with a sensitive nonspecific protein stain, e.g., Coomassie Brilliant Blue, SYPRO Ruby or Silver Staining. By revealing all of the proteins in a preparation, one can ascertain the relative amounts of AAV capsid proteins (VP1, VP2 and VP3) compared to any contaminating proteins that might be present. A highly purified AAV production should have only the three capsid proteins visible in the gel, or at least have them staining substantially more than other non-capsid protein bands. An example is given in Fig. 3a, where preparation 1 has

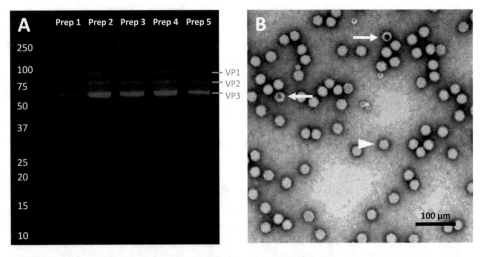

Fig. 3 Quality control for AAV preparations. (**a**) An SDS-PAGE gel stained with SYPRO Ruby for five AAV preparations. Preparations 2–5 have three strong bands corresponding to capsid proteins VP1, VP2, and VP3. Preparation 1 has multiple non-capsid protein bands and is considered a low purity production. (**b**) Electron microscopy can be used to quantify the full to empty virus particle ratio. White arrows indicate the only two AAV particles in the image field that are empty, while all the remaining viruses are packaged (single white arrowhead given as example)

only weak capsid proteins and significant contaminating bands (a low purity vector), whereas preparations 2–5 have predominantly VP1, VP2, and VP3 (high purity vectors). Further confirmation of the identity of the capsid proteins can be achieved by immunoblotting using an antibody against the VP antigen. While SDS-PAGE can give a qualitative picture of the level of purity, it does not inform on the nature of the impurities. One specific concern regarding purity is the presence of agents that would cause inflammatory reactions. Fortunately, a number of endotoxin assays are available commercially to determine the gram-negative bacterial endotoxin level in AAV preparations.

6.4   Electron Microscopy for Full: Empty Particle Ratio

While the analytical techniques described to this point can be used to characterize viral vectors according to their properties, transmission electron microscopy, allows a direct visualization of the actual AAV particle. The glow discharge method is commonly used to make the support film on the grid more hydrophilic [98]. The process of passing electrical current through argon gas produces a glow of ionized argon atoms that bombard the surface of the plastic film, resulting in a clean, negatively charged hydrophilic surface. This process is advantageous in the case of AAV because it discourages artificial aggregation of particles on the surface of the grid and results in a more even stain with fewer gas bubble artifacts upon drying. The main value of electron microscopy is that it allows determination of the full to empty particle ratio. Empty particles have a contrasted dot (or hole) in the center of the 20 nm particles, whereas full particles are opaque. Figure 3b gives an example of a preparation with a >95% full:empty ratio, with 2 empty particles (arrows) shown in the field. Empty particles are incapable of transduction and are therefore undesirable, however, it is not uncommon that full:empty ratios of preps fall as low as 10%.

7   In Vivo Delivery

Following production and quality control, the opsin expressing AAV vector is now ready to be introduced in vivo for optogenetic studies. Direct stereotaxic injection of AAV into the brain provides a rapid, inexpensive and powerful method to induce opsin expression in a region-specific manner. Cell type-specificity can be further increased through the use of projection targeting methods, Cre-driver transgenic mouse lines, or combinatorial approaches [99]. An example of a typical in vivo experiment would be the injection of 500 nL of 2×10^{12} vg/mL (1×109 vg total) of AAV2 expressing ChR2-YFP under a CaMKIIα promoter into the prefrontal cortex of an 8-week-old wild-type mouse (Fig. 4). In this case, the choice of AAV serotype with low diffusional spread allowed opsin expression to be nicely confined to the excitatory neurons of the prelimbic

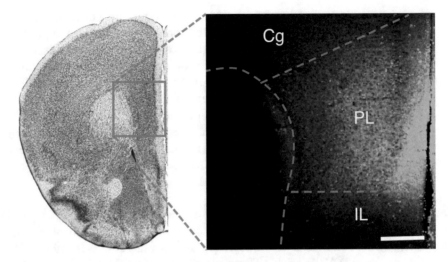

Fig. 4 Anticipated results following in vivo targeting by stereotaxic microinjection. Stereotaxic guidance was used to precisely inject 500 nL of AAV2-CaMKIIα-ChR2-YFP into the prelimbic region (PL) of the medial prefrontal cortex of an 8-week-old mouse at coordinates anterior/posterior = 1.8, medial/lateral = 0.35, dorsal/ventral = −1.65 (coordinates relative to bregma). After 4 weeks of viral incubation, histology was performed to confirm expression and proper anatomical targeting. The choice of AAV2 serotype and injection volume allowed the expression to remain confined to the PL target region without spreading into the neighboring cingulate cortex (Cg) or infralimbic (IL) regions. Scale bar = 250 μm

region, without spreading into neighboring cortical areas. In addition to stereotaxic injection into the brain, AAV vectors have also been used to deliver opsin proteins to the motor neurons [41] and sensory neurons [38] of the peripheral nervous system.

8   Notes

The previous methodological chapter has described a protocol for production of AAV vectors using calcium phosphate triple transfection following by iodixanol ultracentrifugation. The most common problem associated with this protocol and other AAV protocols is low viral titer. Decreased viral yields are typically a result of low calcium phosphate transfection efficiency. Always test transfection solutions for their efficiencies before carrying out large-scale vector purifications. In addition, all transfection material should be at room temperature before transfection. Another cause of low viral titers is the use of 293 cells that have reached high passage numbers. It is recommended to maintain a cell bank with vials of low passage cells that can be thawed regularly to avoid the use of older cell cultures. If transfection efficiency and cell age cannot account for the low viral yield, samples can be taken at each step in the production protocol (e.g., crude cell lysate, post-iodixanol, post-concentration) and measured for viral titer using quantitative PCR. This will help ascertain at which step in the process the viral loss is occurring.

9 Outlook

Viral vectors have greatly facilitated the implementation of optogenetics in awake, behaving animals. Now the challenge is to develop improved vectors with unique properties that can address increasingly sophisticated scientific questions. Towards this end, molecular engineering efforts are underway to generate new viral variants by rational design, which relies on current knowledge of viral biology (e.g., molecular interactions between capsid and receptors) to alter tropism [24, 100], facilitate the expression of larger genomes [101], enhance potency [83], and improve cell health for longer-term study [84]. Another powerful approach being used to derive novel viral variants with unique properties is directed evolution [102]. It involves the generation of an initial library that has high capsid complexity followed by cycles of selection to progressively enrich for the desired target-specific variants. This approach has yielded recombinant AAV serotypes with improved potency of infection [34], enhanced diffusional properties [103], and retrograde functionality [104], and may ultimately yield highly cell type-specific tropisms.

The implementation of viruses in experimental research provides an important basis to assess the potential of viral vectors for use in the clinic. Over the last decade there has been an explosion of clinical trials that use viral vectors to deliver gene therapies to treat numerous human disorders. Thus, the expression of opsins in human neurons may not be so far off into the future. Indeed, in 2015 the FDA granted approval for retinal injections of AAV expressing channelrhodopsin to treat retinitis pigmentosa and the first optogenetic protein was delivered to a human in March 2016. It will be exciting to see whether viral-mediated gene delivery of optogenetic tools is merely a useful research approach to facilitate the study of neural systems, or whether there is real value for clinical applications.

References

1. Simmich J, Staykov E, Scott E (2012) Zebrafish as an appealing model for optogenetic studies. Prog Brain Res 196:145–162

2. Zhang F, Gradinaru V, Adamantidis AR et al (2010) Optogenetic interrogation of neural circuits: technology for probing mammalian brain structures. Nat Protoc 5(3):439–456

3. Naldini L (1998) Lentiviruses as gene transfer agents for delivery to non-dividing cells. Curr Opin Biotechnol 9(5):457–463

4. Parr-Brownlie LC, Bosch-Bouju C, Schoderboeck L et al (2015) Lentiviral vectors as tools to understand central nervous system biology in mammalian model organisms. Front Mol Neurosci 8:14

5. Stroh A, Tsai HC, Wang LP et al (2011) Tracking stem cell differentiation in the setting of automated optogenetic stimulation. Stem Cells 29(1):78–88

6. Bartholomae CC, Arens A, Balaggan KS et al (2011) Lentiviral vector integration profiles differ in rodent postmitotic tissues. Mol Ther 19(4):703–710

7. Abordo-Adesida E, Follenzi A, Barcia C et al (2005) Stability of lentiviral vector-mediated transgene expression in the brain in the presence of systemic antivector immune responses. Hum Gene Ther 16(6):741–751

8. Annoni A, Battaglia M, Follenzi A et al (2007) The immune response to lentiviral-delivered transgene is modulated in vivo by transgene-expressing antigen-presenting cells but not by CD4+CD25+ regulatory T cells. Blood 110(6):1788–1796

9. Blomer U, Naldini L, Kafri T et al (1997) Highly efficient and sustained gene transfer in adult neurons with a lentivirus vector. J Virol 71(9):6641–6649

10. Cronin J, Zhang XY, Reiser J (2005) Altering the tropism of lentiviral vectors through pseudotyping. Curr Gene Ther 5(4):387–398

11. Jakobsson J, Ericson C, Jansson M et al (2003) Targeted transgene expression in rat brain using lentiviral vectors. J Neurosci Res 73(6):876–885

12. Kumar M, Keller B, Makalou N et al (2001) Systematic determination of the packaging limit of lentiviral vectors. Hum Gene Ther 12(15):1893–1905

13. Cetin A, Komai S, Eliava M et al (2006) Stereotaxic gene delivery in the rodent brain. Nat Protoc 1(6):3166–3173

14. Lerchner W, Corgiat B, Der Minassian V et al (2014) Injection parameters and virus dependent choice of promoters to improve neuron targeting in the nonhuman primate brain. Gene Ther 21(3):233–241

15. Boyden ES, Zhang F, Bamberg E et al (2005) Millisecond-timescale, genetically targeted optical control of neural activity. Nat Neurosci 8(9):1263–1268

16. Yizhar O, Fenno LE, Davidson TJ et al (2011) Optogenetics in neural systems. Neuron 71(1):9–34

17. Diester I, Kaufman MT, Mogri M et al (2011) An optogenetic toolbox designed for primates. Nat Neurosci 14(3):387–397

18. Han X (2012) Optogenetics in the nonhuman primate. Prog Brain Res 196:215–233

19. Nassi JJ, Cepko CL, Born RT et al (2015) Neuroanatomy goes viral! Front Neuroanat 9:80

20. Burger C, Gorbatyuk OS, Velardo MJ et al (2004) Recombinant AAV viral vectors pseudotyped with viral capsids from serotypes 1, 2, and 5 display differential efficiency and cell tropism after delivery to different regions of the central nervous system. Mol Ther 10(2):302–317

21. Towne C, Schneider BL, Kieran D et al (2010) Efficient transduction of non-human primate motor neurons after intramuscular delivery of recombinant AAV serotype 6. Gene Ther 17(1):141–146

22. van den Pol AN, Ozduman K, Wollmann G et al (2009) Viral strategies for studying the brain, including a replication-restricted self-amplifying delta-G vesicular stomatis virus that rapidly expresses transgenes in brain and can generate a multicolor golgi-like expression. J Comp Neurol 516(6):456–481

23. Betley JN, Sternson SM (2011) Adeno-associated viral vectors for mapping, monitoring, and manipulating neural circuits. Hum Gene Ther 22(6):669–677

24. Asokan A, Schaffer DV, Samulski RJ (2012) The AAV vector toolkit: poised at the clinical crossroads. Mol Ther 20(4):699–708

25. Kaplitt MG, Feigin A, Tang C et al (2007) Safety and tolerability of gene therapy with an adeno-associated virus (AAV) borne GAD gene for Parkinson's disease: an open label, phase I trial. Lancet 369(9579):2097–2105

26. Gray SJ, Foti SB, Schwartz JW et al (2011) Optimizing promoters for recombinant adeno-associated virus-mediated gene expression in the peripheral and central nervous system using self-complementary vectors. Hum Gene Ther 22(9):1143–1153

27. McCarty DM, Monahan PE, Samulski RJ (2001) Self-complementary recombinant adeno-associated virus (scAAV) vectors promote efficient transduction independently of DNA synthesis. Gene Ther 8(16):1248–1254

28. Gao G, Vandenberghe LH, Wilson JM (2005) New recombinant serotypes of AAV vectors. Curr Gene Ther 5(3):285–297

29. Choi VW, McCarty DM, Samulski RJ (2005) AAV hybrid serotypes: improved vectors for gene delivery. Curr Gene Ther 5(3):299–310

30. Liu X, Ramirez S, Pang PT et al (2012) Optogenetic stimulation of a hippocampal engram activates fear memory recall. Nature 484(7394):381–385

31. Jego S, Glasgow SD, Herrera CG et al (2013) Optogenetic identification of a rapid eye movement sleep modulatory circuit in the hypothalamus. Nat Neurosci 16(11):1637–1643

32. Fenno LE, Mattis J, Ramakrishnan C et al (2014) Targeting cells with single vectors using multiple-feature Boolean logic. Nat Methods 11(7):763–772

33. Rothwell PE, Hayton SJ, Sun GL et al (2015) Input- and output-specific regulation of serial order performance by corticostriatal circuits. Neuron 88(2):345–356

34. Deverman BE, Pravdo PL, Simpson BP et al (2016) Cre-dependent selection yields AAV variants for widespread gene transfer to the adult brain. Nat Biotechnol 34(2):204–209

35. Bi A, Cui J, Ma YP et al (2006) Ectopic expression of a microbial-type rhodopsin restores visual responses in mice with photoreceptor degeneration. Neuron 50(1):23–33

36. Zhang Y, Ivanova E, Bi A et al (2009) Ectopic expression of multiple microbial rhodopsins restores ON and OFF light responses in retinas with photoreceptor degeneration. J Neurosci 29(29):9186–9196

37. Tomita H, Sugano E, Isago H et al (2010) Channelrhodopsin-2 gene transduced into retinal ganglion cells restores functional vision in genetically blind rats. Exp Eye Res 90 (3):429–436

38. Iyer SM, Montgomery KL, Towne C et al (2014) Virally mediated optogenetic excitation and inhibition of pain in freely moving nontransgenic mice. Nat Biotechnol 32 (3):274–278

39. Boada MD, Martin TJ, Peters CM et al (2014) Fast-conducting mechanoreceptors contribute to withdrawal behavior in normal and nerve injured rats. Pain 155 (12):2646–2655

40. Hernandez VH, Gehrt A, Jing Z et al (2014) Optogenetic stimulation of the auditory nerve. J Vis Exp 92:e52069

41. Towne C, Montgomery KL, Iyer SM et al (2013) Optogenetic control of targeted peripheral axons in freely moving animals. PLoS One 8(8):e72691

42. Ambrosi CM, Entcheva E (2014) Optogenetic control of cardiomyocytes via viral delivery. Methods Mol Biol 1181:215–228

43. Vogt CC, Bruegmann T, Malan D et al (2015) Systemic gene transfer enables optogenetic pacing of mouse hearts. Cardiovasc Res 106(2):338–343

44. Nussinovitch U, Gepstein L (2015) Optogenetics for in vivo cardiac pacing and resynchronization therapies. Nat Biotechnol 33 (7):750–754

45. Bruegmann T, van Bremen T, Vogt CC et al (2015) Optogenetic control of contractile function in skeletal muscle. Nat Commun 6:7153

46. Gerits A, Farivar R, Rosen BR et al (2012) Optogenetically induced behavioral and functional network changes in primates. Curr Biol 22(18):1722–1726

47. Jazayeri M, Lindbloom-Brown Z, Horwitz GD (2012) Saccadic eye movements evoked by optogenetic activation of primate V1. Nat Neurosci 15(10):1368–1370

48. Cavanaugh J, Monosov IE, McAlonan K et al (2012) Optogenetic inactivation modifies monkey visuomotor behavior. Neuron 76 (5):901–907

49. Foust KD, Nurre E, Montgomery CL et al (2009) Intravascular AAV9 preferentially targets neonatal neurons and adult astrocytes. Nat Biotechnol 27(1):59–65

50. Maguire CA, Ramirez SH, Merkel SF et al (2014) Gene therapy for the nervous system: challenges and new strategies. Neurotherapeutics 11(4):817–839

51. Gray SJ, Matagne V, Bachaboina L et al (2011) Preclinical differences of intravascular AAV9 delivery to neurons and glia: a comparative study of adult mice and nonhuman primates. Mol Ther 19(6):1058–1069

52. Grimm D, Lee JS, Wang L et al (2008) In vitro and in vivo gene therapy vector evolution via multispecies interbreeding and retargeting of adeno-associated viruses. J Virol 82 (12):5887–5911

53. Seiler MP, Cerullo V, Lee B (2007) Immune response to helper dependent adenoviral mediated liver gene therapy: challenges and prospects. Curr Gene Ther 7 (5):297–305

54. Kremer EJ, Boutin S, Chillon M et al (2000) Canine adenovirus vectors: an alternative for adenovirus-mediated gene transfer. J Virol 74 (1):505–512

55. Kissa K, Mordelet E, Soudais C et al (2002) In vivo neuronal tracing with GFP-TTC gene delivery. Mol Cell Neurosci 20(4):627–637

56. Peltekian E, Garcia L, Danos O (2002) Neurotropism and retrograde axonal transport of a canine adenoviral vector: a tool for targeting key structures undergoing neurodegenerative processes. Mol Ther 5(1):25–32

57. Soudais C, Laplace-Builhe C, Kissa K et al (2001) Preferential transduction of neurons by canine adenovirus vectors and their efficient retrograde transport in vivo. FASEB J 15(12):2283–2285

58. Li Y, Hickey L, Perrins R et al (2016) Retrograde optogenetic characterization of the pontospinal module of the locus coeruleus with a canine adenoviral vector. Brain Res 1641(Pt B):274–290

59. Rajasethupathy P, Sankaran S, Marshel JH et al (2015) Projections from neocortex mediate top-down control of memory retrieval. Nature 526(7575):653–659

60. Lilley CE, Groutsi F, Han Z et al (2001) Multiple immediate-early gene-deficient herpes simplex virus vectors allowing efficient gene delivery to neurons in culture and widespread gene delivery to the central nervous system in vivo. J Virol 75(9):4343–4356

61. Neve RL (2012) Overview of gene delivery into cells using HSV-1-based vectors. Curr Protoc Neurosci Chapter 4:Unit 4 12

62. Zou M, De Koninck P, Neve RL et al (2014) Fast gene transfer into the adult zebrafish brain by herpes simplex virus 1 (HSV-1) and electroporation: methods and optogenetic applications. Front Neural Circuits 8:41

63. Fink DJ, DeLuca NA, Goins WF et al (1996) Gene transfer to neurons using herpes simplex virus-based vectors. Annu Rev Neurosci 19:265–287

64. Covington HE 3rd, Lobo MK, Maze I et al (2010) Antidepressant effect of optogenetic stimulation of the medial prefrontal cortex. J Neurosci 30(48):16082–16090

65. Lobo MK, Covington HE 3rd, Chaudhury D et al (2010) Cell type-specific loss of BDNF signaling mimics optogenetic control of cocaine reward. Science 330(6002):385–390

66. Lima SQ, Hromadka T, Znamenskiy P et al (2009) PINP: a new method of tagging neuronal populations for identification during in vivo electrophysiological recording. PLoS One 4(7):e6099

67. Taber KH, Strick PL, Hurley RA (2005) Rabies and the cerebellum: new methods for tracing circuits in the brain. J Neuropsychiatry Clin Neurosci 17(2):133–139

68. Callaway EM (2008) Transneuronal circuit tracing with neurotropic viruses. Curr Opin Neurobiol 18(6):617–623

69. Ugolini G (2011) Rabies virus as a transneuronal tracer of neuronal connections. Adv Virus Res 79:165–202

70. Bauer A, Nolden T, Schroter J et al (2014) Anterograde glycoprotein-dependent transport of newly generated rabies virus in dorsal root ganglion neurons. J Virol 88 (24):14172–14183

71. Tsiang H, Lycke E, Ceccaldi PE et al (1989) The anterograde transport of rabies virus in rat sensory dorsal root ganglia neurons. J Gen Virol 70(Pt 8):2075–2085

72. Lammel S, Lim BK, Ran C et al (2012) Input-specific control of reward and aversion in the ventral tegmental area. Nature 491 (7423):212–217

73. Callaway EM, Luo L (2015) Monosynaptic circuit tracing with glycoprotein-deleted rabies viruses. J Neurosci 35(24):8979–8985

74. Wickersham IR, Lyon DC, Barnard RJ et al (2007) Monosynaptic restriction of transsynaptic tracing from single, genetically targeted neurons. Neuron 53(5):639–647

75. Krashes MJ, Shah BP, Madara JC et al (2014) An excitatory paraventricular nucleus to AgRP neuron circuit that drives hunger. Nature 507(7491):238–242

76. Lerner TN, Shilyansky C, Davidson TJ et al (2015) Intact-brain analyses reveal distinct information carried by SNc dopamine subcircuits. Cell 162(3):635–647

77. Ogawa SK, Cohen JY, Hwang D et al (2014) Organization of monosynaptic inputs to the serotonin and dopamine neuromodulatory systems. Cell Rep 8(4):1105–1118

78. Watabe-Uchida M, Zhu L, Ogawa SK et al (2012) Whole-brain mapping of direct inputs to midbrain dopamine neurons. Neuron 74 (5):858–873

79. Osakada F, Callaway EM (2013) Design and generation of recombinant rabies virus vectors. Nat Protoc 8(8):1583–1601

80. Osakada F, Mori T, Cetin AH et al (2011) New rabies virus variants for monitoring and manipulating activity and gene expression in defined neural circuits. Neuron 71 (4):617–631

81. Beier KT, Steinberg EE, DeLoach KE et al (2015) Circuit architecture of VTA dopamine neurons revealed by systematic input-output mapping. Cell 162(3):622–634

82. Schwarz LA, Miyamichi K, Gao XJ et al (2015) Viral-genetic tracing of the input-output organization of a central noradrenaline circuit. Nature 524(7563):88–92

83. Kim EJ, Jacobs MW, Ito-Cole T et al (2016) Improved monosynaptic neural circuit tracing using engineered rabies virus glycoproteins. Cell Rep. doi:10.1016/j.celrep.2016.03.067

84. Reardon TR, Murray AJ, Turi GF et al (2016) Rabies virus CVS-N2c(DeltaG) strain enhances retrograde synaptic transfer and neuronal viability. Neuron 89(4):711–724

85. Dong B, Nakai H, Xiao W (2010) Characterization of genome integrity for oversized recombinant AAV vector. Mol Ther 18 (1):87–92

86. Dong JY, Fan PD, Frizzell RA (1996) Quantitative analysis of the packaging capacity of recombinant adeno-associated virus. Hum Gene Ther 7(17):2101–2112

87. Goshen I, Brodsky M, Prakash R et al (2011) Dynamics of retrieval strategies for remote memories. Cell 147(3):678–689

88. Zhu P, Narita Y, Bundschuh ST et al (2009) Optogenetic dissection of neuronal circuits in zebrafish using viral gene transfer and the Tet system. Front Neural Circuits 3:21

89. Miyashita T, Shao YR, Chung J et al (2013) Long-term channelrhodopsin-2 (ChR2) expression can induce abnormal axonal morphology and targeting in cerebral cortex. Front Neural Circuits 7:8

90. Matsushita T, Elliger S, Elliger C et al (1998) Adeno-associated virus vectors can be efficiently produced without helper virus. Gene Ther 5(7):938–945

91. Xiao X, Li J, Samulski RJ (1998) Production of high-titer recombinant adeno-associated virus vectors in the absence of helper adenovirus. J Virol 72(3):2224–2232

92. McClure C, Cole KL, Wulff P et al (2011) Production and titering of recombinant adeno-associated viral vectors. J Vis Exp 57: e3348

93. Burova E, Ioffe E (2005) Chromatographic purification of recombinant adenoviral and adeno-associated viral vectors: methods and implications. Gene Ther 12(Suppl 1):S5–17

94. Strobel B, Miller FD, Rist W et al (2015) Comparative analysis of cesium chloride- and iodixanol-based purification of recombinant adeno-associated viral vectors for preclinical applications. Hum Gene Ther Methods 26 (4):147–157

95. Towne C, Aebischer P (2009) Lentiviral and adeno-associated vector-based therapy for motor neuron disease through RNAi. Methods Mol Biol 555:87–108

96. Ayuso E, Mingozzi F, Montane J et al (2010) High AAV vector purity results in serotype- and tissue-independent enhancement of transduction efficiency. Gene Ther 17 (4):503–510

97. Wright JF (2008) Manufacturing and characterizing AAV-based vectors for use in clinical studies. Gene Ther 15(11):840–848

98. Grieger JC, Choi VW, Samulski RJ (2006) Production and characterization of adeno-associated viral vectors. Nat Protoc 1 (3):1412–1428

99. Tye KM, Deisseroth K (2012) Optogenetic investigation of neural circuits underlying brain disease in animal models. Nat Rev Neurosci 13(4):251–266

100. Buning H, Huber A, Zhang L et al (2015) Engineering the AAV capsid to optimize vector-host-interactions. Curr Opin Pharmacol 24:94–104

101. Hirsch ML, Wolf SJ, Samulski RJ (2016) Delivering transgenic DNA exceeding the carrying capacity of AAV vectors. Methods Mol Biol 1382:21–39

102. Maheshri N, Koerber JT, Kaspar BK et al (2006) Directed evolution of adeno-associated virus yields enhanced gene delivery vectors. Nat Biotechnol 24(2):198–204

103. Dalkara D, Byrne LC, Klimczak RR et al (2013) In vivo-directed evolution of a new adeno-associated virus for therapeutic outer retinal gene delivery from the vitreous. Sci Transl Med 5(189):189ra176

104. Tervo DG, Hwang BY, Viswanathan S et al (2016) A Designer AAV Variant Permits Efficient Retrograde Access to Projection Neurons. Neuron 92(2):372–382

Chapter 2

Cell Type-Specific Targeting Strategies for Optogenetics

Ofer Yizhar and Antoine Adamantidis

Abstract

Optogenetic techniques allow versatile, cell type-specific light-based control of cellular activity in diverse set of cells, circuits, and brain structures. Optogenetic actuators are genetically encoded light-sensitive membrane proteins that can be selectively introduced into cellular circuits in the living brain using a variety of genetic approaches. Gene targeting approaches used in optogenetic studies vary greatly in their specificity, their spatial coverage, the level of transgene expression and their potential adverse effects on neuronal cell health. Here, we describe the major gene targeting approaches utilized in optogenetics and provide a simple set of guidelines through which these approaches can be evaluated when designing an in vitro or in vivo optogenetic study.

Key words Optogenetics, Viral vector, Promoter, Enhancer, Gene expression, Targeting

1 Introduction

Optogenetic techniques utilize a wide range of light-sensitive proteins known from a wide variety of organisms. When heterologously expressed in cells of interest, these proteins are capable of producing light-evoked modulations of various physiological functions, ranging from changes in excitability [1–4] to activation or inhibition of distinct biochemical pathways [5–9], gene expression [10, 11] and enzymatic activity [12, 13]. Since the discovery of channelrhodopsin [14, 15] and the first application of this microbial opsin for activating neurons [1, 4], microbial opsins have been extensively used in experimental neuroscience applications including the functional mapping of neural circuit connectivity and dynamics in the brain and the dissection of neural circuits underlying integrated brain functions and behaviors [16, 17]. These studies all capitalize on the single-component nature of the optogenetic effectors, which allows the use of gene transfer technology for their introduction into post-mitotic neurons.

In optogenetic experiments, light is used to transiently and reversibly modulate the physiological properties of defined cells,

Albrecht Stroh (ed.), *Optogenetics: A Roadmap*, Neuromethods, vol. 133,
DOI 10.1007/978-1-4939-7417-7_2, © Springer Science+Business Media LLC 2018

typically in the context of an active neural circuit in vitro or in vivo. To assure that the desired physiological effect is achieved in such settings, several key factors should be taken into account, including:

1. *Specificity & selectivity of expression*: optogenetic tool expression should be restricted to the desired neuronal population, with minimal leak to nontargeted cell populations.

2. *Robustness of expression*: The optogenetic actuator should be expressed at sufficient levels to allow modulation with moderate light power, avoiding phototoxicity.

3. *Cytotoxicity and other adverse effects*: The method used to express the selected tool should be well-tolerated and nontoxic for the host cells over the entire duration of the experimental period (and ideally well beyond this time).

The gene targeting methods described below all differ in the degree to which they optimize each of these parameters. Further information about optogenetic technology can be found in many other excellent reviews [17–20].

2   Promoter-Based Specificity in Transgenic Expression of Optogenetic Tools

In optogenetic experiments, a precise assessment of the efficacy of targeting, i.e. the percentage of transduced cells among a genetically-, circuit-, or activity-defined population (see below and Chap. 8), provides an important estimate before in vitro or in vivo experimentation. The efficacy of gene expression is dictated by the genetic regulatory elements under which the optogenetic tools are expressed in the targeted cell population. Gene expression is regulated by a large array of noncoding DNA sequences that contain recognition sequences for binding of specific transcription factors, chromatin remodeling proteins and other regulatory elements [21]. Specific genomic promoters allow gene expression selectively within cell types that possess these regulatory proteins. Genomic promoter sequences can span between several hundred bases and several thousands of kilobases (kb). It is therefore often impossible to package complete promoter sequences into a viral vector backbone since viral vectors are limited in their genomic payload size [22, 23] (see also Chap. 1). Many of the commonly used optogenetic viral vectors utilize minimal promoter sequences (0.2–1.5 kb); these are truncated segments of much longer promoters, or repeated sequences of specific transcription factor recognition sites, and are sufficient for specifically targeting a defined population of neurons. Only a few promoters have been identified that can be truncated in this way while retaining sufficient cell type specificity (see [24–27]).

Lentiviral vectors, due to their larger payload size compared with AAVs, can carry larger minimal promoters and are therefore effective in some cases where a minimal promoter sequence exceeds the 1.5–2 kb size, but <3.5 kb [24, 28]. Notably, although minimal promoter sequences can allow specific expression of transgenes, some promoters produce very weak expression of the transgene, thereby limiting their utility in viral vector-based optogenetics since in most cases very strong expression is required to achieve effective light-based control over the targeted neurons. In such cases, alternative methods exist that include classical transgenic mouse engineering (see above). The use of transgenic (multiple random insertion) or knock-in (locus-specific single insertion) recombinant technologies allow one to target the expression of opsins to a specific class of cells from the early embryonic stages of a transgenic animal. For both these approaches, the size of the promoter is less limiting than for the viral targeting. Due to the locus-specific single insertion of the transgene, the knock-in approach is generally preferred for specificity and stability reasons. However, the low single-channel conductance of some optogenetic tools (e.g. channelrhodopsin, halorhodopsin, and archerhodopsin), a single copy of the gene encoding them is typically not sufficient for robust optogenetic modulation. Transgenic expression leads to multiple copies of the opsin gene and therefore permits higher photocurrent sizes in targeted neurons and therefore facilitates optogenetic modulation. The first transgenic mouse model with pan-neuronal expression of ChR2 under the Thy1 promoter was reported in 2010 [29]. Since then, several additional mouse lines were generated, expressing ChR2 in GABAergic neurons (VGAT-ChR2), cholinergic neurons (ChAT-ChR2) and others (see www.jax.org for a list of available and currently generated transgenes). However, there are several limitations inherent to these transgenic approaches. First, despite the use of long promoter sequences, the expression of the transgene can be too low for proper activation/inhibition of cell bodies, but more importantly for stimulation of distant synaptic terminals, which have been shown to require higher light power for efficient optogenetic stimulation [30]. Second, knock-in strategies may results in haplo-insufficiency (a reduction in overall expression of the gene into which the transgene has been targeted) and perturb the expression of the endogenous gene, leading to a molecular phenotype that can have significant synaptic, physiologic and behavioral consequences [31]. Third, transgenic strategies may result in the expression of the transgene in multiple neural circuits in the brain, which strongly hamper the selectivity of the optical manipulation. For instance, VGAT-ChR2 animals express ChR2 in all neurons expressing the VGAT gene (i.e., most of GABAergic cells). Although optical stimulation can be restricted to small brain nuclei, it will activate cell bodies, axons and fiber of passage as well as terminals in the vicinity of the tip of the optical fiber, which often would potentially decrease the specificity of the manipulation. To overcome these limitations, recombinase-based methods or multi-virus circuit/connectivity-based targeting can be used (see below).

3 Viral Vector-Mediated Expression of Optogenetic Tools

Viral vectors are the most popular means of delivering optogenetic tools to the adult brain. Viral vectors are in essence genetically engineered viruses in which a minimal set of viral genes has been retained to allow host cell entry, transport to the nucleus and expression of the transgene while eliminating virulence functions such as replication and cytotoxicity [32]. Lentiviral (LV) vectors [22] and adeno-associated viral (AAV) vectors [33] have both been widely utilized for introducing optogenetic transgenes into post-mitotic neurons [34]. There are several important differences between these vectors that should be considered when designing an optogenetic study. Recombinant AAV (rAAV) vectors are considered safer than LV since the currently available strains do not broadly integrate into the host genome but rather remain inside the nucleus as episomes [35]. AAV-based expression vectors display lower immunogenicity [36], and in many cases allow larger trans-duction volumes than LV [37]. Cell type specificity, the topic of the current chapter, can be achieved using both LV and AAV vectors using cell type-specific promoters [24–26], and both vector types support pseudotyping techniques, which in principle enable a wide range of cell type tropisms and transduction mechanisms [38, 39] (see Chap. 1).

3.1 Combined Promoter- and Recombinase-Based Specificity

When the required cell type-specific minimal promoter fragment cannot be packaged into the viral genome while retaining cell type specificity and adequate expression levels, one can utilize transgenic or knock-in mice expressing a recombinase (e.g., Cre, Flp) [40] under the genomic cell type-specific promoter for the same target population. Recombinase-dependent viral vectors allow specific expression only in cells that contain a specific recombinase protein. For example, a viral vector carrying a Cre-dependent expression cassette will be expressed only in inhibitory neurons of a mouse that expresses Cre under the control of the parvalbumin promoter, which is specific to a population of fast-spiking inhibitory neurons [41, 42]. This approach can be extended to mouse lines carrying the Flp recombinase, which recognizes sequence elements that are incompatible with the Cre recombinase. More complex "boolean logic" gates have been described, utilizing both Cre- and Flpo-dependent expression cassettes that allow Cre-on, Flp-on, Cre-off, Flp-off, and all combinations of dual-recombinase logic [34]. Apart from the obvious advantage of utilizing a large genomic promoter to generate Cre driver lines, the use of recombinase-dependent expression vectors allows the uncoupling of expression specificity from transgene levels in the targeted cells which result, for instance, in better membrane expression of ChR2 in axons allowing the activation of a subset of targeted cells by restricting the optical stimulus to the cell terminals rather than the soma (see below and Fig. 1). The recombinase-dependent approach is quite versatile and economical since a single Cre-driver strain allows: (1) the targeting of different circuits by restricting the virus injection to specific brain nuclei/circuits (e.g., basal forebrain or brainstem cholinergic neurons in ChAT-Cre animal); (2) utilization of many different types of Cre-dependent viral vectors encoding excitatory, inhibitory, and other optogenetic tools.

3.2 Target Volume Considerations

The size of the target brain structure is a major consideration in the design of an optogenetic study. Experiments targeting small nuclei such as the mouse amygdala [43, 44] require different targeting strategies than those involving larger brain regions such as the primate neocortex [45–48]. Therefore, each experimental design necessitates a proper adjustment of the viral delivery methods (e.g. glass pipette, metal needle, *convection-enhanced methods* [49]; see also Chap. 9 for viral delivery methods), and viral vector type to allow efficient transduction of the target region. Furthermore, the choice of optogenetic tool and illumination method is critical to assure that the desired effect is achieved in the targeted cells. The spatial distribution of viral vector particles strongly depends on the targeted brain region. Restricted transduction of smaller brain regions can be achieved by choosing the appropriate viral vector and injection volume. For example, LV and AAV2 injection results in expression patterns that are more localized compared with the pseudotyped rAAV2/1, rAAV2/5, rAAV2/8, or rAAV2/9 (see also Chap. 1 Table 2). AAV2 and LV are therefore well-suited for local expression in volumes smaller than 1 mm^3 [37]. Although viral titer can be reduced in order to decrease the size of the transduced volume, lower-titer injections are also likely to influence the number of genome copies in transduced cells, leading to lower expression levels of the transgene in individual cells within the target region [37]. Compared with AAV2 transduction, LV transduction is more spatially restricted when injected in vivo and thus, can be used to target smaller structures [43]. However, LV has been reported to exhibit a bias towards excitatory neurons in cortex [26], an effect which is likely also region-specific since other more specialized cell types have been successfully targeted with lentiviral vectors [24, 28, 50]. Although such control of viral transduction volume can be achieved with the choice of viral vector, transduction of larger volumes can simply be obtained by performing multiple injections covering a large area. This strategy is commonly used in primate studies, and has also been used successfully in the rodent brain [51].

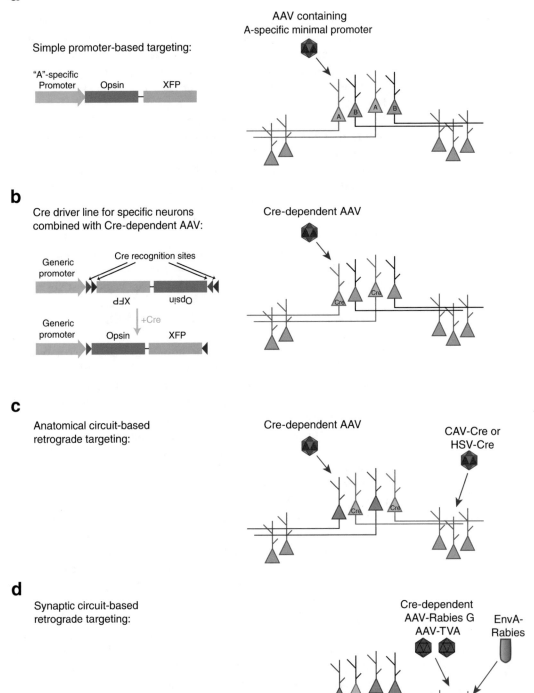

Fig. 1 Schematic of genetic targeting of neural circuits. (**a**) Simple promoter-based targeting relies on a single viral vector (AAV or lentiviral) carrying the opsin gene under the control of a cell type-specific minimal promoter fragment. Expression of the opsin in the soma and axonal efferents is guided by the activity of the

3.3 Circuit-Based Expression of Optogenetic Tools

Neural circuit dissection is one of the most widely used applications of optogenetic techniques. Optogenetic activation combined with electrophysiological recording allows functional anterograde circuit mapping [52, 53] (see also Chap. 8). Introduction of fluorescently tagged channelrhodopsins to the membrane of specific neuronal populations in a defined brain region allows visualization and subsequent photoactivation of long-range axonal connections throughout the brain. Simultaneous electrophysiological recording at the projection site allows the identification of specific post-synaptic components of the circuit both in vivo and in the acute brain slice preparation [44, 54, 55]. While it provides important information regarding the functional properties of specific antero-grade projections, it is hardly scalable due to the need to perform electrophysiological recordings at each target site. Circuit-based expression tools utilize neurotropic viruses for tagging neurons based on their connectivity pattern with identified neurons or macroscopic anatomical projection patterns.

Circuit-based expression methods can be divided to two types: those based on the anatomical location of presynaptic terminals (anatomical circuit-based targeting, Fig. 1c) and those that allow targeting of neurons based on specific synaptic connectivity (mono-synaptic circuit-based targeting; Fig. 1d). Anatomical circuit-based targeting can be achieved using a variety of viral vectors that are capable of transducing neurons through their axons or presynaptic terminals [56–59]. The herpex simplex virus (HSV) has been uti-lized in a variety of optogenetic experiments to label neurons projecting to a specific location in the brain, to attain optogenetic modulation of these cells [60, 61] or to "tag" these neurons using combined electrophysiological recording and photostimulation [57]. The type 2 canine adenovirus (CAV2; [58]) has been used for a similar purpose and is perhaps even more efficient than HSV in retrograde targeting. Although these two viruses are both

←_____

Fig. 1 (continued) promoter. (**b**) Cre-dependent expression using viral vectors. In a mouse that expresses Cre under a cell type-specific promoter, a Cre-dependent AAV is injected to the target circuit. The opsin and fluorescent protein ("XFP") genes are positioned in an inverted orientation with regard to the promoter. Cre-mediated recombination "flips" the opsin and fluorophore genes into the forward orientation, permitting the expression of the opsin-reporter fusion protein only in cells expressing Cre. (**c**) Anatomical circuit-based targeting is carried out by injecting a recombinase-expressing viral vector with the capacity to undergo retrograde transport through transduction of presynaptic nerve terminals. A Cre-dependent opsin-expressing viral vector (as in **b**) is injected at the site of the presynaptic cell bodies, where Cre is expressed only in neurons projecting to the site injected with the Cre-carrying vector. (**d**) Synaptic circuit-based targeting. In this approach, a pseudotyped rabies virus is injected following expression of two Cre-dependent expression vectors encoding the rabies glycoprotein (Rabies G) and the avian receptor TVA. The Env-A pseudotyped rabies virus can only transduce TVA-expressing cells, and requires the Rabies G protein to exit the cell and perform retrograde trans-synaptic transport to its presynaptic partners. Expression of the opsin from the rabies genome (*green*) occurs only in cells that provide monosynaptic input to the first-order neurons (*red*)

considered retrograde-labeling, some studies have mentioned that these two vectors transduce somewhat nonoverlapping neuronal populations [62] and that CAV2 might also be capable of transducing axons of passage [63]. It is therefore important to keep these differences in mind when designing and interpreting circuit-based expression experiments, and to conduct proper anatomical controls.

Monosynaptic circuit-based targeting (Fig. 1d) capitalizes on the exquisite capability of the rabies virus to transport its genetic material across synaptic contacts. This approach utilizes a glycoprotein-deleted variant of the rabies SAD B19 strain, SADΔG [59]. The rabies virus glycoprotein (G), which is embedded in the viral membrane, is required for trans-synaptic spread [64]. By introducing the glycoprotein gene in neurons prior to infection of the G-deleted mutant virus, the virus spreads to presynaptic neurons and is restricted from further spread due to the lack of this complementary glycoprotein in newly transduced neurons. This enables the dissection of direct connections originating from a population of defined neurons, or even from a single primary infected neuron [65, 66].

While this approach provides much more refined selectivity of retrogradely targeted neurons, it still lacks specificity due to the difficulty in targeting the rabies glycoprotein to the primary neurons. To achieve more refined specificity of the primary viral transduction event, the rabies vector can be directed to genetically defined post-synaptic neuronal subtypes by using the avian receptor TVA system. In this approach, the SADΔG rabies variant is pseudotyped with an envelope protein from the avian sarcoma and leukosis virus (ASLV). The avian-specific TVA receptor, which is required for infection by the pseudotyped rabies virus, is then expressed in the cells to be targeted for infection by SADΔG along with the rabies glycoprotein. This allows the virus to spread trans-synaptically from only the TVA-expressing cells to their presynaptic partners. The TVA receptor, along with the rabies G protein gene, can be delivered using AAV to specific neurons using the double-floxed Cre-based expression system [54, 67]. Under this configuration, only Cre-expressing cells will express the proteins required for both uptake of the pseudotyped rabies virus and monosynaptic retrograde transport. The advantages of using rabies-based circuit tracing techniques are its efficient unidirectional retrograde transport and its rapid onset of expression. Unfortunately, the time course of survival of SAD B19-transduced neurons is limited to approximately 2 weeks [59], suggesting that other systems might be required for experiments requiring long-term survival of transduced neurons.

3.4 Activity-Based Tagging and Optogenetic Control

Cell type specificity, the topic of this chapter and one of the key advantages of optogenetics over more classical methods of experimental manipulation in neural circuits, relies on the inherent assumption that cells with distinct gene expression properties perform defined function in neural circuits. Yet, from a systems neuroscience perspective, this assumption is inherently flawed since neuronal ensembles can form purely from processes of synaptic activity and in a way that is at least partially independent of genetic "identity". Can we therefore target ensembles of neurons based on their "assembly/ensemble activity" in a particular behavioral paradigm or neural representation of environmental stimuli? For decades, immediate early genes have been used to represent reliable marker of cell activity, though their expression varies between and within cell population. These genes thus allow the identification of neurons that have been active over a short period of time (minutes to hours) and provided functional tagging of such activity-modulated gene promoters for activity-based expression of optogenetic actuators.

The promoter for c-Fos, an activity-dependent immediate early gene, can be used in combination with the rapidly inducible TRE-ttA expression system in order to achieve expression of ChR2 in hippocampal neurons activated during aversive learning [68] and appetitive experience [69]. In this experimental configuration, the tetracycline transactivator (tTA) is expressed under the control of the c-Fos promoter in a transgenic mouse. An adeno-associated viral vector is then introduced into the hippocampus, expressing ChR2 under the control of the tet-response element (TRE), which under this configuration allows expression of ChR2 in the presence of tTA and in the absence of doxycycline [68]. This system allows selective activation of memory engrams in various paradigms [69], but it requires the constant administration of doxycycline except during the experiment in which the cells are to be labeled and is therefore referred to as "tet-OFF". The tet-ON system, while potentially easier to apply since doxycycline should only be administered during the labeling experiment, is more "leaky" and is therefore potentially less useful for such experiments as they rely on specific optogenetic modulation of cells activated during a strictly defined time window. An alternative approach has been developed which uses the expression of an inducible form of the Cre recombinase (Cre^{ERT2}), expressed transgenically under the control of the promoter for the immediate-early gene Arc [70].

A similar approach, utilizing either the c-Fos or Arc promoters with the Cre^{ERT2} transgene, has been used to generate the Fos-TRAP and Arc-TRAP mice, a general resource for targeting expression of any transgene to recently activated ("TRAPped") neurons [71]. Finally, work from the Bito lab has led to the development of the E-SARE vectors [72], which utilize tandem repeats of the Arc enhancer sequence to generate a viral vector that expresses an activity-dependent form of the Cre^{ERT2} protein. When injected into the brain this viral vector expresses Cre^{ERT2} in a manner that allows activity-dependent expression of Cre-dependent transgenes. The E-SARE approach has the advantage of allowing activity-dependent modulation in nontransgenic animals, but does not allow the brain-wide screening possible with the Fos/Arc-TRAP mice. Importantly, all of these approaches rely on activity-dependent transcription of immediate-early genes, but they vary with regard to the temporal integration time of activity-dependent expression, in its efficacy and in the level of baseline expression levels. Experiments utilizing these constructs for expression of optogenetic tools should be preceded by detailed characterization of these parameters in the particular cell type, region and behavioral paradigm used.

4   Light Delivery in the Animal Brain

After the proper targeting of optogenetic tools to neural circuit(s) of interest, the next step consists of designing an optical neural interface for in vitro/vivo light delivery into (deep) brain structures or brain slices, respectively. In vitro whole cell recordings are frequently used to verify the biological functions of the opsins, as well as a first step towards deconstruction of neural circuits. In this case, a light source can be coupled to the objective of the microscope and controlled by integrated TTL generator/electrophysiology stimulator. In the case of a two-photon microscope set up, a laser beam of small diameter (few dozens of microns) can be focused on smaller targets for high-precision synaptic physiology [52, 73–75]. Sculpted light holds great promises in shaping 3D light stimulation in in vivo preparation [76–79].

Similar to deep brain electrical stimulation that uses metal electrodes to nonselectively activate cells in brain structure (e.g., self-electrical stimulation paradigms, Parkinson disease), optogenetic configuration requires optical fibers to deliver sufficient light to shallow or deep brain targets (see Chap. 13 for methods estimating light distribution in tissue) (Fig. 2a).

4.1 Optical Fiber-Based Light Delivery System to Deep Brain Structures

Typical light sources include high-power diode pumped solid-state lasers (DPSSLs) or light-emitted diodes (LEDs) that are controlled by a waveform generator and commercially available optical shutters. Glass or plastic optical fibers are used for connecting light sources to in vivo preparation (see below). Noninvasive optical fibers or light-emitting diodes can be used with fiber implants or cranial windows for optical stimulation of neuronal networks located in the superficial layers of the cortex or in deep targets,

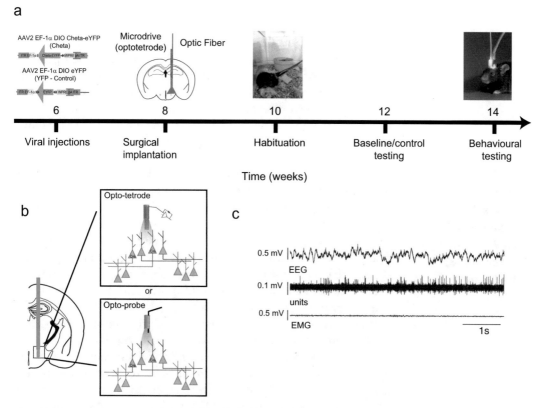

Fig. 2 Representative timeline for opto-tetrodes recording experiment. (**a**) Mice are stereotactically injected with ChETA-eYFP, ArchT-eYFP, or control AAV at 6-weeks old before chronic implantation of optical fiber implants or opto-tetrodes, EEG, LFPs, and EMG (shown in **b**). Schematic of the Cre-inducible AAV vector backbone is shown. Fifteen day after injection, mice will be habituated to recording conditions before baseline recording and actual stimulation/recording experiments at 12-weeks old. Novel Object Recognition Task and Fear conditioning will be conducted 2 and 4 weeks later, respectively. (**c**) Illustration of recording sheme for opto-tetrode (*up*) and -probe (*bottom*) recording in freely moving rodents. (**d**) Example traces of cortical EEG (*top*), multiunits (*middle*), and EMG (*bottom*) recordings in the LH area in freely moving mouse

respectively [80]. Deeper targets may require the use of optical fiber implants [81] (see also Chap. 13) that are may be also chronically implanted and connected to an optical tether for longitudinal experimental strategies. In vivo optogenetic studies have used a variety of multimode fibers that have larger core-size than single-mode fibers and thus, higher numerical apertures and increased "light-gathering" capacity. Light pulses propagate down the fiber-optic by total internal reflection, reaching the fiber tip with minimal power loss. However, it is important to note that the desired experimental application will determine the number of fibers, their shape, length and diameter. Additionally, it is necessary to render the optical fibers opaque (using dark coating or black furcation tubing), since even a small amount of light diffraction through the optical tether can cause sensory stimuli during behavioral testing, particularly in dark environments.

Factors to consider when delivering light to brain structures include (1) the use of appropriate wavelength to activate opsin; (2) the necessity of using one or several optical fiber implants to optimize light delivery to the entire target area (e.g., unilateral vs. bilateral, multisite fibers); (3) sufficient light power through the use of either high power lasers or LEDs; (4) the use of optical swivel to allow free movement of the tethered animals.

4.2 Emerging Techniques: Noninvasive Optical Stimulation (Red-Shifted Opsins, Step-Function Opsins, Nanoparticles)

Alternative to the invasive use of optical fiber implant include red-shifted opsins, step-function opsins and nanoparticles. The blue light wavelength used for activation of channelrhodopsin is both highly absorbed and scattered when it penetrates through the brain tissue, leading to an exponential decay in light power density with distance from the tip of the optical fiber or light source [17, 82] (Chap. 13). This is minimized by the use of red and far-red wavelength and red-shifted channelrhodopsins. Although these wavelengths are more prone to generate heat (that must be assessed with proper control conditions), they allow optical control of cortical circuits through thin/polished skull preparations [83, 84].

Furthermore, the use of channelrhodopsin requires constant optical stimulation at different frequencies to elicit action potentials. An extraordinary alternative consists in using step-function opsins (SFO)—turn ON and OFF with single pulse of blue and yellow light, respectively. Not only do SFOs avoid imposing hypersynchrony and nonnatural firing of action potentials, they have been successfully used to turn ON and OFF cortical circuits remotely, through thinned skull, instead of invasive optical fiber implants [17, 81]. A potential limitation, however, is the need to carefully titrate the amount of light delivered to activate these opsins, since excessive depolarization by these opsins could potentially lead to depolarization block and effective silencing of the targeted neurons. An alternative approach to avoiding hypersynchrony is to use sinusoidal optical stimulus instead of pulses, gradually increasing light intensity. Such an approach would capitalize on the heterogeneity of neuronal intrinsic properties and opsin expression levels and lead to more heterogeneous spike times across the network.

5   Recording Light Evoked Neuronal Activity

A major strength of the optogenetics technology is its compatibility with fast in vitro/vivo electrophysiological/optical/chemical read-out methods [85–87] (see also Chaps. 8, 9, and 11) (Fig. 2b, c). Indeed, if an electrophysiological/optical/chemical probe (tetrodes, glass pipette, dialysis probe, fiber photometry, etc.) is implanted in the vicinity of the targeted neurons (Fig. 2), it offers a direct confirmation of optical modulation. First, it allows one to verify that optogenetic that the optogenetic manipulations work as intended. Although the causality of light-evoked neuronal activity is somewhat debatable, it remains a relatively important verifying step in an experimental procedure. Second, it can be used to characterize the spontaneous activity of these cells during specific behaviors. Defining the precise pattern of firing of opto-tagged cells is particularly important when conducting an optogenetic experiment since it allows one to define optical stimulation parameters that remain within a physiological range for the particular neuronal population targeted. Third, if the probe is implanted in the vicinity of the terminals, rather than the soma, of targeted neurons, functional circuit mapping experiments can be conducted in vivo to reveal a direct and temporally precise readout of circuit modulation before, during and after optogenetic manipulations.

To obtain deeper mechanistic insight into the functional properties of neurons within the circuit, their connectivity and the effects of optogenetic modulation on local-circuit elements, in vitro assays of opsin function have been used to study neurotransmitter release from a variety of cell types, including dopaminergic, cholinergic, noradrenergic, hypocretins/orexins neurons, and MCH (review in [88]).

This rapid feedback from in vivo electrophysiology or electrochemical detection is valuable for measuring the light-evoked response of neuronal activation/silencing in an in vivo preparation, besides the use of classical immunohistochemistry and in vitro electrophysiological procedures. Importantly, it allows the fine-tuning of optical stimulation parameters for efficient control of circuit activity in a physiologically relevant range.

6 Outlook

In this chapter, we focused on the use of cell type-specific targeting strategies for optogenetic modulation of neural activity. Over the last decade, the advent of multiple genetically encoded tools for imaging and manipulation of neuronal activity have expanded the repertoire of techniques for studying the cellular substrates of brain functions and the mechanisms underlying innate, acquired, and pathological behaviors. The field of optogenetics had developed rapidly since its first in vivo application [24, 89]. Current progress in protein engineering is expected to lead to the discovery of novel light-sensitive membrane, cytoplasmic or nuclear proteins for remote control of selective ionic flow, cellular signaling, and gene regulation, while the concurrent development of genetic targeting strategies will allow refined selection of circuit elements for manipulation. The growing application of optogenetic techniques holds great promises for improving our understanding of mammalian brain functions, and identification of novel therapeutical strategies.

7 Notes

- Many gene targeting approaches are available for delivering optogenetic tools to defined neuron populations. Perhaps the most important aspect of gene targeting is specificity, which should be evaluated using immunohistochemistry in the targeted circuit. Robust expression is required for efficient optogenetic modulation, and this is best tested with electrophysiological recording in vivo or ex vivo.

- Viral vector tropism (see Chap. 1) can pose challenges in targeting unique neuron populations, and different serotypes should be tested for optimal expression in a new neuronal population (or organism) with unknown serotype preference.

  A good rule of thumb when examining brain sections expressing viral vectors is that if the fluorescence of the opsin-attached fluorophore (GFP/YFP/mCherry) is visible under a standard fluorescence microscope without the need for antibody staining, the expression levels should be high enough for efficient modulation.

Acknowledgments

We thank all members of the Yizhar and Tidis labs for their helpful comments on the manuscript. We thank K. Deisseroth and L. de Lecea for mentoring our early steps in optogenetics. O.Y. is supported by grants from the Israel Science Foundation (#1351/12), the European Research Commission (ERC starting grant #337637 and Marie Curie Actions grant #321919) and the Israeli Ministry of Science, and Technology and Space (grant #10373). A.A. is supported by the Human Frontier Science Program (RGY0076/2012), the Swiss National Science Foundation (grant #156156), the Inselspital, the University of Bern and the European Research Commission (ERC Consolidator grant).

References

1. Li X, Gutierrez DV, Hanson MG, Han J, Mark MD, Chiel H, Hegemann P, Landmesser LT, Herlitze S (2005) Fast noninvasive activation and inhibition of neural and network activity by vertebrate rhodopsin and green algae channelrhodopsin. Proc Natl Acad Sci U S A 102:17816–17821

2. Zhang F, Wang L-P, Brauner M, Liewald JF, Kay K, Watzke N, Wood PG, Bamberg E, Nagel G, Gottschalk A, Deisseroth K (2007) Multimodal fast optical interrogation of neural circuitry. Nature 446:633–639

3. Chow BY, Han X, Dobry AS, Qian X, Chuong AS, Li M, Henninger MA, Belfort GM, Lin Y, Monahan PE, Boyden ES (2010) High-

performance genetically targetable optical neural silencing by light-driven proton pumps. Nature 463:98–102

4. Boyden ES, Zhang F, Bamberg E, Nagel G, Deisseroth K (2005) Millisecond-timescale, genetically targeted optical control of neural activity. Nat Neurosci 8:1263–1268

5. Airan RD, Thompson KR, Fenno LE, Bernstein H, Deisseroth K (2009) Temporally precise in vivo control of intracellular signalling. Nature 458:1025–1029

6. Siuda ER, McCall JG, Al-Hasani R, Shin G, Park Il S, Schmidt MJ, Anderson SL, Planer WJ, Rogers JA, Bruchas MR (2015) Optodynamic simulation of β-adrenergic receptor signalling. Nat Commun 6:8480

7. Siuda ER, Copits BA, Schmidt MJ, Baird MA, Al-Hasani R, Planer WJ, Funderburk SC, McCall JG, Gereau RW, Bruchas MR (2015) Spatiotemporal control of opioid signaling and behavior. Neuron 86:923–935

8. Masseck OA, Spoida K, Dalkara D, Maejima T, Rubelowski JM, Wallhorn L, Deneris ES, Herlitze S (2014) Vertebrate cone opsins enable sustained and highly sensitive rapid control of Gi/o signaling in anxiety circuitry. Neuron 81:1263–1273

9. Grusch M, Schelch K, Riedler R, Reichhart E, Differ C, Berger W, Inglés-Prieto Á, Janovjak H (2014) Spatio-temporally precise activation of engineered receptor tyrosine kinases by light. EMBO J 33:1713–1726

10. Konermann S, Brigham MD, Trevino AE, Hsu PD, Heidenreich M, Cong L, Platt RJ, Scott DA, Church GM, Zhang F (2013) Optical control of mammalian endogenous transcription and epigenetic states. Nature 500:472–476

11. Wang X, Chen X, Yang Y (2012) Spatiotemporal control of gene expression by a light-switchable transgene system. Nat Methods 9:266–269

12. Taslimi A, Zoltowski B, Miranda JG, Pathak GP, Hughes RM, Tucker CL (2016) Optimized second-generation CRY2-CIB dimerizers and photoactivatable Cre recombinase. Nat Chem Biol 12:425–430

13. Zhou XX, Chung HK, Lam AJ, Lin MZ (2012) Optical control of protein activity by fluorescent protein domains. Science 338:810–814

14. Nagel G, Ollig D, Fuhrmann M, Kateriya S, Musti AM, Bamberg E, Hegemann P (2002) Channelrhodopsin-1: a light-gated proton channel in green algae. Science 296:2395–2398

15. Nagel G, Szellas T, Huhn W, Kateriya S, Adeishvili N, Berthold P, Ollig D, Hegemann P, Bamberg E (2003) Channelrhodopsin-2, a directly light-gated cation-selective membrane channel. Proc Natl Acad Sci U S A 100:13940–13945

16. Zhang F, Vierock J, Yizhar O, Fenno LE, Tsunoda S, Kianianmomeni A, Prigge M, Berndt A, Cushman J, Polle J, Magnuson J, Hegemann P, Deisseroth K (2011) The microbial opsin family of optogenetic tools. Cell 147:1446–1457

17. Yizhar O, Fenno LE, Davidson TJ, Mogri M, Deisseroth K (2011) Optogenetics in neural systems. Neuron 71:9–34

18. Miesenböck G (2009) The optogenetic catechism. Science 326:395–399

19. Zhang F, Wang LP, Boyden ES, Deisseroth K (2006) Channelrhodopsin-2 and optical control of excitable cells. Nat Methods 3:785–792

20. Scanziani M, Häusser M (2009) Electrophysiology in the age of light. Nature 461:930–939

21. Lemon B, Tjian R (2000) Orchestrated response: a symphony of transcription factors for gene control. Genes Dev 14:2551–2569

22. Naldini L, Blömer U, Gallay P, Ory D, Mulligan R, Gage FH, Verma IM, Trono D (1996) In vivo gene delivery and stable transduction of nondividing cells by a lentiviral vector. Science 272:263–267

23. Wu Z, Yang H, Colosi P (2010) Effect of genome size on AAV vector packaging. Mol Ther 18:80–86

24. Adamantidis AR, Zhang F, Aravanis AM, Deisseroth K, de Lecea L (2007) Neural substrates of awakening probed with optogenetic control of hypocretin neurons. Nature 450:420–424

25. Dittgen T, Nimmerjahn A, Komai S, Licznerski P, Waters J, Margrie TW, Helmchen F, Denk W, Brecht M, Osten P (2004) Lentivirus-based genetic manipulations of cortical neurons and their optical and electrophysiological monitoring in vivo. Proc Natl Acad Sci U S A 101:18206–18211

26. Nathanson JL, Yanagawa Y, Obata K, Callaway EM (2009) Preferential labeling of inhibitory and excitatory cortical neurons by endogenous tropism of adeno-associated virus and lentivirus vectors. Neuroscience 161:441–450

27. Rahim AA, Wong AMS, Howe SJ, Buckley SMK, Acosta-Saltos AD, Elston KE, Ward NJ, Philpott NJ, Cooper JD, Anderson PN, Waddington SN, Thrasher AJ, Raivich G (2009) Efficient gene delivery to the adult and fetal CNS using pseudotyped non-integrating lentiviral vectors. Gene Ther 16:509–520

28. Jasnow AM, Rainnie DG, Maguschak KA, Chhatwal JP, Ressler KJ (2009) Construction of cell-type specific promoter lentiviruses for

optically guiding electrophysiological recordings and for targeted gene delivery. Methods Mol Biol 515:199–213

29. Thyagarajan S, van Wyk M, Lehmann K, Löwel S, Feng G, Wässle H (2010) Visual function in mice with photoreceptor degeneration and transgenic expression of channelrhodopsin 2 in ganglion cells. J Neurosci 30:8745–8758

30. Jackman SL, Beneduce BM, Drew IR, Regehr WG (2014) Achieving high-frequency optical control of synaptic transmission. J Neurosci 34:7704–7714

31. Kolisnyk B, Guzman MS, Raulic S, Fan J, Magalhães AC, Feng G, Gros R, Prado VF, Prado MA (2013) ChAT-ChR2-EYFP mice have enhanced motor endurance but show deficits in attention and several additional cognitive domains. J Neurosci 33:10427–10438

32. Verma IM, Weitzman MD (2005) Gene therapy: twenty-first century medicine. Annu Rev Biochem 74:711–738

33. Walsh CE, Liu JM, Xiao X, Young NS, Nienhuis AW, Samulski RJ (1992) Regulated high level expression of a human gamma-globin gene introduced into erythroid cells by an adeno-associated virus vector. Proc Natl Acad Sci U S A 89:7257–7261

34. Fenno LE, Mattis J, Ramakrishnan C, Hyun M, Lee SY, He M, Tucciarone J, Selimbeyoglu A, Berndt A, Grosenick L, Zalocusky KA, Bernstein H, Swanson H, Perry C, Diester I, Boyce FM, Bass CE, Neve R, Huang ZJ, Deisseroth K (2014) Targeting cells with single vectors using multiple-feature Boolean logic. Nat Methods 11:763–772

35. Duan D, Yan Z, Yue Y, Engelhardt JF (1999) Structural analysis of adeno-associated virus transduction circular intermediates. Virology 261:8–14

36. Toromanoff A, Adjali O, Larcher T, Hill M, Guigand L, Chenuaud P, Deschamps J-Y, Gauthier O, Blancho G, Vanhove B, Rolling F, Chérel Y, Moullier P, Anegon I, Le Guiner C (2010) Lack of immunotoxicity after regional intravenous (RI) delivery of rAAV to nonhuman primate skeletal muscle. Mol Ther 18:151–160

37. Burger C, Gorbatyuk OS, Velardo MJ, Peden CS, Williams P, Zolotukhin S, Reier PJ, Mandel RJ, Muzyczka N (2004) Recombinant AAV viral vectors pseudotyped with viral capsids from serotypes 1, 2, and 5 display differential efficiency and cell tropism after delivery to different regions of the central nervous system. Mol Ther 10:302–317

38. Kato S, Kobayashi K, Inoue K-I, Kuramochi M, Okada T, Yaginuma H, Morimoto K, Shimada T, Takada M, Kobayashi K (2011) A lentiviral strategy for highly efficient retrograde gene transfer by pseudotyping with fusion envelope glycoprotein. Hum Gene Ther 22:197–206

39. Auricchio A, Kobinger G, Anand V, Hildinger M, O'Connor E, Maguire AM, Wilson JM, Bennett J (2001) Exchange of surface proteins impacts on viral vector cellular specificity and transduction characteristics: the retina as a model. Hum Mol Genet 10:3075–3081

40. Taniguchi H, He M, Wu P, Kim S, Paik R, Sugino K, Kvitsani D, Fu Y, Lu J, Lin Y, Miyoshi G, Shima Y, Fishell G, Nelson SB, Huang ZJ (2011) NeuroResource. Neuron 71:995–1013

41. Cardin JA, Carlén M, Meletis K, Knoblich U, Zhang F, Deisseroth K, Tsai L-H, Moore CI (2009) Driving fast-spiking cells induces gamma rhythm and controls sensory responses. Nature 459:663–667

42. Sohal VS, Zhang F, Yizhar O, Deisseroth K (2009) Parvalbumin neurons and gamma rhythms enhance cortical circuit performance. Nature 459:698–702

43. Do-Monte FH, Manzano-Nieves G, Quiñones-Laracuente K, Ramos-Medina L, Quirk GJ (2015) Revisiting the role of infralimbic cortex in fear extinction with optogenetics. J Neurosci 35:3607–3615

44. Tye KM, Prakash R, Kim S-Y, Fenno LE, Grosenick L, Zarabi H, Thompson KR, Gradinaru V, Ramakrishnan C, Deisseroth K (2011) Amygdala circuitry mediating reversible and bidirectional control of anxiety. Nature 471:358–362

45. Cavanaugh J, Monosov IE, McAlonan K, Berman R, Smith MK, Cao V, Wang KH, Boyden ES, Wurtz RH (2012) Optogenetic inactivation modifies monkey visuomotor behavior. Neuron 76:901–907

46. Diester I, Kaufman MT, Mogri M, Pashaie R, Goo W, Yizhar O, Ramakrishnan C, Deisseroth K, Shenoy KV (2011) An optogenetic toolbox designed for primates. Nat Neurosci 14:387–397

47. Ohayon S, Grimaldi P, Schweers N, Tsao DY (2013) Saccade modulation by optical and electrical stimulation in the macaque frontal eye field. J Neurosci 33:16684–16697

48. Gerits A, Farivar R, Rosen BR, Wald LL, Boyden ES, Vanduffel W (2012) Optogenetically induced behavioral and functional network changes in primates. Curr Biol 22:1722–1726

49. Hadaczek P, Kohutnicka M, Krauze MT, Bringas J, Pivirotto P, Cunningham J, Bankiewicz K (2006) Convection-enhanced delivery of adeno-associated virus type 2 (AAV2) into

the striatum and transport of AAV2 within monkey brain. Hum Gene Ther 17:291–302

50. Benzekhroufa K, Liu B, Tang F, Teschemacher AG, Kasparov S (2009) Adenoviral vectors for highly selective gene expression in central serotonergic neurons reveal quantal characteristics of serotonin release in the rat brain. BMC Biotechnol 9:23

51. Goshen I, Brodsky M, Prakash R, Wallace J, Gradinaru V, Ramakrishnan C, Deisseroth K (2011) Dynamics of retrieval strategies for remote memories. Cell 147:678–689

52. Petreanu L, Huber D, Sobczyk A, Svoboda K (2007) Channelrhodopsin-2-assisted circuit mapping of long-range callosal projections. Nat Neurosci 10:663–668

53. Arenkiel BR, Klein ME, Davison IG, Katz LC, Ehlers MD (2008) Genetic control of neuronal activity in mice conditionally expressing TRPV1. Nat Methods 5:299–302

54. Lammel S, Lim BK, Ran C, Huang KW, Betley MJ, Tye KM, Deisseroth K, Malenka RC (2012) Input-specific control of reward and aversion in the ventral tegmental area. Nature 491:212–217

55. Scott N, Prigge M, Yizhar O, Kimchi T (2015) A sexually dimorphic hypothalamic circuit controls maternal care and oxytocin secretion. Nature 525:519–522

56. Fink DJ, Glorioso JC (1997) Engineering herpes simplex virus vectors for gene transfer to neurons. Nat Med 3:357–359

57. Lima SQ, Hromádka T, Znamenskiy P, Zador AM (2009) PINP: a new method of tagging neuronal populations for identification during in vivo electrophysiological recording. PLoS One 4:e6099

58. Soudais C, Laplace-Builhe C, Kissa K, Kremer EJ (2001) Preferential transduction of neurons by canine adenovirus vectors and their efficient retrograde transport in vivo. FASEB J 15:2283–2285

59. Wickersham IR, Lyon DC, Barnard RJO, Mori T, Finke S, Conzelmann K-K, Young JAT, Callaway EM (2007) Monosynaptic restriction of transsynaptic tracing from single, genetically targeted neurons. Neuron 53:639–647

60. Xiong Q, Znamenskiy P, Zador AM (2015) Selective corticostriatal plasticity during acquisition of an auditory discrimination task. Nature 521:348–351

61. Znamenskiy P, Zador AM (2013) Corticostriatal neurons in auditory cortex drive decisions during auditory discrimination. Nature 497:482–485

62. Senn V, Wolff SBE, Herry C, Grenier F, Ehrlich I, Gründemann J, Fadok JP, Müller C,

Letzkus JJ, Lüthi A (2014) Long-range connectivity defines behavioral specificity of amygdala neurons. Neuron 81:428–437

63. Schwarz LA, Miyamichi K, Gao XJ, Beier KT, Weissbourd B, DeLoach KE, Ren J, Ibanes S, Malenka RC, Kremer EJ, Luo L (2015) Viral-genetic tracing of the input-output organization of a central noradrenaline circuit. Nature 524:88–92

64. Etessami R, Conzelmann KK, Fadai-Ghotbi B, Natelson B, Tsiang H, Ceccaldi PE (2000) Spread and pathogenic characteristics of a G-deficient rabies virus recombinant: an in vitro and in vivo study. J Gen Virol 81:2147–2153

65. Marshel JH, Mori T, Nielsen KJ, Callaway EM (2010) Targeting single neuronal networks for gene expression and cell labeling in vivo. Neuron 67:562–574

66. Wertz A, Trenholm S, Yonehara K, Hillier D, Raics Z, Leinweber M, Szalay G, Ghanem A, Keller G, Rózsa B, Conzelmann K-K, Roska B (2015) PRESYNAPTIC NETWORKS. Single-cell-initiated monosynaptic tracing reveals layer-specific cortical network modules. Science 349:70–74

67. Pollak Dorocic I, Fürth D, Xuan Y, Johansson Y, Pozzi L, Silberberg G, Carlén M, Meletis K (2014) A whole-brain atlas of inputs to serotonergic neurons of the dorsal and median raphe nuclei. Neuron 83:663–678

68. Liu X, Ramirez S, Pang PT, Puryear CB, Govindarajan A, Deisseroth K, Tonegawa S (2012) Optogenetic stimulation of a hippocampal engram activates fear memory recall. Nature 484:381–385

69. Ramirez S, Liu X, Lin P-A, Suh J, Pignatelli M, Redondo RL, Ryan TJ, Tonegawa S (2013) Creating a false memory in the hippocampus. Science 341:387–391

70. Root CM, Denny CA, Hen R, Axel R (2014) The participation of cortical amygdala in innate, odour-driven behaviour. Nature 515:269–273

71. Guenthner CJ, Miyamichi K, Yang HH, Heller HC, Luo L (2013) Permanent genetic access to transiently active neurons via TRAP: targeted recombination in active populations. Neuron 78:773–784

72. Kawashima T, Kitamura K, Suzuki K, Nonaka M, Kamijo S, Takemoto-Kimura S, Kano M, Okuno H, Ohki K, Bito H (2013) Functional labeling of neurons and their projections using the synthetic activity-dependent promoter E-SARE. Nat Methods 10:889–895

73. Atasoy D, Betley JN, Su HH, Sternson SM (2012) Deconstruction of a neural circuit for hunger. Nature 488:172–177

74. Prakash R, Yizhar O, Grewe B, Ramakrishnan C, Wang N, Goshen I, Packer AM, Peterka DS, Yuste R, Schnitzer MJ, Deisseroth K (2012) Two-photon optogenetic toolbox for fast inhibition, excitation and bistable modulation. Nat Methods 9:1171–1179

75. Packer AM, Peterka DS, Hirtz JJ, Prakash R, Deisseroth K, Yuste R (2012) Two-photon optogenetics of dendritic spines and neural circuits. Nat Methods 9:1202–1205

76. Favre-Bulle IA, Preece D, Nieminen TA, Heap LA, Scott EK, Rubinsztein-Dunlop H (2015) Scattering of sculpted light in intact brain tissue, with implications for optogenetics. Sci Rep 5:11501

77. Vaziri A, Emiliani V (2012) Reshaping the optical dimension in optogenetics. Curr Opin Neurobiol 22:128–137

78. Schrödel T, Prevedel R, Aumayr K, Zimmer M, Vaziri A (2013) Brain-wide 3D imaging of neuronal activity in Caenorhabditis Elegans with sculpted light. Nat Methods 10:1013–1020

79. Baker CA, Elyada YM, Parra A, Bolton MM (2016) Cellular resolution circuit mapping with temporal-focused excitation of soma-targeted channelrhodopsin. elife 5:11981

80. Yizhar O, Fenno LE, Prigge M, Schneider F, Davidson TJ, O'Shea DJ, Sohal VS, Goshen I, Finkelstein J, Paz JT, Stehfest K, Fudim R, Ramakrishnan C, Huguenard JR, Hegemann P, Deisseroth K (2011) Neocortical excitation/inhibition balance in information processing and social dysfunction. Nature 477:171–178

81. Sparta DR, Stamatakis AM, Phillips JL, Hovelsø N, van Zessen R, Stuber GD (2012) Construction of implantable optical fibers for long-term optogenetic manipulation of neural circuits. Nat Methods 7:12–23

82. Zhang F, Gradinaru V, Adamantidis AR, Durand R, Airan RD, de Lecea L, Deisseroth K (2010) Optogenetic interrogation of neural circuits: technology for probing mammalian brain structures. Nat Protoc 5:439–456

83. Chuong AS, Miri ML, Busskamp V, Matthews GAC, Acker LC, Sørensen AT, Young A, Klapoetke NC, Henninger MA, Kodandaramaiah SB, Ogawa M, Ramanlal SB, Bandler RC, Allen BD, Forest CR, Chow BY, Han X, Lin Y, Tye KM, Roska B, Cardin JA, Boyden ES (2014) Noninvasive optical inhibition with a red-shifted microbial rhodopsin. Nat Neurosci 17:1123–1129

84. Lin JY, Knutsen PM, Muller A, Kleinfeld D, Tsien RY (2013) ReaChR: a red-shifted variant of channelrhodopsin enables deep transcranial optogenetic excitation. Nat Neurosci 16:1499–1508

85. Gradinaru V, Thompson KR, Deisseroth K (2008) eNpHR: a Natronomonas halorhodopsin enhanced for optogenetic applications. Brain Cell Biol 36:129–139

86. Dufour S, Lavertu G, Dufour-Beauséjour S, Juneau-Fecteau A, Calakos N, Deschênes M, Vallée R, De Koninck Y (2013) A multimodal micro-optrode combining field and single unit recording, multispectral detection and photolabeling capabilities. PLoS One 8:e57703

87. Canales A, Jia X, Froriep UP, Koppes RA, Tringides CM, Selvidge J, Lu C, Hou C, Wei L, Fink Y, Anikeeva P (2015) Multifunctional fibers for simultaneous optical, electrical and chemical interrogation of neural circuits in vivo. Nat Biotechnol 33:277–284

88. Weber F, Dan Y (2016) Circuit-based interrogation of sleep control. Nature 538:51–59

89. Huber D, Petreanu L, Ghitani N, Ranade S, Hromádka T, Mainen Z, Svoboda K (2008) Sparse optical microstimulation in barrel cortex drives learned behaviour in freely moving mice. Nature 451:61–64

Chapter 3

Molecular Engineering of Channelrhodopsins for Enhanced Control over the Nervous System

André Berndt

Abstract

Light-activated proteins such as channelrhodopsins are powerful tools which reveal the function of neuronal networks with unprecedented precision. The strength of optogenetic applications emerges directly from the unique properties of the utilized proteins. Consequently, modifying the properties of channelrhodopsins extends our research capabilities even further. In this chapter, I describe how targeted protein engineering results in enhanced optogenetic applications. One key element is to align the molecular function of channelrhodopsins with the physiological properties of neuronal circuits. As shown on two examples, understanding protein function as well as the intended host system provides a significant advantage in protein engineering.

Key words Optogenetics, Channelrhodopsin, Protein engineering, Neuronal inhibition

1 Introduction

During all of human history, the acquisition of knowledge and the development of tools have been inextricably intertwined with each other. Particularly, modern neuroscience has tremendously benefited from advances in molecular biology, electrophysiology, and microscopy. The development and usage of new tools has led to amazing interdisciplinary approaches by incorporating aspects from almost every science discipline. It is inevitable that more progress will be made by further extending our toolsets for example by using optogenetics [1]. In a nutshell, optogenetics is a technique that uses light-activated proteins primarily to identify and probe functional connections in brains of vertebrate and invertebrates. One of the major tools in optogenetics is light-triggered ion channels known as channelrhodopsins [2]. Light-induced action potentials in channelrhodopsin-expressing neurons trigger an immediate response in connected downstream neurons. Depending on the type of synapses, downstream neurons can either get excited or

Albrecht Stroh (ed.), *Optogenetics: A Roadmap*, Neuromethods, vol. 133,
DOI 10.1007/978-1-4939-7417-7_3, © Springer Science+Business Media LLC 2018

inhibited at high temporal resolution which reveals existing functional connections [3].

Optogenetics is particularly powerful because it can restrict channelrhodopsin expression to specific neuronal subpopulations, for instance by using cell-specific promoters, transgenic animals or recombinase-driven expression strategies [4] (see also Chaps. 1 and 2). To truly realize why this such a crucial benefit, imagine that one cubic millimeter of mammalian cortex contains more than 100,000 neurons of different subtypes which are all heavily connected with each other [5]. Traditional methods such electrophysiology and microscopy of fixed tissue are very limited in untangling the complexity of neuronal connections. In contrast, optogenetics enables us to effectively control the function of predetermined neurons in freely moving animals during behaviors such as learning, motor control, or reward [6–9]. As a result, optogenetics has helped to identify neurons involved in fear responses, social behavior, and addiction [10–12]. Additionally, it further deepened our understanding of neurological diseases such as depression and epilepsy [13, 14].

The impact of optogenetics in neuroscience and the resulting surge of publications is comparable to the introduction of in vivo and patch clamp electrophysiology several decades ago. However, electrical stimulation and monitoring are unspecific whereas optogenetics allows targeting neuronal subpopulations directly. Therefore, the use of light-activated proteins offers a more efficient way to control neurons and neuronal circuits and to identify their function.

A major goal in neuroscience is to understand the encoding of sensory information and complex behavior in neuronal networks. Optogenetics will play a crucial role here because it has the potential to affect neuronal activity in the brain at high temporal and spatial resolution. For example, channelrhodopsin activation evokes single action potentials in response to single blue light pulses whereas the activation of the light-triggered chloride pump halorhodopsin inhibits action potential firing for the length of a green/yellow light application [2, 15]. However, excitatory and inhibitory neuronal signals are extremely variable in their strength, frequency, and length. Optogenetic tools must be able to cover their full dynamic range at high precision in order to induce natural activity patterns. Modifying the molecular function of the utilized proteins will bring us closer to this goal.

In practical terms, protein engineering means replacing amino acids in the protein sequence by mutating the coding DNA sequence. In general, there are two major approaches: (1) Introduction of random mutations and testing of hundreds or thousands of constructs [16]. Here, promising candidates are identified by high-throughput screening and the underlying mutations are determined in a subsequent step. The advantage is that only minimal knowledge of the structure and molecular properties of the target protein is necessary. However, while the generation of large randomized mutant libraries is relatively simple, high-throughput methods are only available for a limited number of proteins. (2) In structure-guided or targeted engineering, individual amino acids are selected and replaced by point mutations [17]. Consequently, the number of mutated constructs is significantly lower. This approach requires a significant knowledge of the protein structure and function in order to restrict mutation to sites that have a high probability to yield the desired changes in molecular function.

For rhodopsin-engineering, the second approach is more suitable because these proteins carry electrically charged ions over cell membranes which require low-throughput patch clamp electrophysiology to accurately characterize ion transport. While high-throughput imaging techniques can be utilized for screening, they often lack accuracy which increases the chances of false negatives because most individual mutations will cause only subtle changes in molecular function. Furthermore, channelrhodopsins directly affect the electrical properties of neurons which are measured most accurately by patch clamp electrophysiology.

Most rhodopsins comprise around 300 residues in their membrane-bound fraction. Those can be replaced out of a pool of 20 amino acids which hypothetically allows for 20^{300} possible mutant combinations, a number that exceeds both high- and low-throughput screening capabilities by far. Therefore, mutations should be restricted to selected sites and experimental outcomes should be predicted based on available biophysical data to reduce functional screening to reasonable timeframes. Fortunately, there is a vast amount of biophysical data for rhodopsins. In fact, the light-activated proton pump bacteriorhodopsin is one of the most investigated membrane proteins and serves as a paradigm for the rhodopsin class [18, 19]. Furthermore, halorhodopsins, channelrhodopsins, and mutated variants have been the subject to thorough studies. Datasets include molecular X-ray structures, data from time-resolved FTIR and UV/vis spectroscopy and electrophysiology [20–23]. The sum of all biophysical data helps tremendously to understand the underlying molecular mechanism of rhodopsin function and is critical for directed molecular engineering. Consequently, this knowledge allows for more accurate predictions on how specific mutations will change the kinetics of activation and deactivation, absorption spectra, or ion selectivity.

Each engineering approach should include the following aspects (Fig. 1):

1. Understanding the fundamental molecular properties of rhodopsins.

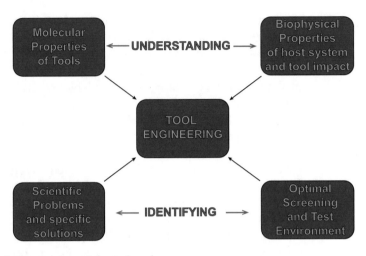

Fig. 1 Conceptual approach to protein engineering

2. Understanding the fundamental physiological properties of the environment in which engineered rhodopsin will be applied. Consideration of how rhodopsin activation will affect those properties.

3. Identifying specific scientific problems and evaluation of how engineered rhodopsins can solve them.

4. Identifying the optimal system for screening and validating the function of engineered rhodopsins.

I will demonstrate how these principles were put into practice by highlighting two recent examples. (1) Acceleration of channelrhodopsin kinetics for enhance precision of optogenetic applications [24]. (2) Conversion of channelrhodopsin into a chloride selective channel for physiologically relevant inhibition of neurons [25, 26].

Example 1: Acceleration of channelrhodopsin kinetics for enhanced precision of optogenetic applications.

1.1 Molecular Properties of Channelrhodopsins

All rhodopsins use a single retinal molecule as chromophore which is buried within the membrane-bound fraction of the protein [19]. This domain consists of seven alpha helices which entirely span through cell membranes. Retinal absorbs energy from photons and transitions into a high-energy configuration which initiates subsequent conformational changes in the surrounding protein environment. Amino acids continue to rearrange while the protein conformation transitions back into the low-energy ground state. During this so called photocycle, different rhodopsins perform different molecular functions. For example, bacteriorhodopsin actively pumps protons out and halorhodopsin actively pumps chloride ions into cells, even against their electrochemical gradients [19, 20]. On the other hand, channelrhodopsins open a pore through which ions passively flow following their electrochemical gradient [27]. Mammalian rhodopsins initiate signal cascades in the photoreceptor cells in the retina which translate to visual information in the visual cortex [28]. Obviously, the difference in molecular function of these proteins is determined by differences in their amino acid composition and their three-dimensional arrangement. In fact, every single biophysical feature of rhodopsins depends on the underlying protein sequence and consequently, can be altered by single or multiple mutations.

One of the first opsins that was mutated for enhanced molecular functions and improved applications in optogenetics was Channelrhodopsin-2 (ChR2). ChR2 originates from the single-cell algae *Chlamydomonas reinhardtii* where it is utilized as a photoreceptor to find optimal light conditions for photosynthesis [29]. Its activation changes the electrical properties of the cell membrane within milliseconds and subsequently affects flagellate motion. Channelrhodopsin-2 is a nonselective light-gated cation channel which forms a pore over cell membranes. Once activated by blue light, the channel pore opens and electrically charged calcium, sodium, potassium, and hydrogen ions flow passively over the cell membrane following their electrochemical gradient. Protons have the highest overall conductivity followed by the monovalent cations Lithium, Sodium, and Potassium [22]. Divalent cations such as Calcium and Magnesium have even higher affinities but lower overall conductivities than monovalent cations. The higher affinity comes at the cost of longer dwell times in the pore which can effectively reduce or block charge transport at high calcium and magnesium ion concentrations [30].

Ions flow through channelrhodopsins in both directions either inside (influx) or outside of cells (efflux) depending on the ion concentration on both sides and the electric field over the membrane (membrane potential). Ions and molecules flow towards lower concentrations, but ions are also affected by electric fields because they are electrically charged. Together, the ion concentration on both sides of the cell membrane and the electric potential form the electrochemical gradient which determines in which direction ions flow. This ion flow equals an electric current which is specifically called photocurrent in the case of rhodopsins. At a specific membrane potential (reversal potential) influx and efflux of ions is equal and observable net current is zero.

At pH 7.2, the ChR2 pore opens and closes with a rate constant of 2 ms (τ-on) and 12 ms (τ-off) respectively [24]. Furthermore, the pH values strongly affect channel kinetics which are faster at low external pH and slower at high external pH values. Additionally, ChR2 has a complex photocycle with a high conducting and low conducting state [31] (see peak and steady state currents, Chap. 13). The high conducting state is populated first, but over time more of the channel molecules transition into the low conducting state. The peak activity is reached at 470 nm, but channelrhodopsin-2 has a relatively broad spectrum and it reaches 50% activity at 410 and 510 nm with significant activity beyond these wavelengths [22].

Replacing single or several amino acids in channelrhodopsins can affect several biophysical properties, for instance: (1) Increase or decrease of speed of the channel opening and closure [24, 32, 33], (2) altered absorption spectra [24], (3) change of the ion selectivity [25, 26] and (4) Ion conductivity, i.e. total number of transported ions in a given period of time [32].

1.2 Molecular Basis of Neuronal Signaling

The basic elements in neuronal activity are action potentials which are strong and fast changes in the membrane potential of neurons. Action potentials travel along the length of the axon and induce neurotransmitter release at the synapse terminals which in turn either stimulates, inhibits, or modulates the activity of connected postsynaptic neurons. The membrane potential of resting neurons lies between −60 and −100 mV, but during action potential firing, it depolarizes (i.e. becomes more positive) to around +50 mV within milliseconds. Subsequently, the membrane potential quickly re-polarizes back to resting state before a new action potential can be generated [34]. Some neurons can fire action potentials at frequencies up to 200 Hz and more, for example certain interneurons, pain and heat receptors and neurons which process auditory inputs.

The shape, size, kinetics and frequency of action potentials is fundamentally dependent on the biophysical properties of the ion channels driving neuronal firing which vary from cell type to cell type [34]. For instance, slow channel kinetics cause slow rise and decay of action potentials which reduces the maximum firing frequencies. Furthermore, the overall conductivity of ion channels determines the extent and size of membrane depolarization during neuronal firing.

Light-triggered ion channels such as ChR2 can manipulate the electrical properties of neurons and thus, can be utilized for action potential generation. As mentioned above channelrhodopsins conduct sodium, potassium, and calcium, but sodium and potassium have almost equal transport rates and can be considered as the same ion species. Under the physiological conditions found in mammalian brains the cytosolic potassium concentration is similar to the external sodium concentration, i.e. from the perspective of channelrhodopsin there is practically no diffusion gradient for monovalent cations across cell membranes. At physiological pH values sodium and potassium constitute the major component of conducted ions [30]. Under these conditions and at the negative resting membrane potential, channelrhodopsin activation induces an influx of sodium which depolarizes cell membranes. However, the overall conductivity is about ten times smaller compared to native sodium channels [35], i.e., channelrhodopsin activation does not generate action potentials directly, but triggers them by activating voltage gated sodium channel. Hence, channelrhodopsins play the role of calcium channels in eliciting action potentials. Consequently, light-triggered action potentials are driven by native ion channels meaning that they are physiologically similar to native signals. Consequently, channelrhodopsins bear the capability to potentially elicit natural activity patterns.

1.3 Improving the Temporal Precision of Optogenetic Applications with Kinetically Improved Channelrhodopsins

Despite being a powerful technique, optogenetics has certain limitations which arise from the biophysical properties of channelrhodopsins and the neurons themselves. First, the size of the photocurrent should be strong enough to exceed the threshold for the activation of voltage gated sodium channels. When a virus is used, channelrhodopsin expression levels can vary and neurons with no or small photocurrent will not get excited by light applications [32]. Furthermore, neuron subtypes that are strongly hyperpolarized or contain a relatively high number of open potassium channels need larger depolarizing photocurrents than neurons that rest near the action potential threshold. Naturally, channelrhodopsins with high conductivity have a higher probability to elicit action potentials. Second, the speed of channelrhodopsin activation and deactivation determines the maximum light-triggered firing frequency. In fact, the kinetics of Channelrhodopsin-2 are significantly slower compared to native channels limiting the maximum light-induced firing frequency to 40 Hz [36]. That is because the channelrhodopsin closure is relatively slow which prolongs the influx of sodium ions and with that the depolarization of neurons. Consequently, re-polarization after action potential generation will be slowed which effectively blocks fast generation of repeated action potentials at frequencies above 40 Hz [24]. Furthermore, prolonged membrane depolarization also increases the probability for double or triple action potentials in response to single light pulses which lowers the overall precision of light stimulation (Fig. 2). To overcome these limitations, one of the first engineering approaches focused on increasing the speed of the opening and closure of channelrhodopsins to reduce the interference of channelrhodopsin

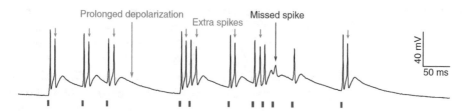

Fig. 2 Whole-cell current clamp recordings from a parvalbumin interneuron strongly expressing wild-type ChR2; note the extra spikes, the missed spike later in the train, and the prolonged depolarizations observed after termination of each 2-ms light pulse (*blue bars*) (*Reproduced from Gunaydin LA, et al. (2010)* [24] *with permission from Macmillan Publishers Limited*)

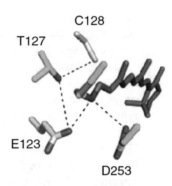

Fig. 3 Homology model of ChR2 based on the bacteriorhodopsin X-ray structure (1KGB, RSCB protein data bank). The retinal is shown in *violet*, conserved residues in *green,* and substitutions in ChR2 in *gray*. Oxygen is *red,* nitrogen is *blue,* and sulfur is *yellow*. Candidate hydrogen bonds are shown as *dotted lines* (*Reproduced from Gunaydin LA, et al. (2010)* [24] *with permission from Macmillan Publishers Limited*)

with native ion channels. Such engineered channelrhodopsins would interfere less with native re-polarization and would enable fast light-triggered signaling at improved precession. The first target for protein engineering was channelrhodopsin-2. Unfortunately, at the time, no molecular structure was available, but a vast amount of data for the structurally related proton pump bacteriorhodopsin. Although the sequence similarity between bacteriorhodopsin and channelrhodopsin is only about 40%, the amino acid composition is most similar in residues near the retinal chromophore and the active center which initiate the photocycle. Hence, the assumption was that the molecular mechanism of channelrhodopsin activation is similar between the two proteins. A residue that plays a crucial role in bacteriorhodopsin activation is aspartate 85 which is homologous to glutamate 123 (E123) in channelrhodopsin. Therefore, this and neighboring residues became the first target for mutations regarding the alteration of channelrhodopsin activation (Fig. 3) [24].

1.3.1 Preparation of ChR2 Mutations

(a) The amino acid sequence of all proteins is coded in DNA. In protein engineering, mutations are directly introduced to the coding DNA sequence which is used for transfecting the respective host systems. Common plasmid vectors are pcDNA3.1 (ThermoFisher, Waltham, MA) for HEK293 and pGEM-HE (Promega, Madison, WI) for oocytes. These plasmids should contain a gene for ampicillin or kanamycin resistance which enables selective DNA amplification in *E. coli* cell cultures. Additionally, vectors that are used for channelrhodopsin expression in mammalian cell cultures or tissue should be co-expressed with a fluorescent protein such as GFP to allow for the identification of transfected cells.

(b) Point mutations are introduced by replacing nucleotide bases using a standard mutagenesis kit such as the QuickChange series (Agilent, Santa Clara, CA). This is basically a PCR reaction using primers that contain the desired mutation.

(c) The PCR product contains mutated plasmids which are used to transform competent *E. coli* cells (high efficiency, XL1-Blue, NEB alpha-5) by heat transformation following standard procedures. The transformed cells are plated on LB agar plates containing ampicillin or kanamycin and incubated overnight at 37 °C.

(d) Colonies are picked the following day and used to inoculate four to six tubes with 5 ml liquid LB medium containing ampicillin or kanamycin to amplify the DNA plasmids.

(e) Vector plasmids are purified the following day using a standard miniprep DNA purification kit (Qiagen Hilden, Germany; ThermoFisher, Waltham, MA; SigmaAldrich, St. Louis, MO, etc.).

(f) For HEK cells, purified miniprep DNA (0.2 μg/ml, ~40 μl) provides enough amount to test up to 10 wells of cell cultures on a standard 24-well cell culture plate. This is sufficient to either confirm or reject a mutation for further consideration by patch clamp electrophysiology.

1.4 Choosing the Best Test Environment for Kinetically Improved Channelrhodopsins

Testing of protein function should always begin in the simplest environment for maximum throughput. In case of channelrhodopsins, simple and accessible host systems that allow electrophysiology are oocytes from *Xenopus laevis* and immortalized human embryonic kidney cells in culture (HEK293). Cell cultures can be produced in larg quantities at low cost. However, patch clamp electrophysiology requires a seal with high resistance between the glass pipettes and cellular membrane. Consequently, this approach is labor intensive and limits the total number of constructs that can be tested in reasonable time frames to only a few dozens.

Transfecting HEK293 cell with purified DNA plasmids is most efficient when a lipophilic transfection agent such as Lipofectamine (ThermoFisher, Waltham, MA) is used. Transfection follows standard procedures at around 40–80% confluency. Cells are kept at 37 °C in DMEM medium (ThermoFisher, Waltham, MA) containing 10% fetal bovine serum and a Penicillin/Streptomycin mix. Incubator atmosphere should contain 5% CO_2 and should be humidified. Patch clamp electrophysiology can be conducted on days 1–3 after transfection.

For Xenopus oocytes, purified DNA must be transcribed into mRNA using an in vitro kit such as mMessageMachine T7 (ThermoFisher, Waltham, MA). The purified mRNA can be directly injected into individual oocytes through micro glass pipettes using a microinjector pump (10–40 nl, 1 µg/µl). Oocytes are kept at 18 °C in Oocyte Ringers Solution containing additional penicillin and streptomycin. Cells can be probed on day 3–7 after injection. The incubator temperature should be 18 °C.

1.4.1 Characterization of Kinetically Improved ChR2 Mutations

Electrophysiology equipment for HEK293 cells (patch clamp) and oocytes (two electrode voltage clamp) is slightly different, but the protocols for rhodopsin testing are very similar. Both techniques provide control over the cell membrane potential and allow to measure ion currents flowing across membranes while the membrane potential is clamped (whole cell mode). Additionally, light has to be applied for channelrhodopsin activation. The most cost effective way is to use single wavelength, high-power LEDs, but depending on the application white-light LEDs or broad spectrum xenon lamps as well as lasers or multi-wavelength LED arrays can be utilized as well.

The goal of this project was to increase the opening and closure speed of the ion conducting pore upon light application [24]. Testing kinetic properties of channelrhodopsin ion transport is simple: Channel opening and closure speeds can be measured in real time by measuring the rise and decay of photocurrent under light applications. The membrane potential of cells expressing channelrhodopsins is clamped between −80 mV and −60 mV, in the range of the resting membrane potential of neurons. There are no other light-activated transporters in oocyte or HEK293 cell membranes, i.e. currents that are measured under light application can be attributed to rhodopsin activity only. The rise and decay of photocurrent in response to light corresponds with the kinetics of channel opening and closure respectively. These, kinetics can be fitted with exponential functions where τ values equal the rate constants of each process, i.e. the time at which $1/e$ of the channels are open and closed respectively. Smaller τ values correspond to faster kinetics and mutant screening aims to identify constructs with the smallest τ values.

Interestingly, mutating ChR2-E123 to threonine and alanine (E123T, E123A) provided the fastest channel kinetics with τ values that are about half compared to wild-type (Fig. 4). Additionally, E123T/A kinetics were independent from the membrane voltage whereas ChR2 channel closure became slower at depolarized membrane potentials. This is a critical disadvantage for ChR2 because its channel kinetics get slower when action potentials depolarize the membrane. In contrast, E123T kinetics stayed constantly fast during the entire action potential cycle which would ensure fast repolarization of neuron membranes and fast light-triggered action potential generation. Therefore, E123T was selected for testing precise light-triggered action potential generation in neurons and named ChETA including the equally fast ChR2-E123A mutation.

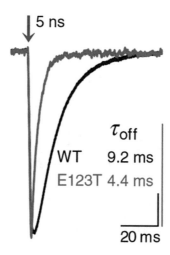

Fig. 4 Oocyte current recordings. Excitation was delivered with a 5-ns laser flash (470 nm, 0.5 mW/cm^2, *blue arrow*) at −100 mV. Photocurrent decay was fitted mono-exponentially with τ values shown for wild-type ChR2 (*black*) and E123T (*red*) and traces were normalized to highlight differences in decay kinetics. Vertical bars correspond to 2.5 nA (wild type, *black*; E123T, *red*) (*Reproduced from Gunaydin LA, et al. (2010)* [24] *with permission from Macmillan Publishers Limited*)

1.4.2 Enhanced Optogenetic Precision in PV+ Interneurons

HEK293 cultures and oocytes are perfectly suited for basic characterization of ion transport, however they do not allow for proper testing of channelrhodopsin function because they do not generate action potentials. Simple neuronal host systems are primary neuron cultures such as from rat brains or organotypical slice cultures. On the other hand, in vivo expression and patch clamp experiments in acute slice enable tests in the environment in which channelrhodopsins are actually being used. However, the preparation requires virus production, injection and appropriate expression time (1–3 months, depending on application). Hence, in vivo and ex vivo characterization of channelrhodopsins should be only considered for the most promising variants.

We aimed to demonstrate that fast ChETA kinetics translate to increased precision and faster stimulation frequencies [24]. The ideal host system for these purposes are fast spiking cortical and hippocampal interneurons which can fire at frequencies up to 200 Hz. This requires viral transduction of channelrhodopsin in living animals and acute slice electrophysiology of mouse brains. Double floxed and inverted ChR2-eYFP wild-type and E123T-eYFP DNA sequences were cloned into an adeno associated virus (AAV) plasmid (see Chaps. 1 and 2). AAV was injected into hippocampus or cortex of transgenic mice that express Cre-recombinase in PV positive (PV+) interneurons. After 1 month, expression levels peak and channelrhodopsin function can be tested by patch clamp

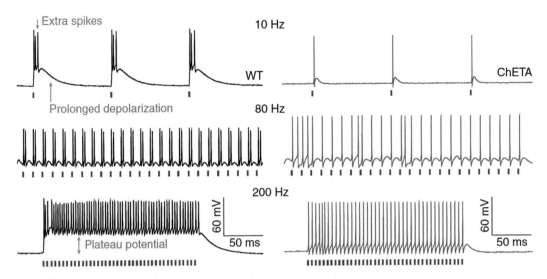

Fig. 5 Whole-cell current clamp recordings from parvalbumin interneurons expressing wild-type ChR2 (*black*) or ChETA (*red*) in response to 10-, 80- and 200-Hz light stimulation (rows 1, 2, and 3, respectively; all 472-nm, 2-ms light pulse widths (*blue bars*)). Note the brisker repolarizations and reduced extra spikes in ChETA-expressing interneurons. Scale bars apply to all traces (*Reproduced from Gunaydin LA, et al. (2010)* [24] *with permission from Macmillan Publishers Limited*)

electrophysiology following standard procedure. Hippocampal or cortical brain areas are cut into 300 μm thick slices and kept in ACSF solution that is constantly being provided with 95% O_2/5% CO_2 gas. Slices are kept at room temperature, but measurements are conducted at 34–37 °C. Following these guidelines, patch clamp experiments can be conducted for up to 14 h after slicing. PV+ cells expressing channelrhodopsin-eYFP can easily be identified by their eYFP fluorescence.

To compare the precision of ChR2 and E123T, PV+ cells were exposed to light pulse trains with increasing frequency and light-induced action potentials were recorded. E123T expressing cells do not exhibit a prolonged depolarization after light-triggered action potential firing due to their fast and voltage-independent channel kinetics. Consequently, the firing probability at fast frequencies above 40 Hz was significantly increased while the probability for a second or third action potential in response to a single light pulse was reduced (Fig. 5).

In conclusion, ChETA constructs are a powerful demonstration of how changed biophysical properties of proteins can open the door to new applications. Furthermore, they highlight the importance of using distinct tools for specific optogenetic applications to overcome experimental limitations.

1.5 Notes

Nevertheless, the search for enhanced tools does not stop here. The fast channel kinetics reduced the overall photocurrent because the accelerated photocycle depopulates the conducting state faster. As a result, a smaller number of channel molecules populate the conducting state. This requires relativity robust expression of ChETA which reduced the number of neurons which could fire light-triggered action potentials in a neuronal population. However, as a result of continuing efforts, today there are more improved channelrhodopsin variants with large photocurrents and fast kinetics available, providing ever improving reliability and precision [32, 37, 38].

Example 2: Physiologically relevant inhibition of neuronal signals.

2 Molecular Properties of Halorhodopsin

Besides excitation by channelrhodopsins, neuronal activity can also be inhibited by using the light-activated chloride pump Halorhodopsin [15]. The opposite mode of action emerges from the fundamentally different molecular function of Halorhodopsin [20].

Halorhodopsin from the halobacteria *Natronobacterium pharaonis* (NpHR) utilizes energy from absorbed photons to actively pump chloride ions across membranes into cells even against electrochemical gradients. NpHR transports only one chloride ion per photon compared to hundreds or thousands in channelrhodopsins making this chloride pump technically less efficient compared to ion channels. Furthermore, the chloride transport is unidirectional, but voltage dependent and thus, photocurrents would theoretically become zero at membrane potentials around -400 mV [39].

The NpHR absorption peak is around 580 nm. However, the absorption spectrum is very broad with 50% activity at approximately 500 nm and 625 nm [40].

In contrast to channelrhodopsins, NpHR activation, deactivation, ion transport and selectivity are more closely linked with each other on the molecular level. Therefore, engineering one of these features will inevitably affect one of the others, often resulting in loss of function. Nevertheless, mutations of NpHR can shift the absorption spectrum without affecting ion transport creating variants with red shifted absorption spectra [40].

2.1 Molecular Basis of Native and Optogenetic Inhibition of Action Potentials

Besides excitation, selective inhibition of electrical signals is a crucial component of neuronal activity. Temporal precise inhibition of neurons ensures that signals are forwarded and processed in synchronized order without creating runaway excitation. The excitation/inhibition balance is mediated by the activity of specific ion channels which either conduct potassium or chloride ions. For instance, the activity of potassium channels primarily hyperpolarizes neuronal membranes whereas chloride channels primarily shunt neurons by decreasing the so-called input resistance of cells. At lower input resistance, larger sodium and calcium currents are necessary to depolarize membrane potentials and trigger action potentials which can effectively inhibit excitatory stimulations. Although the effect on the resting membrane potential can be small or even zero, opening chloride channels makes neurons less excitable because excitatory currents originating from calcium and sodium channels are effectively drained. This is because chloride channels transport negative ions inside whereas sodium and calcium channels transport positive ions inside which generates opposing electrical currents which effectively cancel each other out.

However, inhibition by NpHR works fundamentally different from physiological inhibition. Light activation of chloride pumps has no effect on the input resistance of cells, but instead hyperpolarizes cell membranes strongly, even beyond the level of potassium channel activity. As a result, the strong hyperpolarization can efficiently prevent the generation of action potentials. Therefore, NpHR has become an extremely valuable tool for investigating the causal role of neuronal circuits and has been used to reveal circuits involved in social interaction and anxiety and it has been used to increase the pain tolerance for mice [11, 41, 42].

2.2 Engineering Physiologically Relevant Inhibition

Optogenetic inhibition utilizing NpHR has been successfully applied in several studies. However, there is tremendous potential for improvement because chloride pumps are less efficient compared to channels since they pump only one ion per photocycle. In addition, pumps can strongly hyperpolarize neurons even beyond physiological levels, potentially affecting the physiological state of neurons. This motivated the engineering of a chloride conducting channelrhodopsin which mimics inhibition induced by native chloride conducting channels such as $GABA_A$ receptors.

Therefore, our goal was to convert the ion selectivity of channelrhodopsins from cations to chloride. Fortunately, the X-ray structure of the channelrhodopsin hybrid C1C2 had been published and for the first time, we could identify the residues forming an ion conducting pore [43]. One striking feature here is that the pore contains multiple potentially negatively charged residues such as aspartate or glutamate, which would be ideal binding sites for positive cations. Therefore, our engineering approach aimed to replace these amino acids in C1C2 with neutral or potentially positively charged residues such as histidine, lysine, or arginine to attract negative chloride ions and to repel cations. The strategy was to introduce single mutations and screen for small changes in the ion selectivity and then to combine promising candidates to double, triple and multiple mutations to multiply the effect of individual replacements step by step. The final construct should demonstrate high chloride selectivity and conductivity in response to prolonged blue light pulses.

2.2.1 Preparation of C1C2 Mutations

Individual point mutations were selected based on the C1C2 X-ray structure and were restricted to residues in or near the ion conducting pore. The preparation of mutated constructs followed the steps described in the previous example (1.3.1).

2.3 Choosing the Best Test Environment for Chloride Conducting Channelrhodopsins

The ion selectivity of ion channels is most accurately determined by measuring their reversal potential using patch clamp electrophysiology. As mentioned above, the reversal potential of a channel is the specific membrane potential at which influx and efflux of conducted ions are equal and the effective membrane current is zero. The reversal potential of ion channels depends on the types of conducted ions and the ion concentrations on both sides of the membrane. For instance, C1C2 is a nonselective cation channel with almost equal selectivity for potassium and sodium. Under physiological conditions, cytosolic potassium concentrations almost equal extracellular sodium concentrations and thus the C1C2 reversal potential is near 0 mV. In contrast, chloride conducting channels have negative reversal potentials between -70 and -40 mV in neurons because chloride concentrations are lower on the cytosolic side. Therefore, the screening aimed to identify C1C2 mutations causing small negative shifts in the reversal potential indicating increased chloride selectivity. Measuring the reversal potential of mutated constructs requires patch clamp electrophysiology. Any cell culture system that can be prepared in large quantities at low maintenance cost is suitable such as HEK293 cells, rat primary neuron cultures or oocytes from *Xenopus laevis*.

2.3.1 Protocol for Measuring Reversal Potentials

Electrophysiological experiments are conducted in voltage clamp mode in which membrane currents are measured near the actual reversal potential. The reversal potential is determined by plotting membrane currents against the membrane potential and by reading out the potential at which a connecting line between current values crosses the zero-current mark (Fig. 6). Here, smaller increments of holding potentials increase accuracy.

The second important parameter was the overall conductivity of constructs represented by the size of the current amplitude. For example, constructs generating small currents indicate a low conductivity and were excluded from subsequent multiple mutations.

Single mutations were combined until the reversal potential was more negative than the threshold for action potential generation in neurons. The first construct reaching that benchmark was a ninefold mutation called iC1C2: T98S/E129S/E140S/V156K/E162S/H173R/V281K/T285N/N297Q, ($V_{rev} = -61$ mV under physiological conditions) [25]. As predicted, the final construct contains a number of mutations in which potentially negatively charged residues were replaced by neutral or negatively charged residues. This provides further evidence for the hypothesis

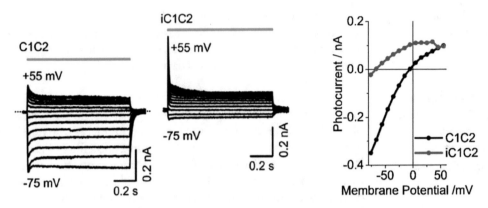

Fig. 6 Representative photocurrents (*left and middle*) and corresponding current-voltage relationships (*right*) of C1C2 (*black*) and iC1C2 (*red*) recorded at membrane potentials from −75 mV to +55 mV upon 475 nm light activation (power density, 5 mW/mm², *blue bars*). The reversal potential is the membrane potential at which observable photocurrent are zero (*Reproduced from Berndt, et al. (2014)* [25] *with permission from the American Association for the Advancement of Science*)

that selectivity in channelrhodopsins is controlled by the electrostatic potential in the ion conducting pore. Overall about 400 mutations were tested in this round.

Interestingly, iC1C2 photocurrents were increased and the reversal potential was more negatively shifted at high external proton concentrations. This indicated that the chloride conductivity and selectivity could be further increased [25].

We introduced more mutations and engineered a second-generation chloride conducting channelrhodopsin iC++ based on the same principles of increasing the positive electrostatic potential in the ion conducting pore. About 200 additional mutations were tested in the second round resulting in the C1C2 multiple mutation T59S/E83N/E90Q/E101S/V117R/E123S/T246N/V242R/N258Q/E273S ($V_{rev} = -78$ mV under physiological conditions, residue numbering has changed due to n-terminal truncation) [26].

Theoretically, the shifted reversal potential could originate from either increased chloride or potassium conductivity. To confirm the ion selectivity, we conducted ion substitution experiments in HEK293 cells. Changing the ion concentrations on either side will affect the reversal potential, but only if those specific ions are conducted. HEK293 is a very robust cell line which can survive in a wide range of different extra- and intracellular ion concentrations and pH. For example, iC++ and iC1C2 have a negative reversal potential when the extracellular solution contained low potassium and high chloride concentrations whereas the intracellular solution contained high potassium and low chloride. This mimics the ionic conditions in neuronal systems. Changing the potassium concentration while keeping chloride constant had no effect on the reversal potential, but changing chloride shifted the reversal potential significantly, thus confirming that iC1C2 and iC++ are chloride conducting channels [25, 26].

2.3.2 Functional Testing of Neuronal Inhibition

The next step aimed to demonstrate that chloride conducting channelrhodopsins effectively inhibit action potentials. One of the simplest systems available is primary neuron cultures from rats. Patch clamp experiments were conducted in current clamp mode. Electrical stimulation through the patch pipette evokes action potentials while blue light application should inhibit the generation of electrically evoked signals. We used two different protocols for action potential generation: (1) Electrical pulses (5–30 ms wide) were applied at a fixed frequency (5–20 Hz). Light was applied for 10 s during the electrical stimulation. Inhibition efficiency was measured as ratio between evoked action potentials and electrical pulses during the duration of the light application (Fig. 7, left). (2) A prolonged electrical stimulus (>10 s) elevates the membrane potential above the threshold for action potential generation and the cells started to generate action potentials at their endogenous firing frequencies (Fig. 7, right). Here, light was also applied for 10 s to inhibit neuronal activity. Inhibition efficiency was measured by comparing the firing frequency pre, post, and during the light application. We applied these protocols to cultured neurons as well as to neurons in acute slice after in vivo expression of iC++ in principal cells of mouse prefrontal cortex.

Most importantly, we aimed to demonstrate that iC++ is more effective in inhibiting neuronal activity than NpHR. We generated action potentials by using extended electrical stimuli (4 s) with increasing current amplitudes. We measured the maximum current amplitude at which NpHR or iC++ cells were still able to inhibit

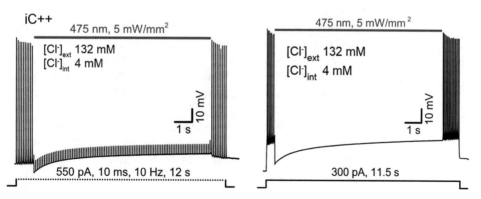

Fig. 7 *Left:* Voltage trace of iC++—expressing neuron showing action potential generation by pulsed current injections (*dotted line*) and inhibition during 10 s light application (*blue bar*). *Right:* Voltage trace of an iC++—expressing neuron showing action potential generation by continuous current injections (*black bar*) and inhibition during 10-s light application (*blue bar*) (*Reproduced from Berndt, et al. (2014)* [26] *with permission from the Proceedings of the National Academy of Sciences*)

action potential generation. NpHR was only able to inhibit action potentials that were generated by electrical stimulation lower than the NpHR photocurrent. On the other hand, iC++ inhibited action potentials generated by electrical stimuli at least 4 times larger than the observable photocurrent, proving that iC++ is more efficient for inhibition than NpHR [26].

These in vitro experiments were a powerful confirmation of iC++ capabilities. However, iC++ was designed with in vivo applications in mind and thus, it was critical to test iC++ in behavioral experiments as well. We choose two experimental settings in which targeted neuronal inhibition has been successfully used to cause behavioral effects: (1) Inhibition of dopaminergic VTA (ventral tegmental area) neurons in mice causing place aversion [44, 45]. (2) Inhibition of neurons involved in memory formation causing loss of fear memory [46]. As expected, iC++ robustly inhibited the targeted cell somata. Surprisingly, the behavioral effects were comparable to NpHR under the same experimental conditions [26]. We assume that inhibition in these specific cell types has a relatively low threshold that is reach by either construct.

2.4 Notes

We demonstrated that iC++ inhibits cell somata in vivo at least as efficient as NpHR, but by utilizing a physiologically relevant effect that bears the potential for even higher inhibition efficiency. In addition, our experiments showed that inhibition triggered by iC++ depends on chloride gradients. iC++ inhibition is most efficient at low intracellular chloride concentration that are found in dendrites and somata (4–10 mM), but decreases above these levels. Therefore, we would expect low inhibition probability or even excitation in axons or synaptic terminals where chloride concentrations can exceed 20 mM [47–49]. Inhibition in these regions is largely mediated by potassium channels such as $GABA_B$ receptors. Therefore, effective optogenetic inhibition of synapses must be provided by light-activated potassium channels or an optogenetic tool that inhibits voltage-gated calcium channels.

Furthermore, expression of light-activated chloride channels such as iC++ or GtACR [50] in specific subcellular compartments (e.g. in somata and/or dendrites) would be advantageous to increase the specificity of chloride-mediated inhibition. To reach this goal, iC++ expression should colocalize with $GABA_A$ receptors by using specific trafficking and protein sorting signals. Both approaches are currently subject to intensive ongoing research efforts.

References

1. Deisseroth K (2014) Circuit dynamics of adaptive and maladaptive behaviour. Nature 505 (7483):309–317

2. Boyden ES, Zhang F, Bamberg E, Nagel G, Deisseroth K (2005) Millisecond-timescale, genetically targeted optical control of neural activity. Nat Neurosci 8(9):1263–1268

3. Petreanu L, Huber D, Sobczyk A, Svoboda K (2007) Channelrhodopsin-2-assisted circuit mapping of long-range callosal projections. Nat Neurosci 10(5):663–668

4. Sohal VS, Zhang F, Yizhar O, Deisseroth K (2009) Parvalbumin neurons and gamma rhythms enhance cortical circuit performance. Nature 459(7247):698–702

5. Carlo CN, Stevens CF (2013) Structural uniformity of neocortex, revisited. Proc Natl Acad Sci U S A 110(4):1488–1493

6. Brown MT et al (2012) Ventral tegmental area GABA projections pause accumbal cholinergic interneurons to enhance associative learning. Nature 492(7429):452–456

7. Tsai HC et al (2009) Phasic firing in dopaminergic neurons is sufficient for behavioral conditioning. Science 324(5930):1080–1084

8. Stuber GD et al (2011) Excitatory transmission from the amygdala to nucleus accumbens facilitates reward seeking. Nature 475 (7356):377–380

9. Kravitz AV et al (2010) Regulation of parkinsonian motor behaviours by optogenetic control of basal ganglia circuitry. Nature 466 (7306):622–626

10. Adhikari A et al (2015) Basomedial amygdala mediates top-down control of anxiety and fear. Nature 527(7577):179–185

11. Gunaydin LA et al (2014) Natural neural projection dynamics underlying social behavior. Cell 157(7):1535–1551

12. Chen BT et al (2013) Rescuing cocaine-induced prefrontal cortex hypoactivity prevents compulsive cocaine seeking. Nature 496 (7445):359–362

13. Warden MR et al (2012) A prefrontal cortex-brainstem neuronal projection that controls response to behavioural challenge. Nature 492(7429):428–432

14. Krook-Magnuson E, Armstrong C, Oijala M, Soltesz I (2013) On-demand optogenetic control of spontaneous seizures in temporal lobe epilepsy. Nat Commun 4:1376

15. Zhang F et al (2007) Multimodal fast optical interrogation of neural circuitry. Nature 446 (7136):633–639

16. Zacharias DA, Tsien RY (2006) Molecular biology and mutation of green fluorescent protein. Methods Biochem Anal 47:83–120

17. Chen TW et al (2013) Ultrasensitive fluorescent proteins for imaging neuronal activity. Nature 499(7458):295–300

18. Oesterhelt D, Stoeckenius W (1973) Functions of a new photoreceptor membrane. Proc Natl Acad Sci U S A 70(10):2853–2857

19. Lanyi JK (1998) Understanding structure and function in the light-driven proton pump bacteriorhodopsin. J Struct Biol 124 (2–3):164–178

20. Kolbe M, Besir H, Essen LO, Oesterhelt D (2000) Structure of the light-driven chloride pump halorhodopsin at 1.8 A resolution. Science 288(5470):1390–1396

21. Rothschild KJ, Bousche O, Braiman MS, Hasselbacher CA, Spudich JL (1988) Fourier transform infrared study of the halorhodopsin pump. Biochemistry 27 (7):2420–2424

22. Nagel G et al (2003) Channelrhodopsin-2, a directly light-gated cation-selective membrane channel. Proc Natl Acad Sci U S A 100 (24):13940–13945

23. Ritter E, Stehfest K, Berndt A, Hegemann P, Bartl FJ (2008) Monitoring light-induced structural changes of Channelrhodopsin-2 by UV-visible and Fourier transform infrared spectroscopy. J Biol Chem 283 (50):35033–35041

24. Gunaydin LA et al (2010) Ultrafast optogenetic control. Nat Neurosci 13(3):387–392

25. Berndt A, Lee SY, Ramakrishnan C, Deisseroth K (2014) Structure-guided transformation of channelrhodopsin into a light-activated chloride channel. Science 344(6182):420–424

26. Berndt A et al (2016) Structural foundations of optogenetics: determinants of channelrhodopsin ion selectivity. Proc Natl Acad Sci U S A 113 (4):822–829

27. Nagel G et al (2002) Channelrhodopsin-1: a light-gated proton channel in green algae. Science 296(5577):2395–2398

28. Scheerer P et al (2008) Crystal structure of opsin in its G-protein-interacting conformation. Nature 455(7212):497–502

29. Nagel G et al (2005) Channelrhodopsins: directly light-gated cation channels. Biochem Soc Trans 33(Pt 4):863–866

30. Berndt A (2011.) Mechanismus und anwendungsbezogene Optimierung von Channelrhodopsin-2. (Mathematisch-Naturwissenschaftliche Fakultät I)

31. Berndt A, Prigge M, Gradmann D, Hegemann P (2010) Two open states with progressive proton selectivities in the branched channelrhodopsin-2 photocycle. Biophys J 98 (5):753–761

32. Berndt A et al (2011) High-efficiency channelrhodopsins for fast neuronal stimulation at low light levels. Proc Natl Acad Sci U S A 108 (18):7595–7600

33. Berndt A, Yizhar O, Gunaydin LA, Hegemann P, Deisseroth K (2009) Bi-stable neural state switches. Nat Neurosci 12(2):229–234

34. Hille B (2001) Ion channels of excitable membranes. Sinauer Associates, Inc, Sunderland, MA

35. Feldbauer K et al (2009) Channelrhodopsin-2 is a leaky proton pump. Proc Natl Acad Sci U S A 106(30):12317–12322

36. Zhang YP, Oertner TG (2007) Optical induction of synaptic plasticity using a light-sensitive channel. Nat Methods 4(2):139–141

37. Tee BC et al (2015) A skin-inspired organic digital mechanoreceptor. Science 350 (6258):313–316

38. Klapoetke NC et al (2014) Independent optical excitation of distinct neural populations. Nat Methods 11(3):338–346

39. Seki A et al (2007) Heterologous expression of Pharaonis halorhodopsin in Xenopus laevis oocytes and electrophysiological characterization of its light-driven Cl- pump activity. Biophys J 92(7):2559–2569

40. Chuong AS et al (2014) Noninvasive optical inhibition with a red-shifted microbial rhodopsin. Nat Neurosci 17(8):1123–1129

41. Kim SY et al (2013) Diverging neural pathways assemble a behavioural state from separable features in anxiety. Nature 496 (7444):219–223

42. Iyer SM et al (2014) Virally mediated optogenetic excitation and inhibition of pain in freely moving nontransgenic mice. Nat Biotechnol 32(3):274–278

43. Kato HE et al (2012) Crystal structure of the channelrhodopsin light-gated cation channel. Nature 482(7385):369–374

44. Ilango A et al (2014) Similar roles of substantia nigra and ventral tegmental dopamine neurons in reward and aversion. J Neurosci 34 (3):817–822

45. Tan KR et al (2012) GABA neurons of the VTA drive conditioned place aversion. Neuron 73 (6):1173–1183

46. Han JH et al (2009) Selective erasure of a fear memory. Science 323(5920):1492–1496

47. Wright R, Raimondo JV, Akerman CJ (2011) Spatial and temporal dynamics in the ionic driving force for GABA(A) receptors. Neural Plast 2011:728395

48. Cossart R, Bernard C, Ben-Ari Y (2005) Multiple facets of GABAergic neurons and synapses: multiple fates of GABA signalling in epilepsies. Trends Neurosci 28(2):108–115

49. Glykys J et al (2009) Differences in cortical versus subcortical GABAergic signaling: a candidate mechanism of electroclinical uncoupling of neonatal seizures. Neuron 63(5):657–672

50. Govorunova EG, Sineshchekov OA, Janz R, Liu X, Spudich JL (2015) NEUROSCIENCE. Natural light-gated anion channels: a family of microbial rhodopsins for advanced optogenetics. Science 349(6248):647–650

Chapter 4

Optogenetic Control of Intracellular Signaling: Class II Opsins

Erik Ellwardt and Raag D. Airan

Abstract

Opsins are classified as either class I (microbial) or II (seven transmembrane) opsins. Class I opsins include channelrhodopsin and halorhodopsin, and are reviewed in previous chapters. Class II opsins are G-protein-coupled receptors (GPCR) and include the vertebrate opsins that underlie mammalian, including human, vision. Chimeras made of class II opsins and other GPCRs allow the precise control of secondary messengers of intracellular signaling like cyclic adenosine monophosphate (cAMP) or the inositol triphosphate (IP$_3$)/calcium system. These allow for optogenetic control of cellular behavior in addition to the excitation/inhibition axis that channelrhodopsin and halorhodopsin offer. The fast kinetics of light transduction within these single-element chimeras allows temporally precise control of GPCR signaling. Spatially precise control can be achieved via small optic fibers and microscopic control of the illumination field. We here give an overview about recent developments of class II opsin/GPCR chimera as promising tools for molecular and behavioral manipulation.

Key words Class II opsins, GPCR, Optogenetics, Viral transduction, Signaling cascade, Light stimulation, Chimeric receptor

1 Introduction

Class II opsins are light-sensitive seven transmembrane proteins which are coupled to G-proteins (GPCR). They are found in vertebrate animals and are usually necessary for vision but as well for circadian rhythm generation. Rhodopsin for instance is present in the mammalian eye and consists of an opsin and 11-cis-retinal, which is tightly bound in a central binding pocket of the opsin protein. Intracellularly, G-proteins transduce the light signal into modulation of the concentration of a variety of secondary messengers. Chimeras between class II opsins and other GPCRs, which have been dubbed the optoXRs, have recently been introduced as additional tools for the optogenetic control of this form of neural signaling.

Albrecht Stroh (ed.), *Optogenetics: A Roadmap*, Neuromethods, vol. 133,
DOI 10.1007/978-1-4939-7417-7_4, © Springer Science+Business Media LLC 2018

1.1 Existing Opsins Linked to G-proteins

Light-sensitive proteins in the mammalian eye are essential for vision and loss of function of those rhodpsins can cause diseases like retinitis pigmentosa. Class II opsins typically consist of seven transmembrane helices which are intracellular coupled to a G-protein. The opsin tightly binds 11-cis-retinal, which after light activation isomerizes to all-trans-retinal. This isomerization leads to a conformational change of the opsin, which then activates its G-protein *transducin* [1]. The alpha subunit of transducin then exchanges GDP for GTP and subsequently activates downstream phosphodiesterases.

Prior structural biology work with GPCR chimeras demonstrated for the rhodopsin-like family of GPCRs—which includes the vast majority of most known and characterized GPCRs—that if a chimeric protein preserved the intracellular domains of the receptor, the activating stimulus could be switched to another GPCR, no matter if the activating stimulus is a drug or light [2, 3]. The idea to use light-induced alteration of intracellular cascades and second messengers for neuroscientific questions became amenable when the first bioengineered rhodopsin/adrenergic receptor chimerae were published as the optoXRs [4]. The fusion of the extracellular and transmembrane domain of rhodopsin with the intracellular domains of adrenergic receptors enabled spatiotemporally precisely control of adrenergic-like intracellular signal transduction with light (Fig. 1).

The canonical function of GPCRs depends upon the G-protein to which they couple. The transducin of rhodopsin transduces the light input signal into phosphodiesterase activity that regulates intracellular cyclic guanosine monophosphate (cGMP) activity. Endogenous β_2-adrenergic receptor (β_2-AR) agonist binding leads to activation of "stimulatory" G-proteins (G_s), which activates

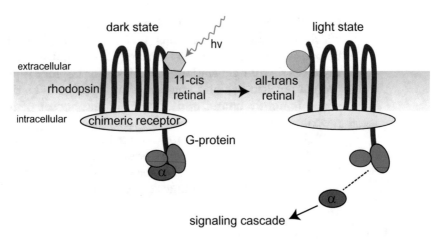

Fig. 1 Class II opsins following light stimulation. Light causes an isomerization of cis-retinal to trans-retinal which activates rhodopsin. The intracellular domain is substituted by a chimeric GPCR which after rhodopsin activation leads to activation of the intracellular signaling cascade of the original GPCR

adenylyl cyclase, which produces cyclic adenosine monophosphate (cAMP) that then, e.g., activates protein kinase A (PKA). In contrast, "inhibitory" G-proteins (G_i), such as those that couple to the α_2-adrenergic receptor (α_2-AR) lead to decreases of cAMP and net downregulation of PKA activity. Finally, activation of α_1-adrenergic receptors (α_1-AR) leads to activation of the G-protein G_q, which leads to activation of phospholipase C (PLC), which cleaves phospholipids to inositol triphosphate (IP_3) and diacyl glycerol (DAG). DAG then activates protein kinase C (PKC) and IP_3 leads to increase of cytosolic calcium through release of organelle calcium stores. In addition, GPCRs may also signal substantially through noncanonical pathways such as β-arrestin, ERK, and MAPK signaling pathways. The creation of a chimera of rhodopsin with the intracellular parts of β_2-AR, called opto-β_2AR, has already proven to retain the functionality of the original β_2-AR following green light stimulation [4–6], in terms of both its canonical and noncanonical signaling pathway activation. Stimulation of this opto-β_2-AR causes light intensity dependent increases of cAMP with similar activation kinetics and internalization to the endogenous beta-2-adrenergic receptor. Viral transduction of opto-β_2-AR via stereotactic injection into specific brain regions already allows studying the effect of elevated cAMP in the living animal [4, 5]. This enables the investigator to elucidate the effect of the cAMP signaling cascade in spatially confined regions under precise temporal control. Chimeras of rhodopsin and α_1-AR, called opto-α_1-AR, exhibit light-induced phospholipase C-mediated increase of inositol triphosphate upon green light stimulation [4]. Both opto-α_1-AR and opto-β_2-AR have been also successfully tested in astrocytes [7], where both lead to an intracellular calcium increase via an autocrine secretion of ATP after G_q or G_s stimulation.

In principle, it is possible to couple any GPCR activity—and in particular the Class A family of GPCRs—to rhodopsin and thereby generate a device for optogenetic control of the intracellular signaling associated with that GPCR. This was recently achieved as well for the G_i-coupled serotonin receptor 5-HT_{1A} [8], called opto-5HT_{1A}, and for μ-opioid signaling via opto-MOR [9]. Those two constructs have been tested in vitro and in vivo and modulate serotonergic and opioid signaling, respectively, via spatiotemporal precise light stimulation. Opto-mGluR6 might be a promising approach for restoring vision after photoreceptor damage in the retina [10]. The metabotropic glutamate receptor 6 is linked to the light-sensitive protein melanopsin which is normally involved in pupillary reflex and circadian rhythm. In contrast to many other opsins, it is very light sensitive and even reacts to daylight [10]. Combining class II opsins with other optical sensors like cAMP sensors to monitor the effect of light stimulations on a molecular level in real time opens new strategies for spatiotemporally precise modulation and monitoring of cellular functions [11]. Although in practice the spectral properties of the excitation and emission wavelengths that are used have to be kept in mind carefully (see the overview of optoXRs in Table 1).

1.2 Other Photoactive Receptors and Enzymes

In addition to class II opsin/GPCR chimeras, other groups have reported the use of similar intracellular signaling control with class I opsin derivatives. The photoactivated adenylyl cyclase from *Beggiatoa* (bPAC) directly produces cAMP after blue light stimulation [13, 14] and shows a 300-fold increase in cAMP production; although it appears to have significant baseline activity without light stimulation. Modulation of tyrosine kinase receptors by light has been described recently [15] as well. Fibroblast growth factor receptor and epidermal growth factor receptor both have been linked to light sensing proteins (opto-RTK), two examples which involve receptor tyrosine kinases.

1.3 Applications for Class II Opsins in Medicine

Promising pioneering experiments have been already conducted which involve a recruitment of anti-tumor cytotoxic T cells into melanoma applying optogenetics [12] in mice using these tools. Their use in, e.g., ophthalmology [10], cardiology, and neurology

Table 1
Overview of existing chimeric Class II opsins

Opsin	Characteristics	Excitation wavelength	References
Opto-β_2-AR	• Bovine rhodopsin and β_2 adrenergic receptor • involves G_s signaling • cAMP increases	504 nm	[4, 5, 7]
Opto-α_1-AR	• Bovine rhodopsin and α_1 adrenergic receptor • involves G_q signaling • IP_3 increases	504 nm	[4, 7]
PA-CXCR4	• Rhodopsin and photoactivatable-chemokine C-X-C motif receptor 4 • involves G_i signaling • Calcium increases	505 nm	[12]
Opto-5-HT	• Vertebrate rhodopsin and serotonergic receptor 5-HT • involves $G_{i/o}$ signaling • K^+ increases intracellular	485 nm	[8]
Opto-MOR	• Rat rhodopsin opioid like receptor • involves $G_{i/o}$ signaling • cAMP decreases	465 nm	[9]
Opto-mGluR6	• Melanopsin and metabotropic glutamate receptor 6 • Closes TRPM1 channels	467 nm	[10]

are also conceivable. The direct optogenetic activation (or inactivation) of neurons, cardiomyocytes or skeletal muscle cells with light activated ion channels like channelrhodopsin-2 (ChR2) has shown promising results in animal or tissue experiments [16]. However, a substantial effort still has to be made for class II opsins to be implemented as therapeutic tools in medical treatments—namely in terms of overcoming the barriers to translate gene therapy. So far, the advantages of temporally and spatially precise control of intracellular signaling cascades have opened new avenues for answering basic scientific questions. The translation from basic neuroscience to clinics seems possible for the future for the application of class II opsins in medicine.

2   Experimental Procedures

2.1   In Vitro

Published constructs are typically available at the researchers' lab who first described the opsin. When the construct/plasmid arrives, normally a clonal bacteria-mediated expansion is necessary. To visualize the protein, constructs are tagged to fluorophores such as mCherry (Fig. 2a) or yellow fluorescent protein (YFP). Make sure that the wavelengths of those reporters do not interfere with the light-sensitive action spectra of the optogenetic tools. For a detailed description of each step for one possible experiment with a class II opsin in HEK293 cells please see below. In principle, every cell type which can be transfected can be used.

1. Transfect cells 24–48 h before light stimulation with the chosen construct, if it is not already available in a frozen cell line. If so, thaw the cells at least 72 h before you want to use them. For transfection you might use lipofectamine (Invitrogen, Germany). For other cell types, electroporation might lead to a higher yield of transfected cells.

2. Differences in cell density can change the result. Check transfection rate before starting the experiment. Always use controls with the exact same treatment.

3. Add 1–10 μM 9-cis-retinal (dissolved in DMSO, shielded from light; Sigma-Aldrich, Germany) to the cells at least 1 h before you start your experiment. You have to try different concentrations. Always shield your cells from light from now on.

4. Determine light intensity with a photometer (Newport, Thorlabs, USA).

5. Check your light stimulation setup before you start. Make sure that you use the correct wavelength in your stimulation setup although many sensors use blue or green light for excitation. You can use DPSS lasers for example. Depending on sensitivity of the construct, you could also use light emitting diodes.

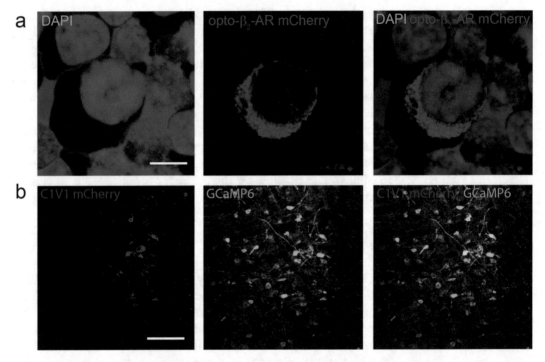

Fig. 2 Transfection/transduction in vitro and in vivo. (**a**) Transfection of primary T cells with the class II opsin opto-β_2-adrenergic receptor (opto-β_2-AR) shows membrane localization and stable transfection. Scale bar 10 μm. (**b**) Transduction of neurons with adeno-associated virus-mediated light-sensitive channel protein C1V1 and GCaMP6 (AAV.CamKII.C1V1-mCherry and AAV.Syn.GCaMP6). Virus linked class II opsins are also available and can be used for in vivo optogenetic experiments. Scale bar 100 μm

6. Ideally, place cells at 37 °C with oxygen and 5% CO_2 supply. Differences in room temperature might cause different activation of the sensor.

7. Stimulate cells with different light intensities and different durations. Keep in mind that sensors might get inactivated or become internalized following activation. Ideally the kinetics have been published already, otherwise titration of best stimulation pattern (maybe repetitive stimulation) has to be determined.

8. Depending on your readout you might have to stop degradation of the product by adding blockers or simply by stopping the reaction with, e.g., HCl. If you use another sensor for the readout simultaneously you have to keep in mind wavelengths so that the sensors do not interfere.

2.2 In Vivo Viral Transduction of Opsins and Imaging

For in vivo experiments, ideally a transgenic animal is used. If one is not available, transduction via lentiviruses or adeno-associated virus (AAV) can be done (see Chaps. 1 and 2). Those viruses, which genetically introduce the opsins are available in several core facilities and can be ordered easily; this means that virus production does not have to be done by your laboratory, which would require additional expertise and hardware. The appropriate titer of the virus, which needs to be injected, has to be determined as each brain region behaves different in terms of expression. Confocal analysis three weeks after injection should be performed to correlate the titer with the expression pattern of the construct (see Fig. 2b for transduction of C1V1 and GCaMP6).

2.2.1 Process for Viral Injection

1. Inject analgesics (e.g. carprofen, 15 µl/g bodyweight) s.c. 30 min prior to first incision.

2. Anesthetize animals with ketamine i.p. or isoflurane via mask and continuous oxygen supply on a warming plate (37 °C). Place animal in a stereotactic apparatus (e.g. Kopf instruments, Tujunga, CA, USA).

3. Check that the animal does not react anymore to external stimuli, otherwise increase anesthesia. Protect eyes by administering ointment (e.g. Bepanthen, Bayer AG, Leverkusen, Germany).

4. Shave the animal where you want to incise the skin. Make an incision with a surgical scissor or scalpel to remove skin. Apply H_2O_2 to remove tissue and for exposure of the skull.

5. Determine with your *x-y-z* controller the target region on the surface based on the position of the bregma ($x = 0, y = 0$).

6. Use an, e.g., dental driller for trepanation of the skull at the target region. Bleeding should be treated with a surgical sponge and 0.9% physiological solution to achieve a fast hemostasis and to remove blood and debris.

7. Under control of your *x-y-z* controller slowly move to the corresponding depth of the target brain region with an, e.g., 26 G (or smaller) Hamilton syringe (Model 701 RN SYR, Hamilton Company, Switzerland) or use micropipettes (e.g. ringcaps from Hirschmann, Germany).

8. Injection of the loaded virus should be performed automatically as slow as possible. Wait 10 min for distribution of the virus within the tissue and slowly remove the syringe.

9. Clip or sew the skin, put the animal into a heating chamber until it awakes and wait 3 weeks for further experiments regarding the construct.

2.2.2 Imaging and Optogenetic Stimulation

1. Repeat steps 1–6 to expose the brain region where the construct was injected 3 weeks earlier.

2. Transfer animal to a two-photon microscope and connect anesthesia and temperature control. Control breathing rate, it should not be less than 90/min (otherwise the animal is anesthetized to deep).

3. Look for expression of the opsin in the epifluorescent mode and start imaging.

4. Light intensities for optogenetics have to be determined with a photometer (Newport, Thorlabs, USA) as stated in the in vitro section. An integrated optogenetic setup (meaning the beam path goes through the objective) should be used ideally with fast closing and opening shutters.

5. When the experiment is finished, transcardially perfuse mice with 4% paraformaldehyde (PFA) and store tissue for further analysis for the expression of the opsin in the tissue.

3   Notes

3.1   In Vitro Experiments

1. Check the transfection rate before you start with your experiment. If possible use transgenic cell lines as variability will decrease.

2. Always use the same distance of the optic fiber to the cells. There is an exponentially decay of light intensity from the tip of the fiber (see Chap. 13). Therefore always measure light intensity using your photometer at the application distance.

3. Use an incubation chamber for experiments as temperature changes can cause significant changes in cellular physiology.

4. Make sure that you shield cells from light once you added the retinal. There is considerable dark activity of some opsins.

5. Add 9-cis-retinal at least 30 min before optogenetic experiments. Vertebrate organisms provide sufficient endogenous cis-retinal, for cell culture experiments it has to be added.

6. Make sure that you chose the correct wavelength for your experiment. If you do not have the appropriate laser, it might still work provided sufficient overlap between the laser wavelength and the excitation spectrum of the opsin. The effect will be lower if you do not use the wavelength for maximal excitation—although opsin action spectra are broad.

7. Control experiments:

    (a) Use transfected cells which will not undergo light stimulation. This is important to rule out a modulation by the transfection itself or dark activity of the opsin.

    (b) Use nontransfected cells which will get light stimulated. This will rule out an effect which might be due to light irradiation such as heat development.

    (c) Use a control plasmid (e.g. GFP under the CAG promoter) without the opsin, which will be light stimulated. Keep in mind that still the properties of the individual plasmid might cause an unspecific change.

(d) Use chemicals (e.g. forskolin or isoprenaline to induce cAMP production) acting on transfected wild-type GPCRs as a positive control.

3.2 In Vivo Experiments

1. Transduction efficacy should be controlled with epifluorescent and confocal microscope analyses to reassure expression of the opsin at the site of injection (target region). Moreover high magnification (63× fold objective) should demonstrate membrane bound expression of the opsins. Class II opsins should be always expressed at the cell surface. Failed transduction might be due to, e.g., defective viruses or leakage of the virus after injection and retrieval of the syringe. Therefore the syringe should be withdrawn after approximately 10 min, which will reduce leakage through the injection canal.

2. The virus titer can be either to low (insufficient expression) or to high (causing toxicity); different titers should be tested.

3. The process of transduction of the target region within the brain can lead to injury of blood vessels and cause stroke symptoms. Those mice have to be excluded from experiments.

4. Control experiments: perform transduction with a virus which carries only a fluorophore to rule out any effect of the transduction process itself.

4 Outlook

Class II opsins exhibit a great potential for neuroscientific analysis as outlined in this chapter. However, its proper use is challenging and experimental conditions (temperature, CO_2, kinetics of constructs, proper light intensities, necessary co factors) need to be optimized carefully to yield proper results. Spatially and temporally precise control of intracellular signaling cascades is feasible with these new tools. Still, at the same time the issue of how to measure these modifications arises and has to be done carefully. Classic techniques like ELISA, quantitative mRNA analysis via qPCR, or western blotting are valid procedures. But, they all bear several limitations in terms of their spatial and temporal resolution. Optical sensors in contrast share properties with class II opsins as they enable us to visualize instant changes in confined areas. The combination of class II opsins and optical sensors for modulation and monitoring of secondary messengers provides an all optical physiology approach. Stimulation of opsins can be conducted in a spatially and temporally precise manner and its intracellular effects could be monitored instantly. For instance, to measure cAMP, several FRET based cAMP sensors have been described already [17] and its combination with optogenetics, although not class II opsins, has been published [11]. However, combining class II

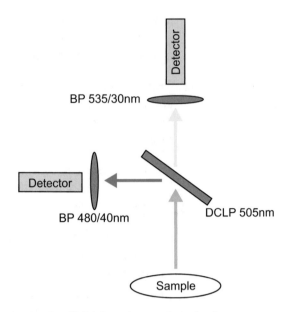

Fig. 3 FRET detection. FRET detection requires simultaneous measurement of acceptor and donor wavelength, e.g., via photomultiplier tubes (PMT) or photodiodes. To measure for instance a CFP/YFP FRET pair, a dichroic long pass (DCLP) filter at 505 nm can be used to separate both beams. *BP* band pass filter

opsins with FRET based sensors generates difficulties and adds requirements to the experimental setup: the potential spectral overlap of excitation wavelengths (of the opsin and FRET sensor) and emission wavelengths (opsin tagged to fluorophores for visibility, donor and acceptor wavelength of FRET sensor) have to be kept in mind. An algorithm for correction of photobleaching should be included for calculation of absolute effects. If a FRET sensor is used, simultaneous detection of acceptor and donor wavelengths has to be ensured (Fig. 3). Other direct possible optical readouts like bioluminescence are available too [5] and show similar temporal resolution. Some groups have detected secondary messengers like cAMP optically indirectly via calcium changes through cAMP gated calcium channels [4]. Calcium sensors can be either genetically encoded proteins (e.g. GCaMP6) or synthetic dyes like Oregon Green BAPTA-1, and are easier to image (just one wavelength for detection).

Although behavioral changes via class II opsins have been demonstrated in vivo [4, 9], simultaneous measurements of the intracellular second messengers on single cell level in vitro and especially in vivo remain challenging and further research needs to be conducted. The combination of class II opsins and optical sensors bears great potential in basic research involving all classes of cells and tissues.

References

1. Grobner G, Burnett IJ, Glaubitz C, Choi G, Mason AJ, Watts A (2000) Observations of light-induced structural changes of retinal within rhodopsin. Nature 405(6788):810–813
2. Kobilka BK, Kobilka TS, Daniel K, Regan JW, Caron MG, Lefkowitz RJ (1988) Chimeric alpha 2-,beta 2-adrenergic receptors: delineation of domains involved in effector coupling and ligand binding specificity. Science 240(4857):1310–1316
3. Kim JM, Hwa J, Garriga P, Reeves PJ, RajBhandary UL, Khorana HG (2005) Light-driven activation of beta 2-adrenergic receptor signaling by a chimeric rhodopsin containing the beta 2-adrenergic receptor cytoplasmic loops. Biochemistry 44(7):2284–2292. doi:10.1021/bi048328i
4. Airan RD, Thompson KR, Fenno LE, Bernstein H, Deisseroth K (2009) Temporally precise in vivo control of intracellular signalling. Nature 458(7241):1025–1029. doi:10.1038/nature07926
5. Siuda ER, McCall JG, Al-Hasani R, Shin G, Il Park S, Schmidt MJ, Anderson SL, Planer WJ, Rogers JA, Bruchas MR (2015) Optodynamic simulation of β-adrenergic receptor signalling. Nat Commun 6:8480. doi:10.1038/ncomms9480. http://www.nature.com/articles/ncomms9480#supplementary-information
6. Siuda ER, Al-Hasani R, McCall JG, Bhatti DL, Bruchas MR (2016) Chemogenetic and optogenetic activation of gαs signaling in the basolateral amygdala induces acute and social anxiety-like states. Neuropsychopharmacology 41(8):2011–2023
7. Figueiredo M, Lane S, Stout RF Jr, Liu B, Parpura V, Teschemacher AG, Kasparov S (2014) Comparative analysis of optogenetic actuators in cultured astrocytes. Cell Calcium 56(3):208–214. doi:10.1016/j.ceca.2014.07.007
8. Oh E, Maejima T, Liu C, Deneris E, Herlitze S (2010) Substitution of 5-HT1A receptor signaling by a light-activated G protein-coupled receptor. J Biol Chem 285(40):30825–30836. doi:10.1074/jbc.M110.147298
9. Siuda Edward R, Copits Bryan A, Schmidt Martin J, Baird Madison A, Al-Hasani R, Planer William J, Funderburk Samuel C, McCall Jordan G, Gereau RW IV, Bruchas Michael R (2015) Spatiotemporal Control of

Opioid Signaling and Behavior. Neuron 86(4):923–935. doi:10.1016/j.neuron.2015.03.066
10. van Wyk M, Pielecka-Fortuna J, Löwel S, Kleinlogel S (2015) Restoring the ON switch in blind retinas: opto-mGluR6, a next-generation, cell-tailored optogenetic tool. PLoS Biol 13(5):e1002143. doi:10.1371/journal.pbio.1002143
11. Karunarathne WKA, Giri L, Kalyanaraman V, Gautam N (2013) Optically triggering spatiotemporally confined GPCR activity in a cell and programming neurite initiation and extension. Proc Natl Acad Sci 110(17):E1565–E1574. doi:10.1073/pnas.1220697110
12. Xu Y, Hyun Y-M, Lim K, Lee H, Cummings RJ, Gerber SA, Bae S, Cho TY, Lord EM, Kim M (2014) Optogenetic control of chemokine receptor signal and T-cell migration. Proc Natl Acad Sci 111(17):6371–6376. doi:10.1073/pnas.1319296111
13. Schroder-Lang S, Schwarzel M, Seifert R, Strunker T, Kateriya S, Looser J, Watanabe M, Kaupp UB, Hegemann P, Nagel G (2007) Fast manipulation of cellular cAMP level by light in vivo. Nat Methods 4(1):39–42. doi:10.1038/nmeth975
14. Stierl M, Stumpf P, Udwari D, Gueta R, Hagedorn R, Losi A, Gärtner W, Petereit L, Efetova M, Schwarzel M, Oertner TG, Nagel G, Hegemann P (2011) Light modulation of cellular camp by a small bacterial photoactivated adenylyl cyclase, bPAC, of the soil bacterium Beggiatoa. J Biol Chem 286(2):1181–1188. doi:10.1074/jbc.M110.185496
15. Grusch M, Schelch K, Riedler R, Reichhart E, Differ C, Berger W, Inglés-Prieto Á, Janovjak H (2014) Spatio-temporally precise activation of engineered receptor tyrosine kinases by light. EMBO J 33(15):1713–1726. doi:10.15252/embj.201387695
16. Bruegmann T, van Bremen T, Vogt CC, Send T, Fleischmann BK, Sasse P (2015) Optogenetic control of contractile function in skeletal muscle. Nat Commun 6:7153. doi:10.1038/ncomms8153
17. Klarenbeek JB, Goedhart J, Hink MA, Gadella TWJ, Jalink K (2011) A mTurquoise-based cAMP sensor for both FLIM and ratiometric read-out has improved dynamic range. PLoS One 6(4):e19170. doi:10.1371/journal.pone.0019170

Chapter 5

Optogenetics in Stem Cell Research: Focus on the Central Nervous System

Johannes Boltze and Albrecht Stroh

Abstract

Stem cell-based therapies of CNS disorders represent a promising approach in translational and regenerative medicine. Stem cell-based tissue replacement and regeneration would, for the first time, offer a causal treatment strategy which is most likely not bound to a specific time window. Therapeutic strategies relying on this paradigm would require administration of exogenous stem cells to the CNS and/or the augmentation of endogenous stem cell capabilities. However, it remains unclear whether tissue replacement or bystander effects are required to induce such effects. Conventional experimental techniques will not be able to causally reveal such information, due to the complexity and coincidence of cellular processes and cell–target interactions, and the inability for longitudinal observations in vivo. Optogenetic approaches allow the targeted activation or inactivation of selected cell types, including stem cells. Optogenetics can therefore help to unravel the major therapeutic mechanism of stem cell therapy in two ways: (1) *to facilitate and improve neuronal differentiation of, e.g., ChR2 expressing stem cells* and (2) *to improve and test for functional integration of stem cell-derived neurons into the endogenous circuitry after transplantation.* Here, we review the state of the art of optogenetic manipulation of stem cells to optimize therapeutic utilization for CNS disorders with a particular focus on endogenous and exogenous stem cell populations.

Key words Embryonic stem cells, Differentiation, Optogenetics, Regenerative medicine, Stem cell transplantation

1 Introduction

Stem cells are undifferentiated, naïve cells being able to give rise to numerous differentiated and functional cell populations. Stem cells exhibit the ability of self-renewal by asymmetric division, resulting in another stem cell and a slightly more differentiated, progenitor cell [1]. The latter divides and differentiates further, giving rise to abundant downstream daughter cells that finally become fully matured and functional.

Stem cells can be classified regarding their potency and occurrence during development. Embryonic stem cells can be derived from the inner cell mass of the blastocyst and are able to

Albrecht Stroh (ed.), *Optogenetics: A Roadmap*, Neuromethods, vol. 133,
DOI 10.1007/978-1-4939-7417-7_5, © Springer Science+Business Media LLC 2018

differentiate into tissues of all three germ layers, ecto-, meso, and endoderm, called pluripotency. Early embryonic stem cells are discussed to be even omnipotent, being able to form extraembryonic tissue such as the trophoblast and therefore being able to give rise to an entire organism [2]. Fetal stem cells are multipotent stem/progenitor cell populations which are primarily found in primordial organs and show a reduced differentiation capacity. The use of embryonic and fetal stem cells is restricted by ethical considerations and the potential risk of uncontrolled growth and formation of disorganized or even tumor tissue. Postnatal stem cell populations are often less potent, being able to give rise to derivates of one germ layer only. However, they can be gained more easily and in higher quantity, while their application is usually not restricted by ethical considerations. Induced pluripotent stem (iPS) cells are a particularly interesting population. These cells are obtained from mature cells such as fibroblasts, which are reprogrammed by viral or mRNA-based transfection with the pluripotency factors c-Myc, Klf-4, Oct-4, and Sox-2 [3] to regain a pluripotent state. Another possible pluripotency factor combination is Oct4, Sox-2, Nanog, and Lin-28 [4] which avoids the use of c-Myc, a potent proto-oncogene. iPS cells show the same pluripotency as embryonic stem cells, and exhibit a similar risk of tissue malformation in vivo. Their main advantage is that they can be derived from mature tissue. Most stem cell populations also exhibit a high secretory activity, enabling them to influence their environment by growth factors, cytokines and immunomodulatory mediators [5]. These effects are frequently observed for post-natal stem cell populations such as mesenchymal stem cells [6] and are often referred to as "bystander" effects. Most stem cells are able of homing to sites of tissue injury by sensing of inflammatory and danger signals, presenting a migratory behavior showing highly interesting similarities to that of leukocytes [7].

Following the discovery of postnatal stem cell populations, it was increasingly recognized that tissue turnover, maintenance, and repair are realized by small endogenous stem cell depots, often embedded in a well-balanced microenvironment called the stem cell niche. Stem cell niches may comprise compartments for hematopoietic stem cells in the bone marrow [8] or depots of cutaneous stem cell population in the hair follicle [9]. Stem cell activity and profound regeneration have been observed and thoroughly studied for decades in tissues and organs showing a naturally rapid turnover and a high regenerative potential such as the blood or the skin. However, stem cell populations were also discovered in organs being much less capable of self-renewal during ageing or following damage.

One of the most prominent examples for such organs is the brain. Stem cell populations in the brain have been discovered mainly in the hippocampus and the subventricular zone. The hippocampus belongs to the medial telencephalon and is a part of the archicortex. It bundles information from different sensory systems and projects to the cerebral cortex. It is a central structure for learning and memory formation (see also Chap. 12). Stem cells residing in the hippocampus are believed to provide the plasticity required to form novel and complex memories. The subventricular zones are thin cellular structures located below the ependymal layer of the cerebral ventricles. As such, they form the outer limit of the former neural tube and play a major role during the ontogenesis of the central nervous system (CNS). The subventricular zones contain mainly astrocytes and astrocyte-like cell populations, but many of those are thought to have stem cell capacities and resemble radial glia cells [10]. The existence of adult cortical stem cells has been proposed repeatedly [11, 12] but their nature as well as the final proof of their existence is subject to further investigation. Indeed, stem cell populations being endogenous to the brain were shown to proliferate and to migrate to sites of injury [13]. Some even differentiate to neuronal or glial cell populations [14]. Since most of these cells do not integrate well into surviving local circuits and degenerate in subsequent stages of tissue reorganization, their overall regenerative benefit is limited. Hence, tissue loss and functional decline in the CNS, for instance occurring after ischemic stroke or during the course of Parkinson's disease (PD), multiple sclerosis (MS) and amyotrophic lateral sclerosis (ALS), are therefore considered almost final once manifested. CNS stem cell populations are believed to be either incapable and/or to lack sufficient numbers to induce real regeneration in the post-lesional mammalian CNS [5].

Stem cell treatments are an emerging paradigm in translational and regenerative medicine. Researchers and clinicians strive to exploit the stem cell's unique potential for future treatments of diseases being characterized by tissue loss and degeneration. Many of those diseases have been considered untreatable in the past, but might be successfully addressed by upcoming therapies allowing tissue replacement and repair [15]. This particularly applies for disorders of the CNS. Stem cell-based tissue replacement and regeneration would, for the first time, offer a causal treatment strategy which is most likely not bound to a specific time window. Therapeutic strategies relying on this paradigm would require administration of exogenous stem cells to the CNS and/or the augmentation of endogenous stem cell capabilities. Profound therapeutic effects have been observed for many CNS disorders including Parkinson's Disease [16], stroke [17, 18], and Multiple Sclerosis [19]. However, it remains unclear whether tissue replacement or bystander effects are required to induce such effects. Conventional experimental setups utilizing molecular biology or histological techniques will not be able to reveal such information, due to the complexity and coincidence of cellular process and cell–target interactions, and the inability for longitudinal observations in vivo.

Optogenetic approaches allow the targeted activation or inactivation of selected cell types and individual cellular functions with high precision and good spatiotemporal discrimination [20]. Optogenetics can therefore help to unravel the major therapeutic mechanism of endogenous cerebral stem cell populations for structural and functional regeneration as well as the role of stem cell transplants [21–23]. Here, we review the state of the art of optogenetic manipulation of stem cells to optimize therapeutic utilization for CNS disorders with a particular focus on endogenous and exogenous stem cell populations.

Importantly, optogenetics can be utilized in two ways: (1) *to facilitate and improve neuronal differentiation of, e.g., ChR2 expressing stem cells* [21] and (2) *to improve and test for functional integration of stem cell-derived neurons into the endogenous circuitry after transplantation* [24]. We will cover both aspects in the following practical sections.

2   Practical Considerations for Choice of Virus and Opsin

Stem cells are by definition dividing cells. Typically, cell division rates range between 8 h for mouse embryonic stem cell to 24 h for human embryonic stem cells. This has important implications for the selection of virus used for the delivery of genes encoding your opsin of choice. In postmitotic neurons, adeno-associated viruses became the favorite choice in the field of optogenetics due to their low immunogenicity and the excellent safety profile (low biosafety level) (see Chap. 1). Moreover, a broad selection of serotype can be chosen to ensure cell-type-specificity (see Chap. 2). However, one reason for the low biosafety level attributed to third-generation AAVs represents the key downside for applications in stem cells: Transduction with AAVs does not lead to the genomic integration of the gene of interest (GOI). The viral plasmid rather represents an episome. This has no practical consequences in postmitotic cells, and expression of, e.g., ChR2 upon AAV-mediated transduction has been reported to remain rather stable for many months [25]. Nevertheless, this represents a key disadvantage in fast dividing cells. The episomal viral plasmid is not replicated during mitosis; therefore, even assuming a multiplicity of infection, the number of viral plasmids is divided into the two daughter cells. Consequently, the expression strength of the opsin—depending on the number of viral plasmids in the nucleus—is reduced by each cell division. Given the necessity of a strong and stable expression of opsins throughout stem cell differentiation and ultimately integration into the neuronal network, within a time frame of weeks and months, AAVs cannot be considered for stem cell-based applications of optogenetics. Therefore, viruses ensuring stable insertion of the GOI into the genome should be employed. Lentiviral vectors, while requiring extended experience in viral production and exhibiting higher immunogenicity and biosafety level, provide genome integration, with the added benefit of a larger packaging capacity (see Chap. 1). The larger packaging capacity might allow for the integration of promoters restricting opsin expression, e.g., to a specific lineage. As mentioned throughout this book, due to the low single-channel conductance of, e.g., ChR2, the expression level has to be high (see Chaps. 3 and 8). This is of particular importance in stem cells. Stem cells are not electrical excitable, they do not fire action potentials, so any effect of optogenetic stimulation solely relies on the conductance of ions modifying signal conductance, chiefly Ca^{2+} [21]. Therefore, channels with increased Ca^{2+} conductance might be advantageous (see Chap. 3). Ideally though, an optogenetic tool should allow for the modification and/or steering of stem cell differentiation *and* for induction of action potentials once the stem cells differentiated into mature neurons.

3   Generating a Stable Opsin Expressing Stem Cell Line

Here, we describe the generation of a stable line of mouse embryonic stem cells expressing high levels of ChR2.

1. Depending on your research question, choose either a promoter which is expressed throughout differentiation, such as elongation factor 1α (EF1α), or a promoter active only in a specific lineage (please see Sect. 4).

2. Choose an opsin directly fused with a fluorophore, to be able to judge membrane-bound location via fluorescence microscopy.

3. Viral transduction of (stem-) cells using viral vectors invariably leads to a heterogeneity of expression strengths due to the multiplicity of infection, leading to the insertion of multiple copies of the GOI into the genome. Therefore, a sorting step selecting only highly expressing—but viable—cells is critical. Cell viability may be assessed by propidium iodide (PI)/annexin V (AnV)-staining using fluorescence-activated cell sorting (FACS) and, ideally, sorting of double-negative populations in the same step.

4. Incubate stem cells with opsin-fluorophore encoding lentivirus for 24 h.

5. Ideally conduct FACS flow cytometry. Select 5% of the population based on the highest expression of opsin-fused fluorophore *and* the negativity of apoptotic (AnV⁺) and necrotic (PI⁺) markers.

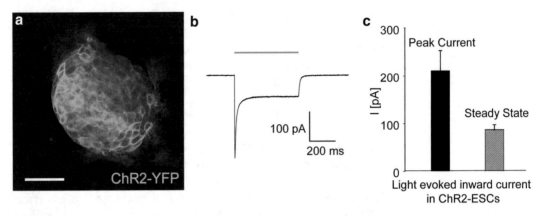

Fig. 1 Functional expression of Channelrhodopsin-2 (ChR2) in embryonic stem cells in vitro. (**a**) Confocal micrograph of ChR2-ESCs revealing membrane localized expression of ChR2-YFP. Scale bar: 50 μm. (**b**) Pulsed light (473 nm) evoked inward currents in a ChR2–YFP ESC in voltage clamp (light pulse indicated by *blue bar*). (**c**) Summary data on evoked photocurrents (mean ± s.e.m., *n* = 15 cells). *Adapted from* [21]

6. Upon sorting, plate a defined number of transduced ESCs and place under a fluorescence microscope with a low-magnification objective. Quantify the fluorescence level, note down all relevant parameters impacting the fluorescence such as laser power, gain, sampling rate. Repeat this measure every other day for two weeks. While an initial dip of fluorescence is unavoidable due to the reduced viability of the highest-expressing cells, the fluorescence level should reach a stable plateau (see Fig. 1). Once this stable plateau is reached, freeze stocks of your now stably opsin expressing stem cell line. Control fluorescence level of the running stem cell line every 2 weeks. Should the fluorescence levels go down significantly, which may occur after several months due to gene silencing, discard and thaw a new stock.

7. For inducing neuronal differentiation, opsin stimulation alone has not been shown to be effective [21]. Optogenetic stimulation can increase neuronal differentiation, in addition to established protocols, such as retinoic acid for differentiation into mainly dopaminergic neurons. The most effective stimulation parameters have not been conclusively explored (Fig. 2). However, be aware of the intrinsic channel kinetics o, e.g., ChR2: The peak current exhibits the by far largest ion flux, but inactivates fast (Fig. 1, see also Chap. 13). Therefore, rather use short pulses of typically 10–50 ms. We suggest employing an automated region-of-interest based stimulation routine (Fig. 2), light intensities should range at 10 mW/mm^2. Illumination of ChR2-expressing mouse ESCs every hour for 10 s/ ROI has shown to significantly increase neuronal marker expression.

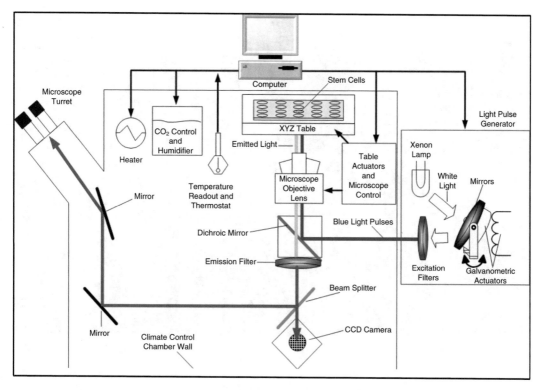

Fig. 2 Spatiotemporally precise long-term optogenetic control of ESCs. Schematic of automated optical stimulation setup. The multiwell plate containing cells is placed on the stage of a fluorescence microscope rig with robotic stage, millisecond-scale optical switching, autofocus, and environmental control of temperature and CO2. Custom software controls all functions of the stage and microscope, as well as the CCD camera and the light source. *Adapted from* [21]

8. During differentiation, expression strength of ChR2—e.g. driven by a ubiquitous promoter such as EF1α—will slowly decrease, not due to gene silencing, but rather most probably due to the overall lower protein expression level in neurons compared to rapidly dividing stem cells.

9. Depending on the differentiation protocol used, transduced ESCs will ultimately differentiate into mature spiking neurons. Use electrophysiological measures, ideally in whole-cell configuration, to demonstrate optogenetically induced action potential firing (Fig. 3). Only then, upon the in vitro prove, it can be assumed, that transplanted pre-differentiated neurons will fire action potentials upon optogenetic modulation.

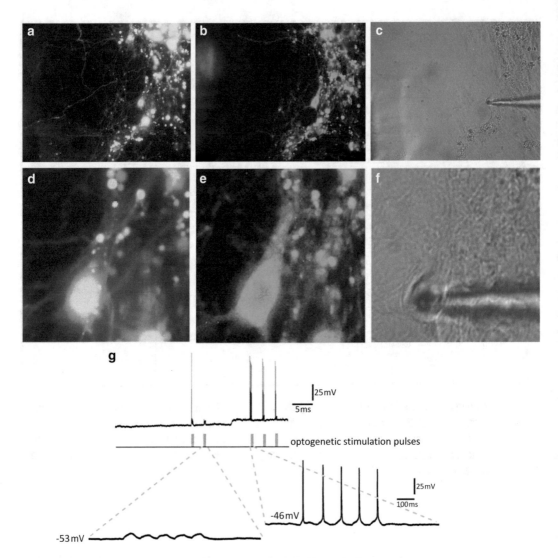

Fig. 3 Optogenetic stimulation of murine adult SEZ neurosphere-derived mature neurons transfected with Ngn2-IRIS-DSRed and transduced with EF1a-ChR2-eYFP lentivirus. (**a**) Expression of Ngn2. (**b**) Coexpression of ChR2-eYFP. (**c**) Patch clamp pipette positioned on a coexpressing stem cell-derived neuron. (**d–f**) High magnification images showing neuronal morphology and smooth membrane-bound expression of opsin. (**g**) Light stimulation leads to action-potential firing of opsin-transduced stem cell-derived neuron. *In cooperation with Benedikt Berninger, Mainz*

4 Notes

1. Choosing a specific promoter only active later in differentiation process might be advantageous for driving specific lineages, but makes it difficult to assess transduction efficacy/sorting in stem cell stage. As packaging capacity of lentiviruses is quite large, a solution would be to include an additional fluorophore under control of a ubiquitous promoter in the plasmid, and using this

color for sorting in ES cell stage, when the specific promoter driving the opsin is not active yet.

2. Red fluorophores such as m-cherry tend to form intracellular aggregates. If possible, use green/yellow fluorophores (GFP or YFP). This will increase the fraction of functional opsin molecules integrated in the membrane.

3. Rather employ culture conditions in which stem cells are adhered to the bottom of the culture plate, avoid free floating cultures. Otherwise, no effective optogenetic modulation of differentiation can be assured, as cells will not be reliably located in the focal plane of the objective of the microscope used for stimulation.

5   Outlook

Numerous cell- and stem cell-based treatments for CNS disorders have been investigated over the past decades [26–28]. Many have shown considerable treatment benefits in relevant preclinical models, and some already advanced to early-stage clinical trials [29–31]. However, the initial enthusiasm regarding stem cell-based regenerative therapies for CNS disorders was mitigated when research results indicated that paracrine mechanisms and factor secretion [32], immunomodulatory properties [33, 34], and other bystander effects rather than brain tissue regeneration are at least partly responsible for the observed treatment effects. Moreover, it became clear that tissue replacement is harder to achieve than originally anticipated. This even accounts for stem cell populations which clearly have the potential to differentiate into functional tissue [5], although spontaneous differentiation and long-term integration of such cells is regularly observed even in human patients [35]. A potential explanation for this observation might be the decline of endogenous cellular guidance structures such as radial glia, which are present during organogenesis and are required for the orchestrated construction of the structurally and functionally highly complex human brain [36]. Artificial replacement of these guidance structures by inert biomaterials has been suggested [37], but those solutions are far from being clinically applicable. The lack of cellular guidance structures might be of particular relevance in CNS disorders leading to massive structural damage such as ischemic stroke or traumatic brain injury whereas diseases causing the loss of selected cell populations such as PD are believed to be easier targets for stem cell-based neurorestorative therapies [38]. Nevertheless, it remains unclear in the field whether or not a certain cellular ability including the differentiation potential into functional tissue is indeed required to induce the therapeutic benefit.

On the other hand, experimental and clinical protocols of future stem cell-based therapies will have to be tailored towards best possible implementation of the most relevant treatment effect. If tissue regeneration was the main therapeutic mechanisms, a more sophisticated, technically challenging and potentially complication-borne local (intracerebral) cell transplantation might be the option of choice, whereas systemic immunomodulation might be best achievable by a systemic intravascular administration. Finally, stem cell therapies come with a number of risks and side effects as it is the case for most treatment paradigms. Those might be related to the naïve nature of the utilized cell populations [39], the chosen transplantation route [6], or potential interactions with concurrent diseases [40]. Hence, the optimum treatment protocol will also require minimization of such potential adverse events, providing the best possible balance of benefit and risk.

Taken together, a clear understanding on the major and most important therapeutic mechanisms is pivotal for the clinical translation of stem cell therapies for CNS disorders. Optogenetics is among the most advanced methods to explore therapeutic effects of stem cell therapies for CNS disorders. For instance, dampening of neuroinflammatory mechanism as a therapeutic effect was observed by utilizing the technology [41]. On the other hand, some studies combined optogenetics with neuronal stem cell transplantation, thereby showing the importance of efficient neural circuit repair for sustained functional improvements after stroke or experimental PD in rodent models [42]. Exact understanding of circuit recovery mechanisms and their impact on functional reconstitution will be pivotal to optimize stem cell therapies while at the same time reducing their risks. Potential directions of such research might include solving the question which afferent and efferent projections emerging from a stem cell graft are required for functional improvements, and which temporal sequence of graft implantation and implementation are best suited to promote those. Moreover, the effects of a stem cell graft-derived signals on "downstream" brain areas has to be explored, including potential detrimental effects such as induction of epileptiform signal patterns. Finally, a thorough understanding of graft-derived circuits and their interaction with preexisting host networks or host circuits being subject to post-injury plastic reorganization will be relevant. Some studies already started to address such questions [43–45], but are far from providing a detailed and final understanding of the complex process of neuronal circuit repair.

Optogenetic techniques hold a great potential not only for the understanding, but also for the optimization therapeutic approaches targeting CNS disorders. Optogenetics has been shown to mitigate seizures in a rat model of ischemic cortical injury [46] and could likewise control post-ischemic neuronal hyperexcitability. Optogenetic manipulation of neuronal circuits could finally be used for targeted stimulation of neuronal networks as well for augmentation of stem cell-based neuronal network formation in order to augment plasticity-based functional recovery [47].

References

1. Horvitz HR, Herskowitz I (1992) Mechanisms of asymmetric cell division: two Bs or not two Bs, that is the question. Cell 68 (2):237–255

2. Ishiuchi T, Enriquez-Gasca R, Mizutani E, Boskovic A, Ziegler-Birling C, Rodriguez-Terrones D, Wakayama T, Vaquerizas JM, Torres-Padilla ME (2015) Early embryonic-like cells are induced by downregulating replication-dependent chromatin assembly. Nat Struct Mol Biol 22(9):662–671. doi:10.1038/nsmb.3066

3. Takahashi K, Yamanaka S (2006) Induction of pluripotent stem cells from mouse embryonic and adult fibroblast cultures by defined factors. Cell 126(4):663–676. doi:10.1016/j.cell.2006.07.024

4. Yu J, Vodyanik MA, Smuga-Otto K, Antosiewicz-Bourget J, Frane JL, Tian S, Nie J, Jonsdottir GA, Ruotti V, Stewart R, Slukvin II, Thomson JA (2007) Induced pluripotent stem cell lines derived from human somatic cells. Science 318(5858):1917–1920. doi:10.1126/science.1151526

5. Janowski M, Wagner DC, Boltze J (2015) Stem cell-based tissue replacement after stroke: factual necessity or notorious fiction? Stroke 46 (8):2354–2363. doi:10.1161/STROKEAHA.114.007803

6. Boltze J, Arnold A, Walczak P, Jolkkonen J, Cui L, Wagner DC (2015) The dark side of the force—constraints and complications of cell therapies for stroke. Front Neurol 6:155. doi:10.3389/fneur.2015.00155

7. Hoehn M, Kustermann E, Blunk J, Wiedermann D, Trapp T, Wecker S, Focking M, Arnold H, Hescheler J, Fleischmann BK, Schwindt W, Buhrle C (2002) Monitoring of implanted stem cell migration in vivo: a highly resolved in vivo magnetic resonance imaging investigation of experimental stroke in rat. Proc Natl Acad Sci U S A 99 (25):16267–16272

8. Hoggatt J, Kfoury Y, Scadden DT (2016) Hematopoietic stem cell niche in health and disease. Annu Rev Pathol 11:555–581. doi:10.1146/annurev-pathol-012615-044414

9. Pasolli HA (2011) The hair follicle bulge: a niche for adult stem cells. Microsc Microanal 17(4):513–519. doi:10.1017/S1431927611000419

10. Hansen DV, Lui JH, Parker PR, Kriegstein AR (2010) Neurogenic radial glia in the outer subventricular zone of human neocortex. Nature 464(7288):554–561. doi:10.1038/nature08845

11. Chen M, Puschmann TB, Wilhelmsson U, Orndal C, Pekna M, Malmgren K, Rydenhag B, Pekny M (2016) Neural progenitor cells in cerebral cortex of epilepsy patients do not originate from astrocytes expressing GLAST. Cereb Cortex. doi:10.1093/cercor/bhw338

12. Vinci L, Ravarino A, Fanos V, Naccarato AG, Senes G, Gerosa C, Bevilacqua G, Faa G, Ambu R (2016) Immunohistochemical markers of neural progenitor cells in the early embryonic human cerebral cortex. Eur J Histochem 60 (1):2563. doi:10.4081/ejh.2016.2563

13. Barkho BZ, Zhao X (2011) Adult neural stem cells: response to stroke injury and potential for therapeutic applications. Curr Stem Cell Res Ther 6(4):327–338

14. Gregoire CA, Goldenstein BL, Floriddia EM, Barnabe-Heider F, Fernandes KJ (2015) Endogenous neural stem cell responses to stroke and spinal cord injury. Glia 63 (8):1469–1482. doi:10.1002/glia.22851

15. Buzhor E, Leshansky L, Blumenthal J, Barash H, Warshawsky D, Mazor Y, Shtrichman R (2014) Cell-based therapy approaches: the hope for incurable diseases. Regen Med 9 (5):649–672. doi:10.2217/rme.14.35

16. Kriks S, Shim JW, Piao J, Ganat YM, Wakeman DR, Xie Z, Carrillo-Reid L, Auyeung G, Antonacci C, Buch A, Yang L, Beal MF, Surmeier DJ, Kordower JH, Tabar V, Studer L (2011) Dopamine neurons derived from human ES cells efficiently engraft in animal models of Parkinson's disease. Nature 480(7378):547–551. doi:10.1038/nature10648

17. Tornero D, Wattananit S, Gronning Madsen M, Koch P, Wood J, Tatarishvili J, Mine Y, Ge R, Monni E, Devaraju K, Hevner RF, Brustle O, Lindvall O, Kokaia Z (2013) Human induced pluripotent stem cell-derived cortical neurons integrate in stroke-injured cortex and improve functional recovery. Brain 136(Pt 12):3561–3577. doi:10.1093/brain/awt278

18. Boltze J, Reich DM, Hau S, Reymann KG, Strassburger M, Lobsien D, Wagner DC, Kamprad M, Stahl T (2012) Assessment of neuroprotective effects of human umbilical cord blood mononuclear cell subpopulations in vitro and in vivo. Cell Transplant 21(4):723–737. doi:10.3727/096368911X586783

19. Grade S, Bernardino L, Malva JO (2013) Oligodendrogenesis from neural stem cells: perspectives for remyelinating strategies. Int J Dev Neurosci 31(7):692–700. doi:10.1016/j.ijdevneu.2013.01.004

20. Azad TD, Veeravagu A, Steinberg GK (2016) Neurorestoration after stroke. Neurosurg Focus 40(5):E2. doi:10.3171/2016.2.FOCUS15637

21. Stroh A, Tsai HC, Wang LP, Zhang F, Kressel J, Aravanis A, Santhanam N, Deisseroth K, Konnerth A, Schneider MB (2011) Tracking stem cell differentiation in the setting of automated optogenetic stimulation. Stem Cells 29 (1):78–88. doi:10.1002/stem.558

22. Avaliani N, Sorensen AT, Ledri M, Bengzon J, Koch P, Brustle O, Deisseroth K, Andersson M, Kokaia M (2014) Optogenetics reveal delayed afferent synaptogenesis on grafted human-induced pluripotent stem cell-derived neural progenitors. Stem Cells 32 (12):3088–3098. doi:10.1002/stem.1823

23. Steinbeck JA, Choi SJ, Mrejeru A, Ganat Y, Deisseroth K, Sulzer D, Mosharov EV, Studer L (2015) Optogenetics enables functional analysis of human embryonic stem cell-derived grafts in a Parkinson's disease model. Nat Biotechnol 33(2):204–209. doi:10.1038/nbt.3124

24. Weitz AJ, Fang Z, Lee HJ, Fisher RS, Smith WC, Choy M, Liu J, Lin P, Rosenberg M, Lee JH (2015) Optogenetic fMRI reveals distinct, frequency-dependent networks recruited by dorsal and intermediate hippocampus stimulations. NeuroImage 107:229–241. doi:10.1016/j.neuroimage.2014.10.039

25. Stroh A, Adelsberger H, Groh A, Ruhlmann C, Fischer S, Schierloh A, Deisseroth K, Konnerth A (2013) Making waves: initiation and propagation of corticothalamic Ca2+ waves in vivo. Neuron 77(6):1136–1150. doi:10.1016/j.neuron.2013.01.031

26. Savitz SI, Chopp M, Deans R, Carmichael T, Phinney D, Wechsler L, Participants S (2011) Stem Cell Therapy as an Emerging Paradigm for Stroke (STEPS) II. Stroke 42(3):825–829. doi:10.1161/STROKEAHA.110.601914

27. Schwarz J, Schwarz SC, Storch A (2006) Developmental perspectives on human midbrain-derived neural stem cells. Neurodegener Dis 3(1–2):45–49. doi:10.1159/000092092

28. Uccelli A, Milanese M, Principato MC, Morando S, Bonifacino T, Vergani L, Giunti D, Voci A, Carminati E, Giribaldi F, Caponnetto C, Bonanno G (2012) Intravenous mesenchymal stem cells improve survival and motor function in experimental amyotrophic lateral sclerosis. Mol Med 18:794–804. doi:10.2119/molmed.2011.00498

29. Petrou P, Gothelf Y, Argov Z, Gotkine M, Levy YS, Kassis I, Vaknin-Dembinsky A, Ben-Hur T, Offen D, Abramsky O, Melamed E, Karussis D (2016) Safety and clinical effects of mesenchymal stem cells secreting neurotrophic factor transplantation in patients with amyotrophic lateral sclerosis: results of phase 1/2 and 2a clinical trials. JAMA Neurol 73(3):337–344. doi:10.1001/jamaneurol.2015.4321

30. Politis M, Wu K, Loane C, Quinn NP, Brooks DJ, Oertel WH, Bjorklund A, Lindvall O, Piccini P (2012) Serotonin neuron loss and non-motor symptoms continue in Parkinson's patients treated with dopamine grafts. Sci Trans Med 4(128):128ra141. doi:10.1126/scitranslmed.3003391

31. Savitz SI, Cramer SC, Wechsler L, Consortium S (2014) Stem cells as an emerging paradigm in stroke 3: enhancing the development of clinical trials. Stroke 45(2):634–639. doi:10.1161/STROKEAHA.113.003379

32. Paczkowska E, Luczkowska K, Piecyk K, Roginska D, Pius-Sadowska E, Ustianowski P, Cecerska E, Dolegowska B, Celewicz Z, Machalinski B (2015) The influence of BDNF on human umbilical cord blood stem/progenitor cells: implications for stem cell-based therapy of neurodegenerative disorders. Acta Neurobiol Exp 75(2):172–191

33. Gao M, Dong Q, Yao H, Zhang Y, Yang Y, Dang Y, Zhang H, Yang Z, Xu M, Xu R (2017) Induced neural stem cells modulate microglia activation states via CXCL12/CXCR4 signaling. Brain Behav Immun 59:288–299. doi:10.1016/j.bbi.2016.09.020

34. Kim KS, Kim HS, Park JM, Kim HW, Park MK, Lee HS, Lim DS, Lee TH, Chopp M, Moon J (2013) Long-term immunomodulatory effect of amniotic stem cells in an Alzheimer's disease model. Neurobiol Aging 34(10):2408–2420. doi:10.1016/j.neurobiolaging.2013.03.029

35. Hallett PJ, Cooper O, Sadi D, Robertson H, Mendez I, Isacson O (2014) Long-term health of dopaminergic neuron transplants in Parkinson's disease patients. Cell Rep 7 (6):1755–1761. doi:10.1016/j.celrep.2014.05.027

36. Sanai N, Nguyen T, Ihrie RA, Mirzadeh Z, Tsai HH, Wong M, Gupta N, Berger MS, Huang E, Garcia-Verdugo JM, Rowitch DH, Alvarez-Buylla A (2011) Corridors of migrating neurons in the human brain and their decline during infancy. Nature 478(7369):382–386. doi:10.1038/nature10487

37. Dihne M, Hartung HP, Seitz RJ (2011) Restoring neuronal function after stroke by cell replacement: anatomic and functional considerations. Stroke 42(8):2342–2350. doi:10.1161/STROKEAHA.111.613422

38. Barker RA, Drouin-Ouellet J, Parmar M (2015) Cell-based therapies for Parkinson disease—past insights and future potential. Nat Rev Neurol 11(9):492–503. doi:10.1038/nrneurol.2015.123

39. Seminatore C, Polentes J, Ellman D, Kozubenko N, Itier V, Tine S, Tritschler L, Brenot M, Guidou E, Blondeau J, Lhuillier M, Bugi A, Aubry L, Jendelova P, Sykova E, Perrier AL, Finsen B, Onteniente B (2010) The postischemic environment differentially impacts teratoma or tumor formation after transplantation of human embryonic stem cell-derived neural progenitors. Stroke 41(1):153–159. doi:10.1161/STROKEAHA.109.563015

40. Chen J, Ye X, Yan T, Zhang C, Yang XP, Cui X, Cui Y, Zacharek A, Roberts C, Liu X, Dai X, Lu M, Chopp M (2011) Adverse effects of bone marrow stromal cell treatment of stroke in diabetic rats. Stroke 42(12):3551–3558. doi:10.1161/STROKEAHA.111.627174

41. Daadi MM, Klausner JQ, Bajar B, Goshen I, Lee-Messer C, Lee SY, Winge MC, Ramakrishnan C, Lo M, Sun G, Deisseroth K, Steinberg GK (2016) Optogenetic stimulation of neural grafts enhances neurotransmission and downregulates the inflammatory response in experimental stroke model. Cell Transplant 25 (7):1371–1380. doi:10.3727/096368915X688533

42. Weitz AJ, Lee JH (2016) Probing neural transplant networks in vivo with optogenetics and optogenetic fMRI. Stem Cells Int 2016:8612751. doi:10.1155/2016/8612751

43. Pina-Crespo JC, Talantova M, Cho EG, Soussou W, Dolatabadi N, Ryan SD, Ambasudhan R, McKercher S, Deisseroth K, Lipton SA (2012) High-frequency hippocampal oscillations activated by optogenetic stimulation of transplanted human ESC-derived neurons. J Neurosci 32(45):15837–15842. doi:10.1523/JNEUROSCI.3735-12.2012

44. Tonnesen J, Parish CL, Sorensen AT, Andersson A, Lundberg C, Deisseroth K, Arenas E, Lindvall O, Kokaia M (2011) Functional integration of grafted neural stem cell-derived dopaminergic neurons monitored by optogenetics in an in vitro Parkinson model. PLoS One 6(3):e17560. doi:10.1371/journal.pone.0017560

45. Weick JP, Liu Y, Zhang SC (2011) Human embryonic stem cell-derived neurons adopt and regulate the activity of an established neural network. Proc Natl Acad Sci U S A 108 (50):20189–20194. doi:10.1073/pnas.1108487108

46. Paz JT, Davidson TJ, Frechette ES, Delord B, Parada I, Peng K, Deisseroth K, Huguenard JR (2013) Closed-loop optogenetic control of thalamus as a tool for interrupting seizures after cortical injury. Nat Neurosci 16 (1):64–70. doi:10.1038/nn.3269

47. Byers B, Lee HJ, Liu J, Weitz AJ, Lin P, Zhang P, Shcheglovitov A, Dolmetsch R, Pera RR, Lee JH (2015) Direct in vivo assessment of human stem cell graft-host neural circuits. NeuroImage 114:328–337. doi:10.1016/j.neuroimage.2015.03.079

Chapter 6

Optogenetic Applications in the Nematode *Caenorhabditis elegans*

Katharina Elisabeth Fischer, Nathalie Alexandra Vladis, and Karl Emanuel Busch

Abstract

The advantages of the nematode *Caenorhabditis elegans*, such as a well-characterized nervous system, complex behavioral patterns, powerful genetics, and experimental tractability, establish this animal as an excellent platform for optogenetic studies and manipulation. A roadmap for conducting optogenetic experiments in *C. elegans* is provided in this chapter. We give advice on the choice of appropriate optogenetic tools, generation of transgenic animals and the preparation of animals for experiments, and describe using the nematode for optogenetic tool engineering by detecting body wall muscle contraction. We also survey specific optogenetic applications in *C. elegans* that give insight in long-term behavior and development; all-optical interrogation, combining optogenetic neuronal manipulation with the simultaneous detection of neural activity by calcium sensors; the optogenetic generation of reactive oxygen species for cell and protein ablation and mutagenesis; optogenetic control of intracellular signaling pathways; and the harnessing of optogenetics for a drug-screening platform by pacing pharyngeal muscle contraction.

Key words *Caenorhabditis elegans*, All-optical interrogation, Genetically encoded photosensitizers, Intracellular signaling pathways, Drug screening, Pharyngeal pumping, Electrophysiology

1 Introduction

Since the nematode *Caenorhabditis elegans* was chosen by Sydney Brenner to study the development of animals four decades ago, it has become one of the most widely used model organisms in biology [1]. Thanks to its powerful genetics, experimental amenability and the extensive resources available in the research community, this organism has enabled numerous discoveries such as of the genes causing apoptosis, and frequently was used to pioneer techniques such as the use of fluorescent proteins. It also has become an important model to elucidate the mechanisms that guide the development or function of nervous systems, and all aspects of its compact nervous system are under intensive investigation [2].

Albrecht Stroh (ed.), *Optogenetics: A Roadmap*, Neuromethods, vol. 133,
DOI 10.1007/978-1-4939-7417-7_6, © Springer Science+Business Media LLC 2018

Optogenetics has rapidly become an indispensable part of neurobiological research, and *C. elegans* has featured in it from the start—it was the first intact animal in which channelrhodopsin (ChR2) was used [3]. Starting with the use of light-gated channels to excite neural activity, optogenetics has expanded into every conceivable direction to yield mechanistic insight in molecular, cellular, and systems neuroscience. *C. elegans* plays a prominent role and holds promise in three areas in particular: the testing and validation of newly developed optogenetic tools; the elucidation of a functional connectome, based on the completely mapped anatomic connectome; and in high-throughput screening approaches.

In this chapter, we will briefly survey the current scope of optogenetics in *C. elegans* for investigating neurobiological questions. The chapter is broadly oriented along an experimental flowchart (Fig. 1) to support designing experiments and to address individual scientific questions. We will describe general considerations and conditions for conducting optogenetic experiments in *C. elegans*, such as how to choose an appropriate tool, starting with a description of how to prepare animals for optogenetic stimulation (Sect. 2). We also give information about how to choose appropriate optogenetic tools and how to improve tool expression in vivo (Sect. 3). We then outline the investigation and validation of optogenetic tools by a simple experimental approach of muscular stimulation, enabling insights into tool characteristics such as kinetics, excitation wavelength, and light sensitivity (Sect. 4).

The next part (Sect. 5) discusses specific applications, giving the researcher the possibility to select from an abundant optogenetic toolkit the tool of choice to address individual scientific questions. A broad spectrum of routes is available for phenotypic readouts in locomotion and other behaviors. We describe selected strategies, namely chronic neuronal manipulation by rhodopsins to investigate locomotion and developmental defects due to long-lasting changes in neuronal circuits (Sect. 5.1). All-optical interrogation of neural circuit function is an exciting new direction of research in neurobiology, where genetically encoded indicators of neural activity are used for functional neuronal imaging coupled with optogenetic stimulation of specific neurons (Sect. 5.2). We describe optogenetic inactivation of proteins or cells (Sect. 5.3) and the optogenetic control of intracellular signaling pathways (Sect. 5.4) for studying neurodegeneration and signaling pathways, respectively. Finally, we introduce a *C. elegans* drug-screening platform based on the optogenetically controlled feeding organ, the pharynx of *C. elegans*, which enables studying aberrations in pharyngeal pumping behavior caused by drug application or genetic defects affecting muscular function (Sect. 5.5).

Several excellent reviews have been published recently that deal with aspects of *C. elegans* optogenetics we have not covered here, and we encourage readers to consult them as well [4–8].

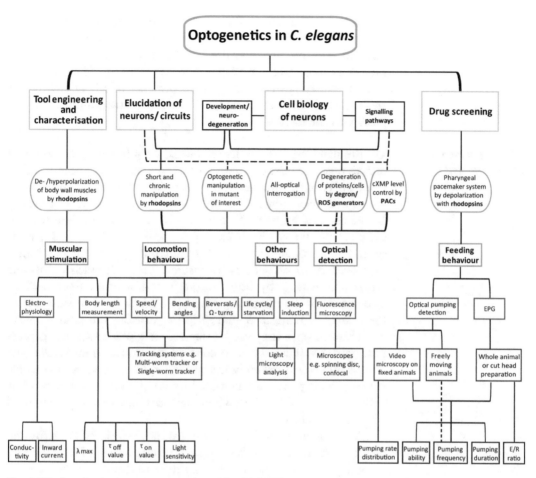

Fig. 1 This flowchart serves as an experimental guideline for optogenetic applications in *C. elegans*. The layers lead the researcher from different research fields (*yellow*) to the corresponding specific optogenetic approaches (*blue*). In the next layer, marked in *green*, an overview about the phenotypic reaction and observation evoked by the chosen optogenetic tool is provided. The *red layer* then gives detailed information about the different possibilities of phenotypic readout. Which experimental setup is suggested for the individual experiment is shown in violet. The *black layer* gives information about the analytical readout of the respective experiment

2   General Considerations for Setting Up Optogenetic Experiments in *C. elegans*

2.1   Preparing Cultivation Dishes

For opsin-based tools, the opsin chromophore all-*trans*-retinal (ATR) must be provided to *C. elegans*, which unlike vertebrates do not produce ATR by themselves. The chromophore is covalently bound to the active site in the channel pore by forming a Schiff base. It changes its conformation to *13-cis*-retinal by illumination with light of channel-specific wavelength, evoking the opening of the channel (*see* Chap. 3), making it light sensitive (*see* **Note 1**). ATR is supplied by feeding of a mixture of *E. coli* OP50, suspended in bacterial growth media, and 0.1 mM ATR (diluted from 100 mM in EtOH). For this, NGM (Nematode Growth Medium) cultivation dishes are seeded with 300–400 µl of the OP50/ATR mixture. Alternatively, ≈30 µl of 5 mM ATR in EtOH can be added to a pre-grown lawn of OP50, as the presence of ethanol inhibits *E. coli* growth. Once the bacteria form a dry lawn, animals can be placed on the dishes.

2.2 Cultivation of C. elegans on ATR Dishes

On the first day, 3–5 adults are placed on a freshly prepared ATR dish to raise the next generation of worms that will appear 2–3 days later. This can be repeated several days in a row to guarantee that a large number of animals is available every day for assays. On the day before assays shall be performed, L4 larvae are transferred to ATR dishes to obtain synchronized young adults for the next day measurement. Particular exposure is required if the optogenetic tool is expressed from an extrachromosomal array. Extrachromosomal arrays are formed by homologous recombination from 100 to 200 copies of the injected plasmid [9]. Since transmission rate of these arrays to the next generation is variable but nearly always <100%, and since they can be silenced or evoke tessellated expression patterns, it is crucial to confirm before the transfer that the gene is indeed expressed in the neuron(s) of interest. As a negative control, transgenic animals raised under the same conditions but without ATR can be used, where bacterial lawns are supplemented with ethanol alone.

An ATR dish can be used for around 3–5 days. To verify the quality of ATR dishes, it is beneficial to run a strain which expresses an optogenetic tool with a clear light-evoked response as an additional ATR control. This can be a strain expressing channelrhodopsin-2 (ChR2) in body wall muscles (BWM) [3], which shows body contraction under a fluorescence microscope upon blue light illumination (*see* also Sect. 4.1). For experiments, it is highly recommended to use ATR dishes that were prepared on the same day or the day before.

2.3 Preparation of Animals for Stimulation

The preparation of the animals for optogenetic manipulation depends on the information that is sought as readout.

To record locomotion, behavioral assays with freely moving animals should be performed on or off food, depending on the selected tracking system. Typically, the animals are placed on the assay dishes and their movement and behavior recorded immediately or after a short acclimatization period of 10–15 min.

For experiments requiring immobilized animals, different setups are possible to use. The simplest method is fixation with polystyrene beads [10]. For that, preparation of 10% agar pads prepared with M9 or similar buffers is needed [11]. After adding 1–2 µl of polystyrene beads to the pad, animals should be transferred to the pad. Then a coverslip is placed on the sample very gently in order not to squeeze the worms. An advantage of this method is that worms can be recovered after the experiment. For that, the coverslip shall be removed gently as well by lifting it up. A droplet of M9 worm buffer can now be added to the worms on the pad and they can be recovered by transferring them with an eyelash pick to a fresh NGM dish.

Microfluidic devices can be used to record fixed animals and expose them to defined sensory stimuli or environmental conditions. The devices are mostly made of polydimethylsiloxane (PDMS), a nontoxic, flexible, gas-permeable, and transparent polymer. PDMS is a material that can partly be combined with other materials such as steel, glass, or silicone, enabling the creation of more complex devices to meet the needs of individual experiments. In a typical PDMS device, worms are moved to the microfluidic chamber slots by pressure control. The animals are pooled in buffer as a source outside the microfluidic device and sucked via a hose into the channels of the device by low pressure. Immobilization can be done by different strategies, such as filling the chambers with the temperature-sensitive PF127 gel, which gelatinizes by temperature increase and so can be controlled by hot water in separate channels covering the animal holding chambers [12]; or by using PDMS membranes deflected by pressure [13]. The most common immobilization strategy is to use microfluidic devices in a tapered shape, where animals are pulled into the tip of the narrowing channel by negative pressure. For behavioral studies on body wall muscles during optogenetic stimulation, Hwang et al. created a device enabling the fixation of the animals on both body ends with pneumatically controlled microvalves [14]. Diverse designs of microfluidic chambers are available for worm culturing [15], worm sorting [16, 17] and microinjection [18] as well as for imaging [19, 20], high-throughput imaging [21], developmental studies [22] and behavioral analysis [23–25]. Also, a microfluidic device for electropharyngeogram (EPG) recordings has been developed [26]. Several of these devices allow the application of drugs, odors or other stimuli during analysis.

Another possibility for immobilization is to glue the animals on an agar pad with tissue glue (e.g., 2-octyl cyanoacrylate) [27]. The required equipment can easily be self-made. For this, a small plastic hose is equipped with a pipet tip on the one side and with a glass capillary on the other side. By holding the capillary tip into a drop of glue and sucking on the pipet tip, the needle can be filled with glue and used to apply the glue to a worm placed on a round Sylgard-covered coverslip (25 mm) in a preparation chamber (here an empty 3 cm cultivation dish can be used), filled with physiological buffer, e.g., M9. Gluing worms requires some practicing and recovery of the worms is not possible, but for experiments requiring strong immobilization, this method can be very beneficial.

For fluorescence imaging, treatment with sodium azide can be more reliable than immobilization with polystyrene beads or gluing, where detection of fluorescence is sometimes impaired by scattered light due to the fixation material. Sodium azide blocks the mitochondrial respiratory chain and leads to paralysis and death at higher concentrations, so worms cannot be recovered unless the concentration of NaN_3 is carefully adjusted. For preparation, after worms are placed on a 2% Agar pad, 1–2 µl of 50 mM NaN_3 in M9 buffer is added to the animals and these are covered with a coverslip. The sample now can be used for up to 30 min to take fluorescence images before necrosis leads to an increase in autofluorescence.

2.4 Microscopy Setups for Optogenetic Stimulation

The basic setup consists of a fluorescence microscope, optionally with a movable stage; a fluorescence light source enabling illumination with the wavelength of interest; and a camera. The fluorescence light can be provided by an LED system or by a mercury lamp. The corresponding filter sets and the light intensity should be adjustable for intensity-dependent activity measurement. In both cases a computer-controlled shutter system is mandatory to provide illumination for distinct periods as well as for repeated illumination, e.g., for quantification of tool recovery during periodic illumination. For locomotion analysis, automated tracking systems have been developed by several labs, so a wide range of tracking systems is available for addressing the individual research question. Available tracking systems are described in detail by Husson et al. and especially for optogenetic applications by Kimura and Busch [4, 28]. Additionally, Kimura and Busch provide a guide to the advantages and disadvantages of different optogenetic assay setups in terms of ease of setting up individual systems.

3 Choosing the Right Optogenetic Tool

3.1 Expression of Optogenetic Tools in C. elegans Neurons

The quality of optogenetic tool expression and functionality is strongly influenced by the specific promotor used for distinct tissue or cell expression (*see* also Chap. 2) and by its codon-dependent strength of expression. Optogenetic genes whose codon usage is not optimized for *C. elegans* frequently have expression levels that are too low to cause a response to optical stimulation. The opposite case, where codon usage is highly optimized and the protein overexpressed, can lead to aggregated structures of the optogenetic protein and to toxicity. To optimize or modulate expression levels in *C. elegans*, Redemann et al. showed that it can be fine-tuned by changing codon usage, with an adaptation index (CAI) that ranges from 1 for highly expressed proteins to 0 for lowest expression rate. The *C. elegans codon* adapter is available as a web tool to choose the best codon usage for the tool of interest [29].

Fusing the tool of interest with a fluorescent protein to detect local tool expression in some cases also induces protein aggregation and reduced functionality by disrupting the correct protein conformation. This problem can be overcome by using bicistronic systems with the *C. elegans*-specific spliced leader RNA 2 (SL2), allowing the parallel expression of the optogenetic tool of interest and a fluorescent protein in the same cells from the same operon if SL2 is placed between the encoding genes [30]. Nonoptimal expression can also occur depending on the promotor used. The pharyngeal myosin promotor *myo-2* is an example of a promotor leading to strong overexpression which frequently results in toxic activity of optogenetic genes. To express optogenetic proteins guided by this and other strong promotors, a balance should be found between functionality and overexpression, where higher concentrations of the injected DNA form larger extrachromosomal arrays and achieve higher expression levels. This has been shown with ChR2(H134R) expression in the *C. elegans* pharynx. Here, injecting the plasmid encoding the gene at 5 ng/µl led to a strong aggregation of the ChR2 protein, where the animals were small and slow-growing (unpublished data), whereas injection at 3 ng/µl showed a regular expression pattern, tool functionality and no abnormal phenotypes [31].

3.2 Standard Optogenetic Tools

The membrane-located, seven transmembrane domain-containing cation channels Channelrhodopsin-1 and -2 (ChR1 and 2) from the green alga *Chlamydomonas* have been first characterized by Nagel et al. as potential optogenetic tools [32, 33] (Fig. 2a). Today, ChR2 with its membrane depolarizing function serves as the most popular photoactivating tool. ChR2 shows maximal activation at a wavelength of 470 nm and channel closure occurs within a few milliseconds after switching off the light (τ_{off} 10 ± 1 ms) [34] (*see* Chap. 3). In 2005, the point-mutated variant ChR2(H134R) was fused to yellow fluorescence protein (YFP) and expressed in *C. elegans* body wall muscles [3] (Fig. 2a, b). ChR2(H134R) benefits from a twofold delayed channel closure (τ_{off} 19 ± 2 ms) [34], enabling an extended cation current and therefore stronger depolarization, and is still one of the best photoactivators available. The functionality of the channel in living *C. elegans* was shown by muscle contraction upon blue light illumination [3]. Meanwhile the arsenal of de- or hyperpolarizing tools has grown immensely and keeps increasing. A number of photoactivating tools offer properties distinct from ChR2(H134R):

- Different excitation/emission wavelength: the red-shifted variants Chrimson [35] or the chimeric C1V1-ET/ET [36] allow experiments requiring spectral separation of illumination.

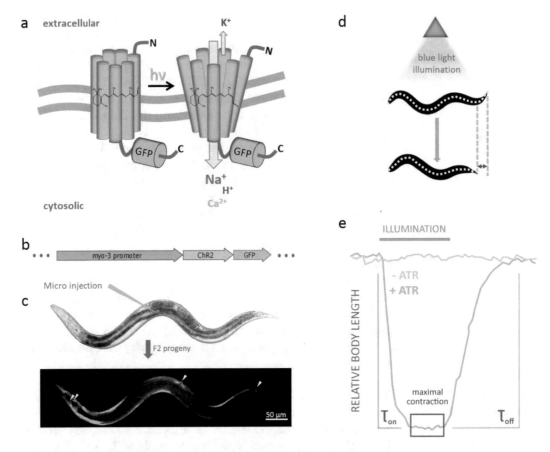

Fig. 2 The experimental scheme of body length measurement to investigate the function and characteristics of novel rhodopsins. The membrane depolarization properties of ChR2 equipped with ATR shall serve as an example (**a**). First, the DNA encoding ChR2, fused to GFP and driven by the myo-3 promoter for expression in body wall muscle (BWM) (**b**), is microinjected followed by verification of expression in the F2 generation via GFP fluorescence (**c**). *White arrowheads* show, apart from BWM expression, anterior fluorescence in muscle arms forming neuromuscular junctions to neurons located in the nerve ring as well as in vulva muscles and posterior in anal muscle. Change of body length (here contraction) is determined by skeletonization of the worm's shape before and after illumination (**d**). (**e**) A typical graph, reflecting the relative body length upon ChR2 stimulation with and without ATR. As described in the text, the curve gives insights in the channel kinetics and the level of maximal contraction

- Light sensitivity and conductivity for specific cations: CatCh (ChR2(L132C)) has higher Ca^{2+} conductivity and is 70-fold more light sensitive compared to ChR2(H134R) [34].

- Duration of open state: ChR2(C128S) shows a strong delay in channel closure with a τ_{off} of 106 ± 9 s and can be used for long-term manipulations (*see* also Sect. 5.1) [37, 38].

    In almost the same manner, variants of photoinhibitory membrane proteins are available. The first tool tested in *C. elegans* was halorhodopsin (NpHR), a yellow light (590 nm)-activated chloride pump found in the halobacterium *Natronomonas pharaonis*, provoking elongation of the animal when expressed in BWMs or cholinergic motor neurons [39]. In comparison, the outward-directed proton pumps archaerhodopsin (Arch) from *Halorubrum sodomense* and Mac from *Leptosphaeria maculans* show a slight blue shift of their maximum activation wavelength, expanded activation spectra and higher photocurrents in *C. elegans* [40–43].

Although there is a wide range of optogenetic tools available, the development of new tools continues to form an important part of optogenetic research. Quality of specific expression, light sensitivity, conductivity, spectral separation, and on/off kinetics remain to be improved. So in *C. elegans* simple body length contraction experiments, first used by Nagel et al. [3] can serve as a basis for tool engineering, since they provide direct information about key channel characteristics such as gating kinetics, light sensitivity, tool recovery and the wavelength of maximal channel activity.

4 Readouts for Validating Optogenetic Tools

4.1 Body Length Measurement

After microinjection and the successful expression of the tool of interest in *C. elegans* BWM is verified by fluorescence microscopy using fluorescent reporters (Fig. 2b, c) or immunocytochemistry, transgenic animals are raised on ATR dishes as described in Sect. 2.1. For measurement, the animals are transferred to unseeded NGM dishes and video recorded, optionally by following the animals with a movable stage. As a target value, 2 s illumination time is sufficient for body length analysis to ensure the maximal tool activity but can be prolonged or reduced contingent on the desired experimental readout. Afterwards the videos are analyzed by extracting the single frames and measurement of the worm length using skeletonization of each worm's shape from head to the tip of the tail for each frame (Fig. 2d). This process has been automated by custom-written scripts in Labview or ImageJ [36, 44–46]. Body length values are normalized to the mean of the period before illumination for each single animal. Body length can then be calculated as the mean of each time point from animals of the same strain. The resulting curve of body length over time can now be used to analyze on/off kinetics (Fig. 2e). The series of data points from the pre-illumination baseline to the state of maximal change of body length, or from the state of maximal change of body length to the part were animals revert to their normal body length after illumination can be fitted with a nonlinear curve, using Origin, Prism, or comparable software. The resulting equation yields the τ_{on}- and τ_{off} –values, respectively, and give insight into the channel kinetics (Fig. 2e). The maximal change of body length result is calculated by averaging the mean of time points 1–2 s after onset of illumination from each single animal (Fig. 2e). To investigate the light sensitivity, sets of animals were measured at different light intensities and the maximal changes of body length calculated. These values were plotted on a logarithmic scale to the corresponding light intensities, fitted with a logarithmic curve and the half-maximal values can be obtained from the equation.

4.2 Electro-physiology

For further tool characterization, electrophysiological patch-clamp measurements on muscle cells can give insight into strength of inward current and conductivity properties for specific ions of the optogenetic tool probed (see Chaps. 3 and 8). For in-depth information on characterizing optogenetic tools using this sophisticated method see references [46–48]. Electrophysiological measurements in *C. elegans* are not widely used in the community because they are laborious, require specialized experience and expensive equipment.

5   Specific Optogenetic Applications

5.1 Long-Term Manipulation of Nervous System Development or Function

Many aspects of development and function of neurons involve long-lasting changes to neuronal properties, such as in long-term memory. To address such questions with optogenetics often necessitates chronic manipulation of neuronal activity. In principle it is possible to use standard optogenetic tools such as ChR2 to activate neurons for several minutes and likely even longer, and they have been used this way in *C. elegans* [25] and other models [49]. Channelrhodopsin has a short-lived high conductivity state which desensitizes to a smaller steady-state conductance in continuous light (*see* also Chap. 3), which can be used for long-term stimulation.

However, a serious drawback of long-term in vivo application of ChR2 is that it requires continuous high-intensity blue light to gate it, causing phototoxic and behavioral avoidance effects. Another aspect to consider is that high-level expression of ChR2 can induce long-term structural changes in neurons, such as abnormal axonal morphology, and very likely functional changes, even when the channel is not activated by light [50]. We speculate that this may be due to chronic low-level permeation of Ca^{2+} through the channel. For experimental considerations, *see* **Note 2**.

Overall more suitable tools for long-term stimulation are bi-stable light-gated neural state switches that operate at different timescales because brief pulses of light cause their long-lasting opening [38]. They offer the key advantage that they require less frequent and less intense light stimulation, thus avoiding phototoxic effects. Schultheis and colleagues demonstrated that slow variants of ChR2 with mutations at cysteine 128 that have delayed off-kinetics can be used in *C. elegans* for chronic or long-lasting stimulation [37]. From the different mutations tested at C128, serine had the best expression levels, the longest opening (with a τ_{off} of 106s) and was the most photosensitive. ChR2(C128S) can be inactivated by green-yellow light, enabling full temporal control as a bi-stable neural state switch. ChR2(C128S) is reliably gated by blue light at an intensity of 0.01 mW/mm^2 to cause body wall muscle contraction, considerably less than ChR2 requires (in the same experiment, 0.08 mW/mm^2 or higher). This slow variant was shown to induce prolonged depolarization of muscles and neurons, and to induce long-lasting muscle contraction and command interneuron-induced long-term behavioral effects, namely excessively long reversals (backward movements). It could be used to manipulate animal development as well: the ASJ sensory neurons serve during larval development to control entry or exit into an alternative developmental path, the dauer larva state. Constant photoactivation of ChR2(C128S) in ASJ with low intensity light (ca. 0.1 µW/mm^2) for 3 days prevented entry into the dauer state, and also promoted exit from this state in dauer larvae [37].

The activity of the C128S variant declines very significantly upon constant illumination, however. After 24 h, responses are down to ≈20% of the maximal response, most likely due to transition to a nonactivatable "lost state."

Other than using channels that use ATR as cofactor, optogenetic tools based on azobenzene-derived chemical photoswitches have the potential to remain in an activated state for long periods of time until they are switched off by a separate light signal [51, 52]. They have not been applied in *C. elegans* yet but constitute promising tools for the long-term control of neurons.

5.2 All-Optical Interrogation of the C. elegans Nervous System

Recent years have seen striking progress in the development of genetically encoded neural activity sensors, optogenetic actuators and microscopy techniques, and an increasing focus has been on using these tools in combination to achieve a deeper understanding of neural circuit function in the living animal. Using such "all-optical" approaches is technically challenging but offers many advantages, namely that they are noninvasive, can target a flexible number of neurons and procedures can be performed in parallel by using discrete wavelengths or spatial targeting.

Three choices have to be made: (1) Which optogenetic control tool; (2) Which genetically encoded activity sensor; and (3) Which microscopy system to use. In our view, the current method of choice is to combine an optogenetic actuator with a genetically encoded calcium sensor which are spectrally separated (*see* also Chap. 9), and target each to a subset of neurons, using specific promoters, in a tracking setup that enables recording in freely moving *C. elegans*. The spectral separation has the benefit, compared to other solutions, that neurons targeted for manipulation and recording can be freely chosen, even if they are close to each other, and that optogenetic stimulation, neural activity recording and behavioral analysis can all be performed simultaneously.

Only a few studies have been performed in this manner in *C. elegans* so far. In a proof-of-principle experiment, Inoue and colleagues expressed ChR2 in the ASH sensory neurons and the red calcium sensor R-CaMP2 in the postsynaptic AVA interneurons. ASH stimulation caused both a Ca^{2+} increase in AVA and transient backward movement [53]. The recording was done on a confocal microscope with laser illumination of R-CaMP2 and blue light stimulation of ChR2 with a mercury lamp. Animals were placed on an agar pad and covered with a coverslip, which is not optimal as it modifies their movement patterns. Tracking was done by manually moving the stage to keep the worm in the field of view.

In the most extensive application of this approach yet, pairs of spectrally separated actuators and calcium sensors (GCaMP3— NpHR and RCaMP—ChR2) were used to dissect the circuit downstream from the AIY first layer interneurons, with the actuators expressed in AIY and the sensors in two different downstream interneurons, RIB and AIZ[54]. For tracking, a motorized stage kept animals in the center of the field of view based on the fluorescent signal from the neuron. Illumination of the sensors required constant illumination with a Xenon light source, while ChR2/ NpHR were stimulated for several seconds by light pulses from a separate LED light source.

5.3 Optogenetic Generation of Reactive Oxygen Species to Ablate Cells, Inactivate Proteins or Mutagenize DNA

5.3.1 Ablating Neurons with Genetically Encoded Photosensitizers

When chromophores absorb photons, they can generate reactive oxygen species (ROS) that rapidly undergo oxidative reactions with nearby molecules, leading to tissue damage. In fluorescent proteins such as GFP, the chromophore is shielded from oxygen in the environment by the β-barrel protein structure, largely avoiding ROS generation upon illumination. However, fluorescent protein variants have been discovered that generate large amounts of reactive oxygen species when illuminated with visible light [55]. Due to their high and rapid reactivity, ROS will have a very localized effect limited to structures very close to the ROS generator. These proteins can therefore be harnessed as optogenetic tools where the highly reactive free radicals they produce are used to kill cells in vivo, to irreversibly inactivate proteins or cellular structures with high temporal and spatial precision, or even to mutagenize genomic DNA. The acute inactivation of proteins tagged with photosensitizers has an advantage over chronic loss-of-function mutations, as the latter will affect development as well and may lead to the neurons and circuits adapting to the absence of the protein, often complicating the interpretation of mutant phenotypes. Compared to ablation with a laser microbeam, using photosensitizers to ablate neurons offers the benefit that no special equipment is needed, that the procedure is easier and less likely to damage neighboring cells, and that larger numbers of worms can be treated, at any developmental stage, whereas laser ablation is typically limited to early larval stages. On the other hand, it critically depends on the availability of strong but specific promoters driving expression of the photosensitizing proteins in the cells of interest.

The genetically encoded photosensitizers KillerRed and miniSOG (mini singlet oxygen generator) are the best-established tools for optogenetically ablating cells in *C. elegans*. The two are not functionally equivalent, as KillerRed almost exclusively produces superoxide radicals ($O_2 \cdot^-$) [56, 57], whereas miniSOG generates primarily singlet oxygen ($^1O_2$) but possibly also other ROS [58]. Superoxide preferentially reacts with iron sulfur centers and nitric oxide and can be converted to hydrogen peroxide (H_2O_2), potentially generating a cascade of different ROS. Singlet oxygen has particularly high reactivity and is believed to react in the immediate vicinity of where it was generated, with no selectivity. It cannot be converted to other ROS.

5.3.2 KillerRed—A Superoxide-Producing Photosensitizer

KillerRed was the first phototoxic fluorescent protein described [59]. It produces superoxide upon illumination. KillerRed has the advantage that it is activated by green light (max. excitation at 585 nm), which is not on its own phototoxic for *C. elegans*, and allows long exposure times with no other adverse effects on the animal. It does not require any cofactors. It has the serious disadvantage, however, that it functions as a homodimer, and thus usually impairs protein function when fused to them. For further experimental considerations, *see* **Note 3**.

KillerRed as a tool to inactivate and kill *C. elegans* cells was investigated extensively by Williams and colleagues and in other studies [60–62]. After illumination, neuronal function was abolished immediately and permanently, and subsequently the cells underwent neurodegeneration and necrotic cell death, with swelling of the cells, vacuolation and fragmentation of the neuronal processes. KillerRed-induced damage was cell-autonomous and did not spread to other nearby cells. Both cytosolic and plasma membrane-targeted KillerRed caused ablation of neurons with similar efficiency; localized activation of KillerRed in axonal processes also triggered neuronal degeneration. Expression and illumination of plasma membrane-targeted KillerRed in the AWA chemosensory neurons largely (but apparently not completely) abolished AWA-dependent chemoattractant responses, but did not impair function of the AWC sensory neurons [60].

As KillerRed functions as a dimer, expressing a tandem dimer version (tdKillerRed) significantly increased killing efficiency and reduced the length of illumination needed from ca. 2 h at 1 mW/mm^2 white light to 5–15 min with green light, although much higher illumination intensities of 5.1 or 46 mW/mm^2 were used. The protein was expressed in a wide variety of neurons and illumination caused the ablation of most of them, but interestingly, certain cells such as the AVM mechanosensory neurons were shown to be refractory to KillerRed. This may be because those neurons have higher expression of proteins such as superoxide dismutase, protecting them from ROS activity. Expression of KillerRed targeted to the outer mitochondrial surface of body wall muscle cells disrupted and fragmented those mitochondria transiently, but did not kill the cells outright. The mitochondria displayed normal morphology before activation of KillerRed. The study also showed that it is possible to combine KillerRed with activation of other optogenetic tools such as channelrhodopsin or miniSOG, as their wavelengths for activation are sufficiently separated [62].

5.3.3 miniSOG—A Singlet Oxygen-Generating Photosensitizer

miniSOG is a fluorescent flavoprotein engineered from the LOV domain of *Arabidopsis* phototropin 2, and generates singlet oxygen and other ROS upon exposure to blue light with a maximum excitation at 448 nm [57]. It is monomeric and with 106 bp less than half the size of GFP, making it superior to KillerRed in this aspect, as it can easily be fused to other proteins without disturbing their normal function. It requires flavin mononucleotide as cofactor, but this is ubiquitously present in *C. elegans* cells.

miniSOG was the first optogenetic tool used for ablating neurons in *C. elegans*. While expressing miniSOG in the cytoplasm had no obvious effect, targeting it to the mitochondrial outer membrane and illuminating it with blue light at 0.57 mW/mm^2 for 30 min killed neurons with high efficiency [63]. Photo-ablation of cells did not appear to occur via apoptosis and was cell-autonomous, no damage to nearby cells or tissues was observed. Rearing miniSOG-expressing worms at normal indoor light levels did not cause obvious behavioral or morphological defects.

Although control animals exposed to the same level of blue light irradiation remained healthy and did not show obvious behavioral defects, the blue light needed for miniSOG activation is problematic, as *C. elegans* can show light avoidance behavior and phototoxic effects cannot be excluded at the long exposure times of 30–90 min used in previous studies. A more efficient variant of miniSOG was described recently that is fused to the Pleckstrin Homology domain from rat PLC-δ, which targets it to the plasma membrane, and which contains the Q103L mutation that increases ROS generation [64]. Together, the two changes make miniSOG at least 5–6× more effective than mitochondrially targeted miniSOG and require much shorter illumination times to achieve permanent miniSOG-mediated ablation. Expressing this variant in cholinergic neurons with 2 min illumination of 2 Hz pulsed (0.25 s on, 0.25 s off) blue light at 2 mW/mm^2 caused fully uncoordinated movement, while mitochondrial miniSOG required at least 15 min for the same effect (*see* **Note 4**). It worked similarly well in neurons as well as in other cell types, acted cell-autonomously and caused no toxic effects in the absence of blue light illumination.

5.3.4 Chromophore-Assisted Light Inactivation (CALI) with Optogenetic Tools

By fusing genetically encoded photosensitizers with other proteins to target them to specific locations in a cell, they can be used to acutely perturb the function of the fused proteins and cellular structures with high temporal and spatial precision. This is because the ROS generated have very short half-lives and usually react with the molecules in the immediate vicinity of the photosensitizer only. miniSOG is the tool of choice as it has a small size and has been shown to not disrupt protein function when fused to them. So far, two studies have demonstrated CALI in *C. elegans* using miniSOG [65, 66]:

Lin and colleagues fused miniSOG with the mammalian pre-synaptic protein synaptobrevin/VAMP2 and expressed it in all neurons of wild-type *C. elegans*, where it correctly localized to presynaptic terminals [65]. Illumination of freely moving L4 larvae with blue light (5.4 mW/mm^2) for 5 min on an inverted compound microscope disrupted locomotion by 80% and frequently induced paralysis; locomotion fully recovered off light after 24 h. Weaker and longer illumination for 25 min at 0.7 mW/mm^2 caused a slightly smaller effect. Synaptic input to muscles, as measured by spontaneous electric postsynaptic current (EPSC) frequency, was reduced by 89% after 3 min illumination. A fusion protein of miniSOG with *C. elegans* synaptotagmin was somewhat less effective.

Wojtovich and colleagues fused miniSOG with the succinate dehydrogenase complex (SDHC) subunit of the mitochondrial respiratory complex II to acutely disrupt complex II activity, whose genetic deletion is lethal [66]. The fusion protein was functional and rescued mutants of the *mev-1* gene encoding the SDHC subunit. Illumination of isolated mitochondria with blue light for 5 min strongly reduced complex II activity but did not eliminate it, while other mitochondrial complexes were not affected. Interestingly, while genetic ablation of *mev-1* is lethal, acute loss of complex II by CALI was survivable; illumination of these worms with blue light (10.2 mW/mm^2 for 5 min) did not kill them but made them susceptible to mild metabolic stress and reduced brood size by selectively impairing spermatozoa.

KillerRed, in comparison, is a poor tool for CALI due to its larger size and functional dimerization, which usually disrupts the function and localization of other proteins when fused to them (*see* **Note 5**). To overcome this limitation, the photosensitizing protein **SuperNova** was derived from KillerRed by random mutagenesis [67]. It is monomeric and appears to form dimers with other proteins that are functional and localize correctly. It has been postulated to produce both superoxide and singlet oxygen. When expressed in *C. elegans* mechanosensory neurons with no photostimulation, KillerRed impaired behavioral responses to gentle touch, but SuperNova showed normal touch responses, suggesting that it may be more tolerable.

5.3.5 Acute Light-Induced Protein Degradation with a Photosensitive Degron

Apart from the undesirable effects of strong blue light illumination, ROS-mediated inactivation of proteins with miniSOG has the disadvantage that it can also inactivate any other protein nearby with no selectivity. To specifically and rapidly remove proteins of interest, a photosensitive degron can be used which has been shown to work in *C. elegans* [68]. The photosensitive degron psd is unmasked by blue light and then targets fusion proteins for proteasomal degradation [69] (*see* also **Note 6**). Psd was c-terminally fused to synaptotagmin, which is required for synaptic vesicle release, expressed pan-neuronally and illuminated with blue light for 1 h at 0.03 mW/mm^2, about an order of magnitude less light than what ROS generators need. This severely reduced swimming locomotion within an hour, an effect comparable to a synaptotagmin mutation or irradiation of miniSOG-synaptotagmin.

5.3.6 Optogenetic Mutagenesis

Remarkably, as reactive oxygen species can also modify and damage DNA, a genetically encoded photosensitizer has been developed to enable forward genetic screening and transgene integration in *C. elegans* by inducing random mutations and chromosome breaks in its genome [70]. This way, no toxic chemicals or special equipment are needed. miniSOG was fused to a histone H3 variant and expressed in the germline. While no obvious defects were seen in nonirradiated worms, illumination of gravid young adults with 30 min of blue LED light (at 2 mW/mm^2) induced scores of mutants with visible phenotypes. Mutation frequency was calculated to be ≈ 0.7 per 1000 haploid genomes—about a quarter of that of standard chemical EMS mutagenesis. The treatment induced a wide range of mutations, including single nucleotide variants and deletions, which differs from the spectrum of mutations induced by chemical or radiation mutagenesis.

5.4 Optogenetic Manipulation of Cellular Signaling Pathways

A number of new optogenetic tools enables the targeting and control of intracellular signaling proteins, providing a window into the pathways underpinning diverse cellular behaviors. Given the generally high conservation of signaling pathways, *C. elegans* provides an ideal testbed for these tools. In the past, chemical and genetic techniques have played a pivotal role in the identification of signaling components involved in cellular regulation. However, there were numerous shortcomings of these methods; these include the difficulty in restricting diffusible factors to particular cellular compartments but also the lack of reversibility. In sum, it is challenging to provide variable inputs or exert precise spatial and temporal control.

On the other hand, optogenetic tools with their high degree of spatiotemporal control and high spectral selectivity can provide a noninvasive and accurate solution. Exploiting these physical properties makes it possible to probe neurotransmission and behavior by directly manipulating cellular signaling.

5.4.1 Optogenetic Control of cAMP Signaling

Weissenberger and colleagues modulated intracellular cAMP levels in the cholinergic neurons of *C. elegans* by using PACα, a photo-activated adenylyl cyclase isolated from the flagellate *Euglena gracilis* [71]. Worms placed on a wormtracker platform were exposed to intense blue light ($25.6 \, \text{mW/mm}^2$ for 25 s). Photoactivation led to an increase in locomotory speed and reduced reversal frequency. Whole-cell patch-clamp electrophysiological recordings in body wall muscle cells showed that spontaneous neurotransmitter release at the neuromuscular junctions increased within less than a second after onset of blue light stimulation. It is likely that the rise in cAMP activated the $G\alpha_s$ pathway which regulates synaptic release. As opposed to pharmacological agonists, effects on cAMP were spatially specific, with activity limited to a subset of neurons by using specific promoters. Responses occurred within seconds of the activation and could be modulated by changing light intensity. Interestingly, previous studies with membrane-permeable cAMP analogues to manipulate *C. elegans* locomotion were unsuccessful [72]. This suggests that the timing and location of cAMP stimulation are crucial for manipulating neuronal responses.

In a more recent study, Ryu and colleagues engineered a light-regulated adenylyl cyclase by fusing a photosensory module from the *Rhodobacter sphaeroides* bacteriophytochrome diguanylate cyclase to an adenylyl cyclase domain from the *Nostoc* sp. CyaB1 protein (IlaC) [73] (*see* Chap. 7). The photosensory module uses biliverdin IXα as chromophore, which is naturally present in *C. elegans* cells and does not need to be supplied. Compared to the blue-light-activated PACα, bacteriophytochromes, which are microbial photoreceptors, are sensitive to the near-infrared light in the spectral region of ~680–880 nm, which can penetrate biological tissues more deeply and efficiently, with the additional advantage of reduced photooxidative damage compared to blue light stimulation. When expressed in *C. elegans* cholinergic neurons, IlaC increased the frequency of body bends upon stimulation with red light for 30 s, presumably through manipulation of cAMP levels. For advantages and disadvantages of using PACα and IlaC in *C. elegans see* **Note** 7.

5.4.2 Optogenetic Manipulation of cGMP Signaling

Gao and colleagues used a rhodopsin fused with a guanylyl cyclase domain from the fungus *Blastocladiella emerdonii* as an optogenetic tool in *C. elegans*, termed BeCyclOp, enabling fast and highly specific light-triggered cyclic guanosine monophosphate (cGMP) release [74] (*see* also Chap. 4). Much like cAMP, cGMP acts as a second messenger. Mostly known in *C. elegans* for its regulatory role in ion channel conductance, cGMP plays a crucial role in the relay and encoding of primary sensory signals. When expressed in the BAG neurons which act as sensors for CO_2 and O_2, blue light photoactivation ($70 \, \mu\text{W/mm}^2$) for as little as 0.5 s induced slowing responses that could be rapidly turned on and off. Compared to other photoactivated guanylyl cyclases such as bPGC (also named BlgC or EROS) described by Ryu et al. [75], BeCyclOp showed higher specificity as no cAMP generation was detected, had a higher turnover on the membrane and exhibited a greater light versus dark ratio. BeCyclOp is mostly sensitive to light in the green spectral range but also responds to violet and red light.

5.4.3 Optogenetic Manipulation Through Diacylglycerol Signaling

Diacylglycerols (DAGs) can trigger the translocation of proteins equipped with a C1 domain towards the plasma membrane and are involved in controlling exocytosis and neuronal activity. In 2016, Frank et al. described a series of photosensitive DAGs (PhoDAGs) as potential tools for the manipulation of intra- and extracellular signaling [76]. The team designed an assay where they exposed animals grown in the presence of trans- and cis-PhoDAG-3 to aldicarb, an acetylcholinesterase inhibitor which paralyses nematodes. Under conditions of excess acetylcholine release, C. elegans can become hypersensitive to the compound and become paralyzed faster. The animals were illuminated with UV-A light (366 nm, 18 μW/mm^2) for the first 5 min after being placed on aldicarb and then again during the last 3 min of each 15-min time interval. Whereas worms exposed to cis-PhoDAG-3 showed a faster onset of aldicarb induced paralysis, trans-PhoDAG-3 animals showed no hypersensitivity to the drug compared to controls. These results suggest that PhoDAGs can increase neurotransmitter release by affecting the presynaptic machinery, thus offering a promising tool for studying exocytosis and neurotransmission. Compared to caged CAGs, they have the advantage that they can be applied with unmatched spatiotemporal precision and be quickly switched off again. However, the currently available PhoDAGs require high-energy UV-A irradiation light for their activation, which is cytotoxic and considerably limits their application in living C. elegans.

5.5 Harnessing Optogenetics for Drug Screening in C. elegans

The method we aim to introduce here was established as a drug-screening platform to mechanistically characterize and treat orthologues of human mutations that cause heritable cardiac arrhythmia, such as the LQT8/Timothy syndrome mutations in L-type voltage-gated calcium channels (Ca$_v$1.2 in mammals, EGL-19 in C. elegans), using the C. elegans pharynx as paradigm for a rhythmic muscle contraction system [31]. However, it can more broadly be used for any pharyngeal muscle related question.

The pharynx is a neuromuscular pump and serves as the feeding organ of C. elegans. It consists of 20 muscle cells, connected by gap junctions and mostly arranged in a tubular, threefold symmetric order starting with the pro corpus, interrupted by the anterior bulb, continuing with the isthmus and ending with the terminal bulb (Fig. 3a). The C. elegans pharyngeal nervous system consists of 20 neurons which act mostly independently from the rest of the nervous system and are required for fast coordinated pumping and

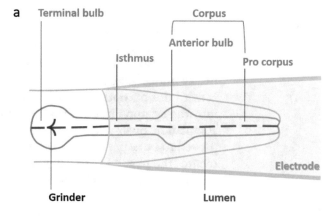

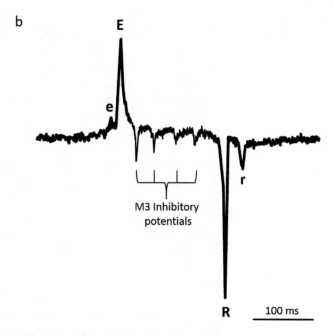

Fig. 3 (**a**) An outline of the *C. elegans* pharynx with a description of its anatomy, where the head is placed in an electrode for electrophysiological recording. An example of a typical electropharyngeogram (EPG) of *C. elegans* is shown in (**b**). The small e peak describes the contraction of the corpus followed by the big E peak that characterizes the contraction of the terminal bulb. The following *baseline* shows M3 neuronal inhibitory potentials, leading to the termination of the EPG signal by relaxation of the corpus (big R peak) and the terminal bulb (small r peak)

for controlling relaxation timing of pharyngeal muscles [77, 78]. With expression of ChR2 in the plasma membrane of the pharyngeal muscle cells, a pacemaker system was created to enable the induction of regular pumping at rates up to 5 Hz in N2 animals, sustained over minutes [31]. In the paced system, hyperpolarization of the cholinergic motor neurons by NpHR did not influence the pacing quality, whereas counting natural pumping in animals on food showed a strong decrease in pumping frequency upon NpHR activation, identifying the pacemaker system as primarily muscular acting approach. For further insight into neuronal activity during optical pacing, the pacemaker system was crossed into the synaptic transmission defective mutant *unc-13(s69)*. The stimulation quality in this system with almost no neuronal input proved to be as good as in wild type animals. This gives the experimenter the possibility to study the influence of drugs or mutations on the muscle-controlled pharyngeal pumping.

5.5.1 Electro-pharyngeographic Measurements of the Optogenetically Stimulated Pharynx

To detect contraction and relaxation processes in the *C. elegans* pharynx, electropharyngeograms (EPGs), similar to electrocardiograms, can be recorded. This method provides general information about pumping duration, pumping frequency, pumping ability and the E/R ratio. To this end excitation currents of the pharyngeal muscle are detected and recorded by two Ag/AgCl electrodes [79]. An EPG starts with the small e peak followed by the big E peak, corresponding to the contraction of the corpus and the terminal bulb, respectively. The consecutive baseline is interrupted by inhibitory potentials evoked by influence from the M3 neuron leading to the end of the EPG cycle with the big and small R/r peaks, indicating the relaxation of corpus and terminal bulb (Fig. 3b). The ratio between E and R gives the scientist insight into the strength of contraction and relaxation process during one pumping event, and characterizes the stability of pumping behavior from an average of E/R ratio values. For EPG analysis the program Auto-EPG [80] is most commonly used. The measurement can be done on whole or cut head-prepared animals. For measurement, the tip of the animals' head is sucked into a borosilicate electrode (inner $\varnothing$ ~ 20 μm) equipped with an Ag/AgCl wire, and animals are placed in a recording chamber filled with EmD50 buffer and equipped with a round Sylgard covered coverslip (25 mm) (Fig. 3a). Here substances of interest can be added during measurement. For cut head preparations the head is cut with a scalpel directly posterior to the terminal bulb on the Sylgard coverslip. This kind of preparation provides EPG traces of more precision, as in intact animals movement can interrupt the signal, but the cutting process will affect the nervous system and the inner cuticle ionic conditions will be changed by buffering. So, for muscular pathway studies e.g. of calcium signaling, which require native ionic intra- and inner-cellular conditions, recording in the whole animal is beneficial. For testing drugs the substances can be added to the recording chamber. Light can be applied during measurement by LED or by a mercury lamp with the corresponding filter sets, and delivered in specific illumination frequencies by a computer-controlled shutter system [31, 36].

EPG measurements deliver detailed and precise information about pharyngeal pumping behavior but require single worm measurements. As a technically easier approach the natural pump frequency on food can be counted by eye with a high magnification microscope. This method is reliable to obtain rudimentary information about the pumping behavior in mutants or animals exposed to drugs, but does not give the amount of information EPGs can provide. For that reason, a method for optical recording of pharyngeal pumping has been developed by Schüler et al. to enable a higher throughput recording of pumping characteristics [31].

The method requires the immobilization of animals expressing the light-gated pacemaker system on polystyrene beads as described in Sect. 2.3. During video recording the animals can be paced with a computer controlled shutter system. Depending on the required experimental readout, the stimulation frequency can be chosen in lower or higher ranges. A synchronization of video start and corresponding light stimulation can be programmed. For unsynchronized measurements the start of illumination can be extracted from the videos by measurement of the whole video in ImageJ, shown as the first increase in the overall image grey value. For video analysis, ImageJ will be used as well. The pumping is defined as grinder or luminal movement during muscle contraction (Fig. 4a, e). The change of grey values during that process is extracted by multiple kymographic traces. To generate multi kymographs, a line scan is set in a chosen area (Fig. 4a). This can be anterior or posterior to the grinder, which is driven by terminal bulb contraction to grind pumped-in bacteria, or longitudinal to the lumen, recording the peristaltic movement of bacteria transport. The kymographs show changes in grey values corresponding to the pumping movements (Fig. 4b). Via line scanning through the kymograph these values are extracted, serving as base for further calculations (Fig. 4c). For the analysis a custom written script in the KNIME software package has been created by Wagner Steuer Costa [31]. The script provides tools to extract and manually correct the start time and the maximal contraction time of each pumping event according to the beginning of a peak and the minimum, respectively. From those values the pumping frequency (peak start to peak start) can be calculated, as well as the contraction duration (peak start to minimum) (Fig. 4d). Furthermore, the aberration from pulse frequency, also described as pumping ability, the ability to follow the pace frequency over time and the pumping rate distribution can be calculated. These data give insight into the pumping behavior of mutated strains or strains under pharmacological influence affecting the pharyngeal muscle cells.

Animals were studied using this paradigm that carried a Timothy syndrome orthologous mutation in the *egl-19* L-type voltage-gated calcium channels, which in humans leads to delayed channel

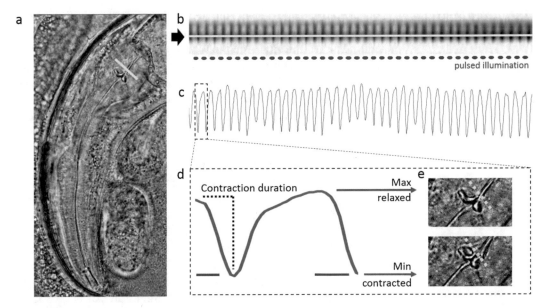

Fig. 4 The analysis of pharyngeal pumping behavior in optogenetically paced immobilized animals expressing ChR2 in the plasma membrane of pharyngeal muscle. From the recorded videos the movement of the grinder, located in the pharyngeal terminal bulb, is detected by a line scan (*yellow*) (**a**) and displayed as a kymograph (**b**). The pump movements, represented as changes of *grey values*, are extracted from the kymograph via line scan (*yellow*) as a graph (**c**). The graph provides information such as the contraction duration, the aberration from pulse frequency, ability to follow the pulse frequency over time and the pumping rate distribution (**d**). Thereby the maxima describe pharyngeal relaxation and the minima contraction

closure and prolonged membrane depolarization. Both in this optical detection method as well as in EPG measurements, the mutants showed a prolonged pump duration. The phenotype indicated here allows comparisons to the extension in QT stretch recorded in Timothy syndrome patients electrocardiograms [31, 81, 82].

6   Notes

1. All-trans-retinal (ATR) is light sensitive, so please avoid as much light exposure as possible during performing all steps that involve ATR.

2. To control for chronic effects of expressing unstimulated ChR2 and other optogenetic tools on the development, morphology and function of neurons in *C. elegans* is relatively straightforward, by using animals expressing the channel as controls which were raised in the absence of the cofactor all-trans-retinal. It appears that reducing channel expression levels can alleviate such effects.

   An important concern is that ChR2 photocurrents substantially diminish during illumination; to get stronger stimulation it is thus preferable to use millisecond-range light-pulsed stimuli (*see* Chap. 3 for details on ChR2 photocycle). Due to ChR(C128S) being activated by even dim light, animals must be grown in the dark. Interestingly, this variant, unlike ChR2, requires a continuous supply of ATR—its functionality halves 4 h after removal from ATR.

3. It is always advised to confirm the actual loss of the cell of interest after KillerRed illumination. This can be done by observing the loss of fluorescent protein expression in the cell, and depending on the cell in question, by recording behavioral changes.

    Neurons refractory to ablation by KillerRed require higher doses of KillerRed illumination. Some neurons were ablated with lower efficiency due to lower expression of KillerRed, which needs to be controlled for.

    The membrane-targeted version of KillerRed is said to be preferable as no phenotypic effect of the expression itself was observed, unlike the mitochondrially targeted version [62].

4. Pulsing the illumination light increased the efficiency of killing by a factor of three at the same total dose of illumination; pulsed light may facilitate the diffusion of oxygen into the active site and produce more ROS [63, 64].

5. The effectiveness of CALI depends strongly on which protein is tagged, and different tags should be tried depending on the actual target.

6. Animals should be kept in the dark before the experiment, as protein degradation is induced by low levels of blue light already. In addition, there is some indication that the psd domain may be partially active even without illumination to degrade the fusion protein.

7. To eliminate the photophobic response of *C. elegans* to the intense blue stimulation light, the experiments with PACα should be conducted in a *lite-1* mutant background, which strongly reduces avoidance of blue light. IlaC stimulation with red light can be done in a wild-type background.

    PACα exhibited an undesired basal activity which potentially restricts its application. This might be mitigated by using lower expression levels. Thus, one should find a balance between an expression low enough to minimize dark activity while being high enough to cause significant effects.

    Naturally occurring cAMP signaling is restricted to small domains close to the plasma membrane [83] which may well influence the way that this signaling molecule affects downstream pathways. Ectopically expressed PACα, however, is neither localized to the membrane nor restricted to small domains. Thus, cAMP produced by PACα may have more diverse and possibly unwanted effects. Modifications of the protein to restrict its subcellular localization would be desirable.

Animals expressing the IlaC light-gated adenylate cyclase showed locomotory hyperactivity already when grown at ambient light levels, and thus should be raised in the dark.

7 Outlook

Optogenetics provides a wide diversity of prospective development and areas of application in *C. elegans* neurobiology, and will continue to play a critical role in it. The tools introduced here and their utilization can be improved in many cases, and we would like to highlight a few desirable future directions:

- Novel tools and strategies to target specific subcellular organelles would provide the possibility to control and study intracellular processes without affecting other cellular functions, as mutations often do by evoking additional phenotypes. This could be combined with optical recording of Ca^{2+} or voltage signals in freely moving animals, giving insights into signaling pathways as well as into neuronal physiology.

- The combination of all-optical approaches to both control and record the activity of neurons with genetically encoded tools in live behaving animals represents the "holy grail" of understanding nervous system function and is still in its infancy. The approach will be beneficial for the elucidation of neural circuit function, especially regarding electrical and chemical synaptic connections. A pan-neuronally expressed optical activity sensor combined with a spectrally separated activation of optogenetic activators targeted to single neurons could provide transformational insight into neural signal processing. The specific targeting of optogenetic tools is still limited by the availability of promotors driving their expression and this goal will require novel tools to drive specific neuronal expression. Also, the existing optical sensors are still too slow for showing many signaling responses, optical sensors that enable the real-time detection of these fast processes need to be developed.

- *C. elegans* uses diverse sensory cues and sophisticated behavioral programs to navigate its environment, where the nervous system processes and integrates sensory information over space and time to guide behavioral strategies. To understand the neuronal computation underpinning navigational behavior, it would be valuable to create virtual environments based on optogenetic stimulation. This could be done by adapting existing solutions for the targeted illumination of optogenetic tools to create spatial or temporal gradients of light, mimicking the nonuniform distribution of sensory cues in the environment. Experiments of this kind are closely linked to the dynamic closed-loop control of optogenetic stimulation based on feedback in real time from the behavior the animal is showing. A few closed-loop optogenetic systems have been established in *C. elegans* so far [84, 85].

Acknowledgments

We thank the Wellcome Trust (109614/Z/15/Z) and the Medical Research Council (MR/N004574/1) for financial support.

References

1. Corsi AK, Wightman B, Chalfie M (2015) A transparent window into biology: a primer on *Caenorhabditis elegans*. Genetics 200:387–407. doi:10.1534/genetics.115. 176099

2. de Bono M, Villu Maricq A (2005) Neuronal substrates of complex behaviors in *C. elegans*. Annu Rev Neurosci 28:451–501. doi:10. 1146/annurev.neuro.27.070203.144259

3. Nagel G, Brauner M, Liewald JF et al (2005) Light activation of channelrhodopsin-2 in excitable cells of *Caenorhabditis elegans* triggers rapid behavioral responses. Curr Biol 15:2279–2284. doi:10.1016/j.cub.2005.11. 032

4. Kimura KD, Busch KE (2017) From Connectome to function: using optogenetics to shed light on the *Caenorhabditis elegans* nervous system. In: Appasani K (ed) Optogenetics: from neuronal function to mapping and disease biology. Cambridge University Press, Cambridge

5. Husson SJ, Gottschalk A, Leifer AM (2013) Optogenetic manipulation of neural activity in *C. elegans*: from synapse to circuits and behaviour. Biol Cell 105:235–250. doi:10.1111/ boc.201200069

6. Tsukada Y, Mori I (2015) Optogenetics in *Caenorhabditis elegans*. In: Yawo H, Kandori H, Koizumi A (eds) Optogenetics. Springer Japan, Tokyo, pp 213–226

7. Fang-Yen C, Alkema MJ, Samuel ADT (2015) Illuminating neural circuits and behaviour in *Caenorhabditis elegans* with optogenetics. Philos Trans R Soc Lond Ser B Biol Sci 370:20140212. doi:10.1098/rstb.2014.0212

8. Glock C, Nagpal J, Gottschalk A (2015) Microbial rhodopsin optogenetic tools: application for analyses of synaptic transmission and of neuronal network activity in behavior. In: Cell senescence. Humana Press, Totowa, NJ, pp 87–103

9. Mello CC, Kramer JM, Stinchcomb D, Ambros V (1991) Efficient gene transfer in *C. elegans*: extrachromosomal maintenance and integration of transforming sequences. EMBO J 10:3959–3970

10. Kim E, Sun L, Gabel CV, Fang-Yen C (2013) Long-term imaging of *Caenorhabditis elegans* using nanoparticle-mediated immobilization. PLoS One 8:e53419–e53416. doi:10.1371/ journal.pone.0053419

11. Stiernagle T (2006) Maintenance of *C. elegans*. WormBook 11:1–11. doi:10.1895/ wormbook.1.101.1

12. Krajniak J, Lu H (2010) Long-term high-resolution imaging and culture of *C. elegans* in chip-gel hybrid microfluidic device for developmental studies. Lab Chip 10:1862–1867. doi:10.1039/c001986k

13. Mondal S, Ahlawat S, Rau K et al (2011) Imaging in vivo neuronal transport in genetic model organisms using microfluidic devices. Traffic 12:372–385. doi:10.1111/j.1600-0854. 2010.01157.x

14. Hwang H, Barnes DE, Matsunaga Y et al (2016) Muscle contraction phenotypic analysis enabled by optogenetics reveals functional relationships of sarcomere components in *Caenorhabditis elegans*. Sci Rep 6:1–10. doi:10.1038/ srep19900

15. Ma H, Jiang L, Shi W et al (2009) A programmable microvalve-based microfluidic array for characterization of neurotoxin-induced responses of individual *C. elegans*.

Biomicrofluidics 3:044114–044118. doi:10.1063/1.3274313

16. Rezai P, Siddiqui A, Selvaganapathy PR, Gupta BP (2010) Electrotaxis of *Caenorhabditis elegans* in a microfluidic environment. Lab Chip 10:220–226. doi:10.1039/B917486A

17. Rezai P, Salam S, Selvaganapathy PR, Gupta BP (2012) Electrical sorting of *Caenorhabditis elegans*. Lab Chip 12:1831–1810. doi:10.1039/c2lc20967e

18. Song P, Dong X, Liu X (2016) A microfluidic device for automated, high-speed microinjection of *Caenorhabditis elegans*. Biomicrofluidics 10:011912–011912. doi:10.1063/1.4941984

19. Miyawaki A (2003) Fluorescence imaging of physiological activity in complex systems using GFP-based probes. Curr Opin Neurobiol 13:591–596. doi:10.1016/j.conb.2003.09.005

20. Allen PB, Sgro AE, Chao DL et al (2008) Single-synapse ablation and long-term imaging in live *C. elegans*. J Neurosci Methods 173:20–26. doi:10.1016/j.jneumeth.2008.05.007

21. Larsch J, Ventimiglia D, Bargmann CI, Albrecht DR (2013) High-throughput imaging of neuronal activity in *Caenorhabditis elegans*. Proc Natl Acad Sci U S A 110:E4266–E4273. doi:10.1073/pnas.1318325110

22. Kopito RB, Levine E (2014) Durable spatiotemporal surveillance of *Caenorhabditis elegans* response to environmental cues. Lab Chip 14:764–770. doi:10.1039/C3LC51061A

23. Chronis N, Zimmer M, Bargmann CI (2007) Microfluidics for in vivo imaging of neuronal and behavioral activity in *Caenorhabditis elegans*. Nat Methods 4:727–731. doi:10.1038/nmeth1075

24. Chalasani SH, Chronis N, Tsunozaki M et al (2007) Dissecting a circuit for olfactory behaviour in *Caenorhabditis elegans*. Nature 450:63–70. doi:10.1038/nature06292

25. Busch KE, Laurent P, Soltesz Z et al (2012) Tonic signaling from O_2 sensors sets neural circuit activity and behavioral state. Nat Neurosci 15:581–591. doi:10.1038/nn.3061

26. Lockery SR, Elizabeth Hulme S, Roberts WM et al (2012) A microfluidic device for whole-animal drug screening using electrophysiological measures in the nematode *C. elegans*. Lab Chip 12:2211–2217. doi:10.1039/c2lc00001f

27. Kerr RA (2006) Imaging the activity of neurons and muscles. WormBook:1–13. doi:10.1895/wormbook.1.113.1

28. Husson SJ, Costa WS, Schmitt C, Gottschalk A (2013) Keeping track of worm trackers. WormBook:1–17. doi:10.1895/wormbook.1.156.1

29. Redemann S, Schloissnig S, Ernst S et al (2011) Codon adaptation–based control of protein expression in *C. elegans*. Nat Methods 8:250–252. doi:10.1038/nmeth.1565

30. Li D, Wang M (2012) Construction of a bicistronic vector for the co-expression of two genes in *Caenorhabditis elegans* using a newly identified IRES. Biotech 52:173–176. doi:10.2144/000113821

31. Schüler C, Fischer E, Shaltiel L et al (2015) Arrhythmogenic effects of mutated L-type Ca^{2+}-channels on an optogenetically paced muscular pump in *Caenorhabditis elegans*. Sci Rep 5:14427. doi:10.1038/srep14427

32. Nagel G, Ollig D, Fuhrmann M et al (2002) Channelrhodopsin-1: a light-gated proton channel in green algae. Science 296:2395–2398. doi:10.1126/science.1072068

33. Nagel G, Szellas T, Huhn W et al (2003) Channelrhodopsin-2, a directly light-gated cation-selective membrane channel. Proc Natl Acad Sci U S A 100:13940–13945. doi:10.1073/pnas.1936192100

34. Kleinlogel S, Feldbauer K, Dempski RE et al (2011) Ultra light-sensitive and fast neuronal activation with the Ca^{2+}-permeable channelrhodopsin CatCh. Nat Neurosci 14:513–518. doi:10.1038/nn.2776

35. Klapoetke NC, Murata Y, Kim SS et al (2014) Independent optical excitation of distinct neural populations. Nat Methods 11:338–346. doi:10.1038/nmeth.2836

36. Erbguth K, Prigge M, Schneider F et al (2012) Bimodal activation of different neuron classes with the spectrally red-shifted Channelrhodopsin chimera C1V1 in *Caenorhabditis elegans*. PLoS One 7:e46827–e46829. doi:10.1371/journal.pone.0046827

37. Schultheis C, Liewald JF, Bamberg E et al (2011) Optogenetic long-term manipulation of behavior and animal development. PLoS One 6:e18766. doi:10.1371/journal.pone.0018766

38. Berndt A, Yizhar O, Gunaydin LA et al (2009) Bi-stable neural state switches. Nat Neurosci 12:229–234. doi:10.1038/nn.2247

39. Zhang F, Wang L-P, Brauner M et al (2007) Multimodal fast optical interrogation of neural circuitry. Nature 446:633–639. doi:10.1038/nature05744

40. Husson SJ, Liewald JF, Schultheis C et al (2012) Microbial light-Activatable proton pumps as neuronal inhibitors to functionally

dissect neuronal networks in *C. elegans*. PLoS One 7:e40937–e40914. doi:10.1371/journal. pone.0040937

41. Chow BY, Han X, Dobry AS et al (2010) High-performance genetically targetable optical neural silencing by light-driven proton pumps. Nature 463:98–102. doi:10.1038/nature08652

42. Stirman JN, Crane MM, Husson SJ et al (2011) Real-time multimodal optical control of neurons and muscles in freely behaving *Caenorhabditis elegans*. Nat Methods 8:153–158. doi:10.1038/nmeth.1555

43. Okazaki A, Sudo Y, Takagi S (2012) Optical silencing of *C. elegans* cells with arch proton pump. PLoS One 7:e35370. doi:10.1371/journal.pone.0035370

44. Stephens GJ, Johnson-Kerner B, Bialek W, Ryu WS (2008) Dimensionality and dynamics in the behavior of *C. elegans*. PLoS Comput Biol 4: e1000028–e1000010. doi:10.1371/journal.pcbi.1000028

45. Stirman JN, Crane MM, Husson SJ et al (2012) A multispectral optical illumination system with precise spatiotemporal control for the manipulation of optogenetic reagents. Nat Protoc 7:207–220. doi:10.1038/nprot.2011.433

46. Liewald JF, Brauner M, Stephens GJ et al (2008) Optogenetic analysis of synaptic function. Nat Methods 5:895–902. doi:10.1038/nmeth.1252

47. Richmond J (2006) Electrophysiological recordings from the neuromuscular junction of *C. elegans*. WormBook 6:1–8. doi:10.1895/wormbook.1.112.1

48. Richmond JE, Jorgensen EM (1999) One GABA and two acetylcholine receptors function at the *C. elegans* neuromuscular junction. Nat Neurosci 2:791–797. doi:10.1038/12160

49. Lignani G, Ferrea E, Difato F et al (2013) Long-term optical stimulation of channelrhodopsin-expressing neurons to study network plasticity. Front Mol Neurosci 6:22. doi:10.3389/fnmol.2013.00022

50. Miyashita T, Shao YR, Chung J et al (2013) Long-term channelrhodopsin-2 (ChR2) expression can induce abnormal axonal morphology and targeting in cerebral cortex. Front Neural Circuits 7:8. doi:10.3389/fncir.2013.00008

51. Gorostiza P, Isacoff EY (2008) Optical switches for remote and noninvasive control of cell signaling. Science 322:395–399. doi:10.1126/science.1166022

52. Mourot A, Tochitsky I, Kramer RH (2013) Light at the end of the channel: optical manipulation of intrinsic neuronal excitability with chemical photoswitches. Front Mol Neurosci 6:5. doi:10.3389/fnmol.2013.00005

53. Inoue M, Takeuchi A, Horigane S-I et al (2015) Rational design of a high-affinity, fast, red calcium indicator R-CaMP2. Nat Methods 12:64–70. doi:10.1038/nmeth.3185

54. Li Z, Liu J, Zheng M, Xu XZS (2014) Encoding of both analog- and digital-like behavioral outputs by one *C. elegans* interneuron. Cell 159:751–765. doi:10.1016/j.cell.2014.09.056

55. Wojtovich AP, Foster TH (2014) Optogenetic control of ROS production. Redox Biol 2:368–376. doi:10.1016/j.redox.2014.01.019

56. Pletnev S, Gurskaya NG, Pletneva NV et al (2009) Structural basis for phototoxicity of the genetically encoded photosensitizer KillerRed. J Biol Chem 284:32028–32039. doi:10.1074/jbc.M109.054973

57. Shu X, Lev-Ram V, Deerinck TJ et al (2011) A genetically encoded tag for correlated light and electron microscopy of intact cells, tissues, and organisms. PLoS Biol 9:e1001041. doi:10.1371/journal.pbio.1001041.g005

58. Pimenta FM, Jensen RL, Breitenbach T et al (2013) Oxygen-dependent photochemistry and photophysics of "MiniSOG," a Protein-Encased Flavin. Photochem Photobiol 89:1116–1126. doi:10.1111/php.12111

59. Bulina ME, Chudakov DM, Britanova OV et al (2006) A genetically encoded photosensitizer. Nat Biotechnol 24:95–99. doi:10.1038/nbt1175

60. Kobayashi J, Shidara H, Morisawa Y et al (2013) A method for selective ablation of neurons in *C. elegans* using the phototoxic fluorescent protein, KillerRed. Neurosci Lett 548:261–264. doi:10.1016/j.neulet.2013.05.053

61. Shibuya T, Tsujimoto Y (2012) Deleterious effects of mitochondrial ROS generated by KillerRed photodynamic action in human cell lines and *C. elegans*. J Photochem Photobiol B Biol 117:1–12. doi:10.1016/j.jphotobiol.2012.08.005

62. Williams DC, Bejjani RE, Ramirez PM et al (2013) Rapid and permanent neuronal inactivation in vivo via subcellular generation of reactive oxygen with the use of KillerRed. Cell Rep 5:553–563. doi:10.1016/j.celrep.2013.09.023

63. Qi YB, Garren EJ, Shu X et al (2012) Photoinducible cell ablation in *Caenorhabditis elegans* using the genetically encoded singlet oxygen generating protein miniSOG. Proc Natl

Acad Sci U S A 109:7499–7504. doi:10.1073/pnas.1204096109

64. Xu S, Chisholm AD (2016) Highly efficient optogenetic cell ablation in *C. elegans* using membrane-targeted miniSOG. Sci Rep 6:21271. doi:10.1038/srep21271

65. Lin JY, Sann SB, Zhou K et al (2013) Optogenetic inhibition of synaptic release with chromophore-assisted light inactivation (CALI). Neuron 79:241–253. doi:10.1016/j.neuron.2013.05.022

66. Wojtovich AP, Wei AY, Sherman TA et al (2016) Chromophore-assisted light inactivation of mitochondrial electron transport chain complex II in *Caenorhabditis elegans*. Sci Rep 6:29695. doi:10.1038/srep29695

67. Takemoto K, Matsuda T, Sakai N et al (2013) SuperNova, a monomeric photosensitizing fluorescent protein for chromophore-assisted light inactivation. Sci Rep 3:2629. doi:10.1038/srep02629

68. Hermann A, Liewald JF, Gottschalk A (2015) A photosensitive degron enables acute light-induced protein degradation in the nervous system. Curr Biol 25:R749–R750. doi:10.1016/j.cub.2015.07.040

69. Renicke C, Schuster D, Usherenko S et al (2013) A LOV2 domain-based Optogenetic tool to control protein degradation and cellular function. Chem Biol 20:619–626. doi:10.1016/j.chembiol.2013.03.005

70. Noma K, Jin Y (2015) Optogenetic mutagenesis in *Caenorhabditis elegans*. Nat Commun 6:8868. doi:10.1038/ncomms9868

71. Weissenberger S, Schultheis C, Liewald JF et al (2011) PACα- an optogenetic tool for in vivo manipulation of cellular cAMP levels, neurotransmitter release, and behavior in *Caenorhabditis elegans*. J Neurochem 116:616–625. doi:10.1111/j.1471-4159.2010.07148.x

72. Schade MA, Reynolds NK, Dollins CM, Miller KG (2005) Mutations that rescue the paralysis of *Caenorhabditis elegans* ric-8 (Synembryn) mutants activate the Gαs pathway and define a third major branch of the synaptic signaling network. Genetics 169:631–649. doi:10.1534/genetics.104.032334

73. Ryu M-H, Kang I-H, Nelson MD et al (2014) Engineering adenylate cyclases regulated by near-infrared window light. Proc Natl Acad Sci U S A 111:10167–10172. doi:10.1073/pnas.1324301111

74. Gao S, Nagpal J, Schneider MW et al (2015) Optogenetic manipulation of cGMP in cells and animals by the tightly light-regulated guanylyl-cyclase opsin CyclOp. Nat Commun 6:8046. doi:10.1038/ncomms9046

75. Ryu M-H, Moskvin OV, Siltberg-Liberles J, Gomelsky M (2010) Natural and engineered photoactivated nucleotidyl cyclases for optogenetic applications. J Biol Chem 285:41501–41508. doi:10.1074/jbc.M110.177600

76. Frank JA, Yushchenko DA, Hodson DJ et al (2016) Photoswitchable diacylglycerols enable optical control of protein kinase C. Nat Chem Biol 12:755–762. doi:10.1038/nchembio.2141

77. Avery L (1993) Motor neuron M3 controls pharyngeal muscle relaxation timing in *Caenorhabditis elegans*. J Exp Biol 175:283–297

78. Avery L, Horvitz HR (1989) Pharyngeal pumping continues after laser killing of the pharyngeal nervous system of *C. elegans*. Neuron 3:473–485. doi:10.1016/0896-6273(89)90206-7

79. Cook A (2006) Electrophysiological recordings from the pharynx. WormBook 17:1–7. doi:10.1895/wormbook.1.110.1

80. Dillon J, Andrianakis I, Bull K et al (2009) AutoEPG: software for the analysis of electrical activity in the microcircuit underpinning feeding behaviour of *Caenorhabditis elegans*. PLoS One 4:e8482–e8413. doi:10.1371/journal.pone.0008482

81. Splawski I, Timothy KW, Sharpe LM et al (2004) CaV1.2 calcium channel dysfunction causes a multisystem disorder including arrhythmia and autism. Cell 119:19–31. doi:10.1016/j.cell.2004.09.011

82. Splawski I, Timothy KW, Decher N et al (2005) Severe arrhythmia disorder caused by cardiac L-type calcium channel mutations. Proc Natl Acad Sci U S A 102:8089–8096. doi:10.1073/pnas.0502506102

83. Beavo JA, Brunton LL (2002) Cyclic nucleotide research – still expanding after half a century. Nat Rev Mol Cell Biol 3:710–718. doi:10.1038/nrm911

84. Kocabas A, Shen C-H, Guo ZV, Ramanathan S (2012) Controlling interneuron activity in *Caenorhabditis elegans* to evoke chemotactic behaviour. Nature 490:273–277. doi:10.1038/nature11431

85. Satoh Y, Sato H, Kunitomo H et al (2014) Regulation of experience-dependent bidirectional Chemotaxis by a neural circuit switch in *Caenorhabditis elegans*. J Neurosci 34:15631–15637. doi:10.1523/JNEUROSCI.1757-14.2014

Chapter 7

Optogenetic Interpellation of Behavior Employing Unrestrained Zebrafish Larvae

Soojin Ryu and Rodrigo J. De Marco

Abstract

The zebrafish larva, *Danio rerio*, provides superb genetic access for studying how systematic variations in behavioral profiles relate to differences in brain activity. Larvae respond predictably to various sensory inputs and their nervous system is readily accessible. Also, their transparent body allows for noninvasive optogenetics and their small size allows for measuring behavior with full environmental control. In tethered larvae, neural activity has been correlated to eye and tail movements. The challenge now is to tackle the building blocks of behavior: internal states (maturation and learning), motivations (drives), reversible phenotypic adaptations (humoral actions), and decision processes (choice and task selection). These phenomena are best addressed through the analysis of freely behaving subjects. This chapter provides the basics for applying optogenetics to the analysis of behavior in freely swimming larvae. As a study case, we offer information from recent tests showing how optogenetic manipulation of hormone-producing cells can be used to address reversible phenotypic adaptations. Because larvae are highly reactive to optic stimuli, light control is pivotal in employing noninvasive optogenetics. This point is covered in detail, starting from the general rules of light delivery and maintenance prior to the tests.

Key words Optogenetics, Larval zebrafish, Behavior, Photoactivated adenylyl cyclases, Stress

1 Introduction

Caenorhabditis elegans and *Danio rerio* are widely used for studying the neuronal bases of behavior. In recent years, research on both organisms has seen an increasing number of optogenetic tools. *C. elegans*, a transparent nematode, has a nervous system of 302 neurons whose wiring diagram has been established [1]. In combination with easily quantifiable movements serving responses to various sensory inputs and simple forms of learning, *C. elegans* provides a productive ground for the analysis of locomotion patterns [2–4]. It was the first organism in which the microbial rhodopsin, ChR2, was implemented for remote locomotion control [5]; numerous photo-switchable actuators and reporters have

Albrecht Stroh (ed.), *Optogenetics: A Roadmap*, Neuromethods, vol. 133,
DOI 10.1007/978-1-4939-7417-7_7, © Springer Science+Business Media LLC 2018

subsequently been used to link the activity of neural circuits to locomotion schemes [6] (*see* Chap. 6).

Due to their genetic amenability and transparent body, larval zebrafish have also become popular for studying the neuronal bases of behavior. They show a range of locomotion patterns and reactions [7–10], visual reflexes [11, 12], and prey capture movements [13–15]. In tethered larvae, brain regions [16–20] and circuit motives regulating locomotion have been identified via genetically encoded calcium indicators (GECIs) [21–23] (*see* also Chap. 9). Photo-switchable actuators have been used to either increase or decrease neuronal activity [24]. Examples include the use of LiGluR or ChR2 in spinal cord neurons [25, 26], halorhodopsin or ChR2 in hindbrain neurons [27–30], and ChR2 in midbrain (nMLF) [31], pretectal, and tectal neurons [32]. In sum, in combination with high-resolution in vivo imaging and optogenetics, tethered larvae have provided a great deal of help to link brain activity to principles of locomotion control. The analysis of complex behavior, however, cannot be accurately carried out employing restrained subjects. Behavioral complexity rests on the joint work of four regulatory components functionally defined: internal states, motivations, adaptive responses to the environment, and decision processes. To tackle these fundamentals, records of freely behaving subjects will be compulsory.

1. *Internal states* are measured through novelty responses. When released into a novel environment, e.g., animals initially show signs of fear and may remain motionless. This is often followed by exploratory movements. The probability that a given novel situation will elicit exploration rather than quiescence depends on the animal's internal state. Animals that have recently had a stressful experience are more likely to be wary of novel environments.

2. *Motivation* is a reversible aspect of an animal's state that plays a causal role in behavior. Such state is made up of both internal and external factors relevant to incipient activities as well as the animal's current behavior. It follows that motivations can only be studied through the analysis of the relative importance of different activities. To determine motivations, therefore, a subject should be free to respond to local environments with a full repertoire of actions.

3. *Humoral actions* may be slow and prolonged, or quick and short lived. For example, hormones have important effects on the ontogeny of reproductive behavior, migration and other seasonal activities. They control many rhythms, such as the sexual and menstrual cycles, and also exert a short-term influence on fear. All in all, multiple measures from freely behaving subjects over multiple time domains are necessary to specify

short- and long-term effects of brain neuropeptides and peripheral hormones.

4. There are many activities in which an animal could engage at any particular time: feeding, courtship, sleep, and so on. It is virtually impossible for an animal to carry out such different activities simultaneously, simply because the movements required are mutually incompatible. There are processes that determine which activity has priority at any particular time. Such processes are termed *decision processes* [33]. To unravel the process of deciding on different activities, an animal must be presented with alternatives and must be free to choose and experience the consequences of its own choice. Experimentally, this demands a focus on freely behaving subjects.

Here, we present general recommendations for applying optogenetics to the analysis of behavior in freely swimming zebrafish larvae. We offer data from a series of recent experiments in which a photoactivated adenylate cyclase (PAC) was used to manipulate the activity of pituitary cells in combination with newly developed assays for measuring goal-directed actions [34]. The pituitary is the major link between nervous and hormonal systems, which allow the brain to generate flexible behavior. Embedded in the hypothalamo-pituitary-adrenal (HPA) axis, pituitary corticotroph cells are known to control the release of glucocorticoids from the adrenal gland into the blood [35], yet their contribution to stress behavior had been difficult to pin down due to the limited accessibility of the hypothalamus and pituitary and the coupled release of hypothalamic and pituitary neuropeptides. The results of our experiments revealed rapid organizing effects of corticotroph cell products on locomotion, avoidance and arousal directly after the onset of stress. PAC contains a "blue light sensor using FAD" (BLUF) domain and, once activated by blue light, generates cAMP [36]. *Beggiatoa* PAC (bPAC) is an improved PAC, as compared to EuPAC [37, 38]. In zebrafish, bPAC had previously been used for analyses of stress reactions and neuronal repair [39, 40, 41, 42] (*see* also Chap. 4).

2 Materials

2.1 Common Buffers and Chemicals

1. E2 medium: 5 mM NaCl, 0.25 mM KCl, 0.5 mM MgSO$_4$, 0.15 mM KH$_2$PO$_4$, 0.05 mM Na$_2$HPO$_4$, 0.66 mM CaCl$_2$, 0.71 mM NaHCO$_3$, pH adjusted to 7.0.

2. 0.2 mM 1-Phenyl-2-thiourea (PTU; Sigma-Aldrich, #P7629).

2.2 Generation of Transgenic Larvae Expressing Photo-Switchable Actuators (PSAs)

1. Wild-type zebrafish: e.g., cross between AB and TL strains.

2. Microinjector for DNA injection into single-cell-stage zebrafish embryos (e.g., Eppendof Femtojet).

3. Plasmid DNA allowing tissue-specific expression of PSAs in larvae.

4. Routine injection reagents: phenol red, injection mold or slide, glass capillary.

5. Fluorescence stereomicroscope to screen for embryos expressing PSAs according to signals from coupled fluorescence proteins (e.g., Leica MZ16F with excitation light source and filters).

6. Vibration-free incubator for raising larvae under controlled temperature and light conditions (e.g., RUMED type 3101).

7. Containers covered with light filters for raising embryos expressing PSAs; 550 nm long-pass filters (Thorlabs, Dachau, Germany) for bPAC.

2.3 Environmental Control

Several aspects of the environment can impinge on freely behaving larvae during a test, which makes it utterly important to control and monitor the global environment of the test chamber. The temperature, composition and motions of the medium as well as its overall level of illumination should be controllable. A straightforward approach consists of a cylindrical chamber equipped with a temperature monitoring sensor and an inlet, outlet and overflow, particularly if optogenetics is to be combined with proxies perfused into the medium and washed away in a highly controlled fashion. Figure 1 presents one such chamber:

1. The exemplary swimming chamber in Fig. 1 (internal diameter: 10 mm, height: 10 mm) has a transparent bottom and two opposite overtures, inlet and outlet (width: 2.5 mm, height: 400 μm). The chamber also had two cylindrical side channels (internal diameter: 400 μm) opposite to each other opening 200 μm above the transparent glass bottom, with their longest axis oriented at an angle of 30° relative to horizontal (**see** also **Note 1**).

2. A peristaltic pump allows the medium inside the chamber to circulate at a constant rate (IPC Ismatec, IDEX Health and Science GmbH, Wertheim, Germany).

3. A thermocouple (TS200, npi electronics GmbH, Tamm, Germany) monitors the temperature inside the chamber and provides feedback to a control system (PTC 20, npi electronics GmbH, Tamm, Germany; Exos-2V2 liquid cooling system, Koolance, Auburn, WA, USA).

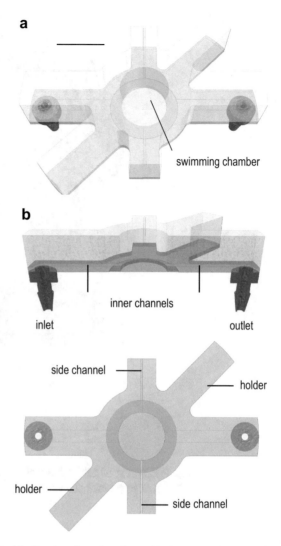

Fig. 1 Cylindrical swimming chamber with transparent bottom (**a**) and inner channels (**b**) allowing the medium to flow at constant or increasing temperature; side channels hold thermocouples constantly monitoring the temperature of the flowing medium. Scale bar, 10 mm. Modified from [34]

4. If experiments require perfusion, a computer-controlled perfusion system (Octaflow, ALA Scientific Instruments, Inc., Farmingdale, NY, USA) can be used to inject known solutions into a mixing compartment (internal diameter: 1 mm) situated 10 mm from the inlet of the swimming chamber. The mixing compartment can be connected to single reservoirs of solutions coupled to computer-controlled solenoid valves via Teflon tubing (internal diameter: 230 μm, outer diameter: 600 μm). TTL signals can trigger the opening and closing of the valves, allowing the solutions to be well mixed with the flowing medium before reaching the inner chamber.

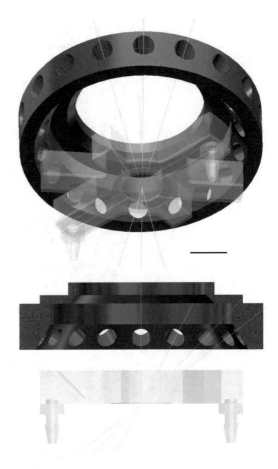

Fig. 2 LED ring on top of the test chamber in Fig. 1 for *blue* and *yellow light* stimulation in combination with video tracking. Scale bar, 10 mm. Modified from refs. 44, 45

2.4 Light Delivery

For the purpose of video recording freely behaving larvae, the investigator must adopt an approach for delivering light that is capable of providing homogeneous illumination within a relatively large chamber, particularly during tests of exploratory activity and goal-directed actions. Figure 2 presents an example based on a ring of LEDs.

1. The incident angle of the LEDs in Fig. 2, each pointing towards the center of the bottom surface of the chamber, allows for a homogeneous illumination of the chamber's inner compartment. Both the wavelength and intensity of the light needed for activating PSAs need to be separately adjustable and empirically determined for each transgenic line. This can be achieved by using distinct LEDs in combination with a control system for light delivery, composed of custom-made drivers for setting the light power, pulse generators for setting

the frequency and duration of a squared pulse of light, and a TTL control box for computer control of the LEDs.

2. A hand-held light power meter (Newport Corp., Irvine, CA, USA) can be used to measure light power (*see* **Note 2**)

2.5 Video-Tracking

Video tracking provides a suitable solution in many situations where a camera can access an entire swimming chamber from above. The essential features of a video-tracking system consist of a solid-state digital camera with a charge-coupled device (CCD) detector and a lens placed above the chamber so that the entire area of interest is in the field of view. A challenge arises from the poor contrast of the semi-transparent larva. Low contrast results in very small differences from background, which are difficult to track. The best solution is to equip the swimming chamber with a translucent bottom surface and provide homogeneous white and IR light from below through an array of LEDs covered by diffusing glass. With such an array, problems arising from object reflectance close to background brightness can be reduced if the LEDs have a wide range of brightness levels. A basic set of elements is as follows:

1. A light-proof enclosure for the entire set up placed on a vibration-free platform (Newport Corp., Irvine, CA, USA).

2. Infrared-sensitive cameras (25 frames*s^{-1}, ICD-49E B/W, Ikegami Tsushinki Co., Ltd. Japan, and 100 frames*s^{-1}, Fire-wire Camera, Noldus Information Technology, Wageningen, Netherlands) and lens (TV Lens, Computer VARI FOCAL H3Z4512 CS-IR, CBC; Commak, NY, USA) with a maximum 3× zoom (*see* also **Note 3**).

3. A source of infrared (IR) illumination opposite to the camera lens—we use a custom-made array of IR-LEDs below the swimming chamber of transparent bottom.

4. We use EthoVision XT software (Noldus Information Technology, Wageningen, Netherlands) for online video-tracking, but *see* also **Note 4**.

3 Experimental Procedures

3.1 Obtaining Transgenic Zebrafish Larvae for Optogenetics (See Note 5)

1. Clone a construct harboring the protein of interest (PSA) under the control of a tissue-specific promoter (*see* Chap. 2) and couple the PSA with a fluorescent protein to aid in subsequent identification/maintenance of the line.

2. Set up mating crosses of wild-type zebrafish the day before and collect embryos after the light onset on the following day.

3. Inject the construct (approximately 50–100 pg) into single-cell-stage embryos in the presence of 0.05% phenol red. If the

construct was generated with the Tol2 kit [43], then incubate with 100 pg Tol2 transposase RNA for 10 min prior to the injection.

4. After injection, maintain embryos inside an incubator at 28 °C on a 12:12 h light-dark cycle. Raise a maximum of 60 larvae in a (10 cm) petri dish with 15 ml of egg water, check at 1 day post fertilization (dpf) and remove dead embryos. At 5 dpf, transfer the larvae to the fish facility and start feeding.

5. Once the fish reach sexual maturity, identify transgenic carriers ("founders") by inter-crossing injected and wild-type fish and screening the progenies for the expression of the marker gene.

6. For transgenic larvae expressing PAC, set up a cross of *Tg (POMC:bPAC-2A-tdTomato)* with wild-type (AB/TL) fish. On the following day, collect the fertilized embryos and place a maximum of 60 embryos into a container with 15 ml of E2 medium. (If the experiment requires PTU, then add 0.2 mM at this point). Cover the container with a light filter (*see* Sect. 2.2) and place it in a temperature-controlled incubator at 28 °C on a 12:12 h light-dark cycle. At 3 or 4 dpf, depending on experimental needs, screen for larvae that express tdTomato in the pituitary under fluorescence microscope. For 4 dpf larvae, tricaine methane sulfonate (MS222, 0.01 mg/ml) may be added before (5 min) the screening to sedate the larvae; 3 dpf larvae are less likely to move and can be screened without MS222. Keep tdTomato-positive vs. tdTomato-negative larvae separately inside the incubator under filtered light as described above (*see* **Notes 6** and **7**).

3.2 Optogenetic Interpellation of Behavior

Larval zebrafish are highly sensitive to optic stimuli [44]. When briefly (~5 min) adapted to darkness, for example, they react to a sudden dark-to-light transition with a brief period of increased locomotion directly after the light onset, followed by reduced locomotion during the light period and increased locomotion after the light offset. Notably, in briefly dark-adapted larvae, we observed that a squared pulse of either blue or yellow light not only elicits locomotor reactions, but also increases whole-body cortisol in a graded fashion, depending on light power and exposure time (Fig. 3) [41]. By virtue of its complexity alone, the regulation of behavior is susceptible to stress. The above finding is relevant because it shows that a light change in itself can be stressful and has, for that reason, the capacity to alter the regulation of a behavioral scheme; future tests on intact larvae must control for such an effect. The simplest way to activate PSAs without causing stress is to modify the wavelength composition of light while maintaining the overall illumination of the test chamber unchanged. Still, the fact that light can be stressful also has an advantage. It can be used in a dual role to understand how distinct modulators influence behavior

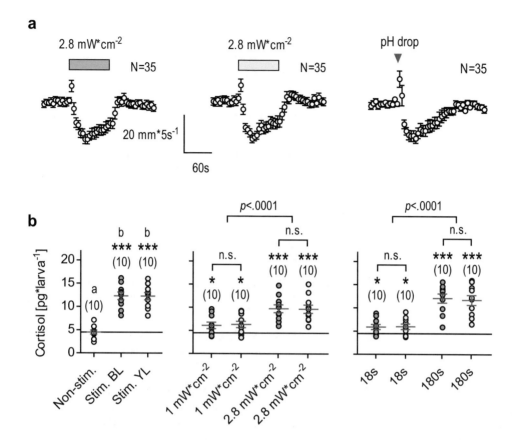

Fig. 3 (**a**) Left, Center, briefly dark-adapted larvae react to a squared pulse of either *blue* (left) or *yellow* (center) light with reduced and increased locomotion after the onset and offset of light, respectively. Right, they react similarly to a pH drop, a known potent stressor in fish. Locomotion measured as swim velocity [mm/$(5 \ s)^{-1}$]. (**b**) A squared pulse of light can increase whole-body cortisol (left, light power: 4.4 mW/cm^{-2}), depending on light power (center) and exposure time (right, light power: 2.8 mW/cm^{-2}). Left, One-Way ANOVA, $F(2,29) = 42.1$, $p < 0.0001$, followed by Bonferroni's tests. Center, Two-Way ANOVA, Wavelength factor: $F(1,36) = 0.01$, $p = 0.93$, Light Power factor: $F(1,36) = 19.8$, $p < 0.0001$, Wavelength × Light Power factor: $F(1,36) = 0.03$, $p = 0.86$, followed by post hoc comparisons. Right, Two-Way ANOVA, Wavelength factor: $F(1,36) = 0.03$, $p = 0.87$, Length factor: $F(1,36) = 52.1$, $p < 0.0001$, Wavelength × Length factor: $F(1,36) = 0.08$, $p = 0.79$, followed by post hoc comparisons. *$p < 0.05$, ***$p < 0.001$ after one sample t-tests against basal cortisol in nonstimulated larvae (*black line*), Non-stim: $t(9) = 0.01$, $p = 0.99$, Stim. BL: $t(9) = 9.6$, $p < 0.0001$, Stim. YL: $t(9) = 9.9$, $p < 0.0001$, 1 mW/cm^{-2} BL: $t(9) = 8.8$, $p < 0.0001$, 1 mW/cm^{-2} YL: $t(9) = 9.3$, $p < 0.0001$, 2.8 mW/cm^{-2} BL: $t(9) = 11.4$, $p < 0.0001$, 2.8 mW/cm^{-2} YL: $t(9) = 10.9$, $p < 0.0001$, 18 s BL, $t(9) = 2.4$, $p = 0.04$, 18 s YL, $t(9) = 2.5$, $p = 0.04$, 180 s BL, $t(9) = 7.7$, $p < 0.0001$, 180 s YL, $t(9) = 6.8$, $p < 0.0001$). Sample size in parentheses. See also De Marco et al. [34]

after the onset of stress. This is exactly what we did to study the role of pituitary corticotroph cells in mediating reversible phenotypic adaptations [34]. In larvae expressing bPAC specifically in pituitary corticotrophs, we showed that blue light can be used as both a potent stressor and a means to enhance corticotroph cell activity directly after the onset of stress, thus prompting locomotion changes, meaningful adjustments in avoidance behaviors, higher levels of whole-body cortisol, and greater arousal. General recommendations from these experiments are listed below (*see* also **Note 8**).

1. Carry out individual video-recordings (i.e., one larva at the time), with different experimental groups intermixed throughout the day.

2. Provide each larva with an adaptation period of no less than 10 min to the test conditions.

3. Use an image spatial resolution of -no less than- 37 or 16 pixels per mm at 100 or 25 Hz, respectively.

4. Systematically use baseline recordings and compare baseline and test data across clutches, developmental stages, individuals and time of the day.

5. If experiments require drug substances using perfusion, use a maximum flowing medium of 200 $\mu l/min^{-1}$ in order to avoid rheotaxis.

6. Confirm that larvae do not show bias swim directions under IR light if experiments require darkness before and after the activation of PSAs.

7. Larvae show low response thresholds to light change. If light of a given frequency or light power is required during tests, locomotion profiles elicited by the selected frequency or power could mask, at least partially, changes related to PSA activation. Use light of different wavelengths as a means for wavelength-dependent optogenetic control of selective cell activity.

4 Notes

1. The level of baseline activity expressed by a larva is very important to consider when designing a swimming chamber. Larvae that move very little in a novel environment may cover only a small area of the chamber, undermining tests by inactivity. The shape and dimensions of the chamber will depend directly upon the type of targeted actions. A test for rheotaxis, e.g., will demand an elongated chamber as well as accurate control of the medium flowing through it. A test for feeding or any other goal-directed action causing a differential usage of space will demand a compartmentalized chamber so that a heterogeneous distribution of prey—or any other goal or condition—can be created (e.g., [34, 45]). Measuring escape reactions demands absent corners so that sustained displacements can occur.

2. Many light sources have a manual dial for setting the current intensity that determine light power, but this is not more than a rough guide to setting a desired power of light. The investigator needs to monitor the light power directly at the tip of the objective or at the bottom of the swimming chamber. The best practice is to determine the actual power of light that can reach a freely behaving larva by monitoring light power at specific wavelengths. For this purpose, place the sensor of the light power meter below a swimming chamber with a fully transparent bottom and measure light power—with and without liquid medium inside the chamber—across currents and distances from the light source.

3. If the lens is X mm above the bottom surface of the chamber and the width of the chamber is Y mm, the tangent of half the angle of view equals $(Y/2)/X$. The required focal length of a lens that can see the entire chamber must be less than or equal to $(X \times CCD)/(Y + CCD)$. The format of the CCD chip is important because the size of the chip and the focal length determine the angle of view. A CCD sensitive to very long wavelength or infrared light (IR) is desirable if optogenetic intervention is to be combined with behavioral recordings carried out in darkness.

4. Goal-directed actions by freely behaving larvae have already been measured using different video-tracking approaches [45, 46]. Open source software has also been published [47]. Commercially available systems provide readily measures of path length, latency until the subject enters specific zones, time and distance travelled in each such zone, number of entries, and so on, with the possibility to organize data on the basis of an overall estimate of mobility level. Usually, these systems locate the larva being tracked using background subtraction. An image is first stored before the larva is introduced into the chamber; it is important that illumination during this step be identical to that used when the larva is present. Once the larva is present, every few milliseconds—depending on the temporal resolution of the camera—another image is saved, and the difference between the new image and the background is determined for every pixel array. The entire difference array of Nx by Ny pixels is scanned and, for pixels where the absolute difference exceeds some preset threshold, the X and Y values for that pixel are added to two counters (Σx, Σy) and the pixel counters (nx, ny) are incremented by one unit. After the scan is complete, the coordinates of the weighted center of the image are found as $\Sigma x/nx$ and $\Sigma y/ny$. A scale factor from an initial calibration of the image is finally used to convert pixels to millimeters and the vectors of scaled X and Y values for each image are used to determine path length and other measures.

Based on similar principles, algorithms to measure the overall motion level of not just one, but a group of larvae have also been implemented ([34, 45]). They can detect the movements of several larvae within one or several areas of interest using the pixel-by-pixel mean squared error (m.s.e.) of gray-scale transformed and adjusted images from consecutive video frames, given by m.s.e $= 1 \ N \ \Sigma$pixel $= 1 \ N$ (image frame, pixel–image frame-1, pixel)^2 where N corresponds to the total number of pixels of each frame. Notably, the m.s.e. index is sensitive to both the amount of motion and the number of swimming larvae, enabling estimates of differential space use or place preference in semi-compartmentalized chambers or in open chambers with varying nearby environments.

5. A list of different steps involved in obtaining transgenic larvae is given here for the sake of providing a complete overview involving the protocol. However, a detailed treatment of zebrafish transgenesis is well beyond the scope of this chapter. A general guideline for zebrafish husbandry and genetic methods can be found in "A Guide for the Laboratory Use of Zebrafish Danio (Brachydanio) rerio" by M. Westerfield, 5th Edition. For detailed instruction on microinjection in zebrafish, see published protocols [48, 49].

6. In studies involving noninvasive optogenetics and freely-swimming larvae, test groups are commonly identified by evaluating the expression of fluorescent proteins. This can be done either before or after the tests. Doing it before the tests aids in adjusting the size of the samples, which is particularly beneficial when the complexity of the tests makes them difficult to carry out and time consuming. By contrast, a blind design (i.e., doing it after the test) may prevent accurate control of the sample size, but it will rule out any possible bias during testing arising from the experimenter's hand. Sometimes a blind design is necessary for another reason: fluorescence proteins cannot be detected before a certain stage of development. Under such circumstances, screening for false negatives at one or more time points after the tests become utterly important.

7. Exposing transgenic larvae expressing optogenetic proteins to excitation lights during screening can activate the protein. New optogenetic tools with improved light sensitivity are activated at extremely low light and some could be activated even at ambient light [50] (see Chaps. 3 and 6). Light exposure before tests can therefore transform experimental outcomes. Transgenic larvae can be evaluated post hoc based on fluorescence.

8. *Do's:* Keep checking if medium temperatures in the testing chamber are constant throughout the whole testing period at

any day. Keep track of the exact time of the onset of light in your data. This is your fixed time point throughout all following data analysis. Depending on the data analysis planned, EthoVision XT allows for separate adjustments for the raw data output to Excel (amounts of data tend to be huge and difficult to compute sometimes). Create one master template per testing protocol each, and adjust specifics on a day-to-day basis in copies accordingly. *Don'ts*: Don't stick to the same recording order of different groups every day. Don't keep supplying tubes and the testing chamber filled with stagnant medium over longer periods of time (more than a few days) to prevent algae from building up and contaminating the system.

5 Outlook

The zebrafish provides a powerful system for optogenetics. In combination with high-resolution in vivo imaging, application of optogenetics in tethered larvae has already provided a great deal of insight to link neural circuit activity to principles of basic behavioral control. The next frontier for zebrafish optogenetics will be the analysis of complex behavior, which rests on joint work of four regulatory components such as internal states, motivations, adaptive responses to the environment, and decision processes. Such complex behavior cannot be accurately carried out employing restrained subjects, and thus the future zebrafish optogenetics research will rest increasingly on combination of precise and sophisticated genetic manipulations combined with exquisite and fine-grained analysis of behavior in freely moving subjects.

References

1. White JG, Southgate E, Thomson JN, Brenner S (1986) The structure of the nervous system of the nematode *Caenorhabditis elegans*. Philos Trans R Soc Lond Ser B Biol Sci 314 (1165):1–340

2. de Bono M, Maricq AV (2005) Neuronal substrates of complex behaviors in *C. elegans*. Annu Rev Neurosci 28:451–501. doi:10. 1146/annurev.neuro.27.070203.144259

3. Sengupta P, Samuel AD (2009) *Caenorhabditis elegans*: a model system for systems neuroscience. Curr Opin Neurobiol 19(6):637–643. doi:10.1016/j.conb.2009.09.009

4. Yemini E, Jucikas T, Grundy LJ, Brown AE, Schafer WR (2013) A database of *Caenorhabditis elegans* behavioral phenotypes. Nat Methods 10(9):877–879. doi:10.1038/nmeth. 2560

5. Nagel G, Brauner M, Liewald JF, Adeishvili N, Bamberg E, Gottschalk A (2005) Light activation of channelrhodopsin-2 in excitable cells of *Caenorhabditis elegans* triggers rapid behavioral responses. Curr Biol 15(24):2279–2284. doi:10.1016/j.cub.2005.11.032

6. de Bono M, Schafer WR, Gottschalk A (2013) Optogenetic actuation, inhibition, modulation and readout for neuronal networks generating behavior in the nematode *Caenorhabditis elegans*. In: Hegemann P, Sigrist SJ (eds) Optogenetics. De Gruyter, Berlin, pp 61–74

7. Fero K, Yokogawa T, Burgess HA (2011) The behavioral repertoire of larval zebrafish. In: Kalueff AV, Cachat JM (eds) Zebrafish models in neurobehavioral research. Springer Science + Business Media, pp 249–291

8. Budick SA, O'Malley DM (2000) Locomotor repertoire of the larval zebrafish: swimming,

turning and prey capture. J Exp Biol 203(Pt 17):2565–2579

9. Kimmel CB, Patterson J, Kimmel RO (1974) The development and behavioral characteristics of the startle response in the zebra fish. Dev Psychobiol 7(1):47–60. doi:10.1002/dev.420070109

10. Burgess HA, Granato M (2007) Sensorimotor gating in larval zebrafish. J Neurosci 27 (18):4984–4994. doi:10.1523/JNEUROSCI.0615-07.2007

11. Neuhauss SC (2003) Behavioral genetic approaches to visual system development and function in zebrafish. J Neurobiol 54 (1):148–160. doi:10.1002/neu.10165

12. Portugues R, Engert F (2009) The neural basis of visual behaviors in the larval zebrafish. Curr Opin Neurobiol 19(6):644–647. doi:10.1016/j.conb.2009.10.007

13. Borla MA, Palecek B, Budick S, O'Malley DM (2002) Prey capture by larval zebrafish: evidence for fine axial motor control. Brain Behav Evol 60(4):207–229. doi:66699

14. McElligott MB, O'Malley DM (2005) Prey tracking by larval zebrafish: axial kinematics and visual control. Brain Behav Evol 66 (3):177–196. doi:10.1159/000087158

15. Bianco IH, Kampff AR, Engert F (2011) Prey capture behavior evoked by simple visual stimuli in larval zebrafish. Front Syst Neurosci 5:101. doi:10.3389/fnsys.2011.00101

16. Dunn TW, Mu Y, Narayan S, Randlett O, Naumann EA, Yang CT, Schier AF, Freeman J, Engert F, Ahrens MB (2016) Brain-wide mapping of neural activity controlling zebrafish exploratory locomotion. Elife 5. doi:10.7554/eLife.12741

17. Ahrens MB, Engert F (2015) Large-scale imaging in small brains. Curr Opin Neurobiol 32:78–86. doi:10.1016/j.conb.2015.01.007

18. Severi KE, Portugues R, Marques JC, O'Malley DM, Orger MB, Engert F (2014) Neural control and modulation of swimming speed in the larval zebrafish. Neuron 83(3):692–707. doi:10.1016/j.neuron.2014.06.032

19. Portugues R, Feierstein CE, Engert F, Orger MB (2014) Whole-brain activity maps reveal stereotyped, distributed networks for visuomotor behavior. Neuron 81(6):1328–1343. doi:10.1016/j.neuron.2014.01.019

20. Ahrens MB, Li JM, Orger MB, Robson DN, Schier AF, Engert F, Portugues R (2012) Brain-wide neuronal dynamics during motor adaptation in zebrafish. Nature 485 (7399):471–477. doi:10.1038/nature11057

21. Renninger SL, Orger MB (2013) Two-photon imaging of neural population activity in zebrafish. Methods 62(3):255–267. doi:10.1016/j.ymeth.2013.05.016

22. Portugues R, Severi KE, Wyart C, Ahrens MB (2013) Optogenetics in a transparent animal: circuit function in the larval zebrafish. Curr Opin Neurobiol 23(1):119–126. doi:10.1016/j.conb.2012.11.001

23. Friedrich RW, Jacobson GA, Zhu P (2010) Circuit neuroscience in zebrafish. Curr Biol 20(8):R371–R381. doi:10.1016/j.cub.2010.02.039

24. Simmich J, Staykov E, Scott E (2012) Zebrafish as an appealing model for optogenetic studies. Prog Brain Res 196:145–162. doi:10.1016/B978-0-444-59426-6.00008-2

25. Wyart C, Del Bene F, Warp E, Scott EK, Trauner D, Baier H, Isacoff EY (2009) Optogenetic dissection of a behavioural module in the vertebrate spinal cord. Nature 461 (7262):407–410. doi:10.1038/nature08323

26. Fidelin K, Djenoune L, Stokes C, Prendergast A, Gomez J, Baradel A, Del Bene F, Wyart C (2015) State-dependent modulation of locomotion by GABAergic spinal sensory neurons. Curr Biol 25(23):3035–3047. doi:10.1016/j.cub.2015.09.070

27. Goncalves PJ, Arrenberg AB, Hablitzel B, Baier H, Machens CK (2014) Optogenetic perturbations reveal the dynamics of an oculomotor integrator. Front Neural Circuits 8:10. doi:10.3389/fncir.2014.00010

28. Miri A, Daie K, Arrenberg AB, Baier H, Aksay E, Tank DW (2011) Spatial gradients and multidimensional dynamics in a neural integrator circuit. Nat Neurosci 14(9):1150–1159. doi:10.1038/nn.2888

29. Arrenberg AB, Del Bene F, Baier H (2009) Optical control of zebrafish behavior with halorhodopsin. Proc Natl Acad Sci U S A 106 (42):17968–17973. doi:10.1073/pnas.0906252106

30. Schoonheim PJ, Arrenberg AB, Del Bene F, Baier H (2010) Optogenetic localization and genetic perturbation of saccade-generating neurons in zebrafish. J Neurosci 30 (20):7111–7120. doi:10.1523/JNEUROSCI.5193-09.2010

31. Thiele TR, Donovan JC, Baier H (2014) Descending control of swim posture by a midbrain nucleus in zebrafish. Neuron 83(3):679–691. doi:10.1016/j.neuron.2014.04.018

32. Kubo F, Hablitzel B, Dal Maschio M, Driever W, Baier H, Arrenberg AB (2014) Functional architecture of an optic flow-responsive area that drives horizontal eye movements in zebrafish. Neuron 81(6):1344–1359. doi:10.1016/j.neuron.2014.02.043

33. McFarland DJ (1977) Decision making in animals. Nature 269:15–21

34. De Marco RJ, Thiemann T, Groneberg AH, Herget U, Ryu S (2016) Optogenetically enhanced pituitary corticotroph cell activity post-stress onset causes rapid organizing effects on behavior. Nat Commun 7:12620. doi:10.1038/ncomms12620

35. Charmandari E, Tsigos C, Chrousos G (2005) Endocrinology of the stress response. Annu Rev Physiol 67:259–284. doi: 10.1146/annurev.physiol.67.040403.120816

36. Iseki M, Matsunaga S, Murakami A, Ohno K, Shiga K, Yoshida K, Sugai M, Takahashi T, Hori T, Watanabe M (2002) A blue-light-activated adenylyl cyclase mediates photoavoidance in Euglena gracilis. Nature 415 (6875):1047–1051. doi: 10.1038/4151047a

37. Stierl M, Stumpf P, Udwari D, Gueta R, Hagedorn R, Losi A, Gartner W, Petereit L, Efetova M, Schwarzel M, Oertner TG, Nagel G, Hegemann P (2011) Light modulation of cellular cAMP by a small bacterial photoactivated adenylyl cyclase, bPAC, of the soil bacterium Beggiatoa. J Biol Chem 286(2):1181–1188. doi: 10.1074/jbc.M110.185496

38. Ryu MH, Moskvin OV, Siltberg-Liberles J, Gomelsky M (2010) Natural and engineered photoactivated nucleotidyl cyclases for optogenetic applications. J Biol Chem 285(53): 41501–41508. doi:10.1074/jbc.M110.177600

39. Pujol-Marti J, Faucherre A, Aziz-Bose R, Asgharsharghi A, Colombelli J, Trapani JG, Lopez-Schier H (2014) Converging axons collectively initiate and maintain synaptic selectivity in a constantly remodeling sensory organ. Curr Biol 24(24):2968–2974. doi: 10.1016/j.cub.2014.11.012

40. Xiao Y, Tian W, Lopez-Schier H (2015) Optogenetic stimulation of neuronal repair. Curr Biol 25(22):R1068–R1069. doi: 10.1016/j.cub.2015.09.038

41. De Marco RJ, Groneberg AH, Yeh CM, Castillo Ramirez LA, Ryu S (2013) Optogenetic elevation of endogenous glucocorticoid level in larval zebrafish. Front Neural Circuits 7:82. doi:10.3389/fncir.2013.00082

42. Gutierrez-Triana JA, Herget U, Castillo-Ramirez LA, Lutz M, Yeh CM, De Marco RJ, Ryu S (2015) Manipulation of Interrenal cell function in developing zebrafish using genetically targeted ablation and an optogenetic tool. Endocrinology 156(9):3394–3401. doi: 10.1210/EN.2015-1021

43. Kwan KM, Fujimoto E, Grabher C, Mangum BD, Hardy ME, Campbell DS, Parant JM, Yost HJ, Kanki JP, Chien CB (2007) The Tol2kit: a multisite gateway-based construction kit for Tol2 transposon transgenesis constructs. Dev Dyn 236(11):3088–3099. doi: 10.1002/dvdy.21343

44. Burgess HA, Granato M (2007) Modulation of locomotor activity in larval zebrafish during light adaptation. J Exp Biol 210(Pt 14):2526–2539. doi: 10.1242/jeb.003939

45. De Marco RJ, Groneberg AH, Yeh CM, Trevino M, Ryu S (2014) The behavior of larval zebrafish reveals stressor-mediated anorexia during early vertebrate development. Front Behav Neurosci 8:367. doi: 10.3389/fnbeh.2014.00367

46. Groneberg AH, Herget U, Ryu S, De Marco RJ (2015) Positive taxis and sustained responsiveness to water motions in larval zebrafish. Front Neural Circuits 9:9. doi: 10.3389/fncir.2015.00009

47. Zhou Y, Cattley RT, Cario CL, Bai Q, Burton EA (2014) Quantification of larval zebrafish motor function in multiwell plates using open-source MATLAB applications. Nat Protoc 9(7):1533–1548. doi:10.1038/nprot.2014.094

48. Rosen JN, Sweeney MF, Mably JD (2009) Microinjection of zebrafish embryos to analyze gene function. J Vis Exp 25. doi: 10.3791/1115

49. Yuan S, Sun Z (2009) Microinjection of mRNA and morpholino antisense oligonucleotides in zebrafish embryos. J Vis Exp 27. doi: 10.3791/1113

50. Dawydow A, Gueta R, Ljaschenko D, Ullrich S, Hermann M, Ehmann N, Gao S, Fiala A, Langenhan T, Nagel G, Kittel RJ (2014) Channelrhodopsin-2-XXL, a powerful optogenetic tool for low-lightapplications. Proc Natl Acad Sci U S A 111(38):13972–13977. doi: 10.1073/pnas.1408269111

Chapter 8

Combining Optogenetics with MEA, Depth-Resolved LFPs and Assessing the Scope of Optogenetic Network Modulation

Jenq-Wei Yang, Pierre-Hugues Prouvot, Albrecht Stroh, and Heiko J. Luhmann

Abstract

The development of multi-electrode arrays enables researchers to record neuronal activity from several to hundreds or even thousands of neurons simultaneously. Optogenetics provides the toolbox to excite or inhibit a genetically defined subpopulation of neurons. Here, we present a detailed description of how to combine multi-electrode array (MEA) recording and optogenetics to study the role of parvalbumin positive interneurons in mouse somatosensory cortical signal processing. We also provide the tools to quantify the density of opsin-expressing cells and estimate the effective illuminated area by optogenetic stimulation, resulting in a quantification of optogenetically modulated neurons.

Key words Optogenetics, ArchT, Electrical recordings, Multi-electrode arrays, Light propagation

1 Introduction

The mammalian brain is comprised of billions of neurons which are spatially organized in different brain structures to process different types of information. One of the goals of system neuroscience is to understand how groups of neurons work together to perform efficient information processing. To reach this goal, simultaneous recordings of multiple neurons are needed. In the past several decades, both scientists and commercial enterprises undertook great efforts to develop the multi-electrode array (MEA). This technique enables researchers to record the local field potential (LFP) and extracellular multiunit activity (MUA) from hundreds or even thousands of neurons simultaneously. Up to date, several types of high-density MEA probes have been developed according

---

Jenq-Wei Yang and Pierre-Hugues Prouvot contributed equally to this work.

Albrecht Stroh and Heiko J. Luhmann are equal contributing last authors.

Albrecht Stroh (ed.), *Optogenetics: A Roadmap*, Neuromethods, vol. 133,
DOI 10.1007/978-1-4939-7417-7_8, © Springer Science+Business Media LLC 2018

to the respective experimental design and have been routinely used for large-scale LFP and unit recording in vivo [1–4]. For example, to record neural activity simultaneously in several barrel-related cortical columns of rats, an eight-shank 128-channel MEA probe (200 μm horizontal shank distance and 75 μm vertical inter-electrode distance) was developed [1]. Another example is an eight-shank 256-channel MEA probe (300 μm horizontal shank distance and 50 μm vertical inter-electrode distance) which can record neural activity covering a large area of the dorsal hippocampus of the rat [3]. The recently developed complementary metal-oxide semiconductor (CMOS) MEA containing 1028 recording sites enable researchers to simultaneously record thalamocortical network activity and select the target recording sites according to the experimental needs [4].

In addition to record the activity of many neurons simultaneously by MEA, the precise and selective manipulation of individual components of the network is critical for a causal understanding of their role in information processing within a given brain circuit. The recent developments of optogenetic techniques meet those needs. Unlike traditional electrophysiological and pharmacological methods, optogenetics allows researchers to optically activate or silence a specific, genetically defined neuronal subtype by expressing a light-sensitive opsin in a given neuronal population (*see* Chaps. 1 and 2). Successfully performing an optogenetic experiment involves various steps and considerations, for example, the selection of target cell type and opsin, the makeup of light delivery system and choosing the right method to evaluate the outcome of the optogenetic experiment [5]. Several laboratories have focused on developing opto-MEA neural interfaces [6–8]. The simplest way is to assemble a conventional MEA with one or several optical fibers to control the neuronal activity in one shank of the MEA containing several recording sites [7]. Recently, a new type of opto-MEA containing multiple μLEDs nearby several recording sites was developed to direct optical stimulation more precisely on target recording sites [8]. Furthermore, another type of opto-MEA was designed to perform simultaneous light delivery and electrical recording in individual recording channels [6].

Here we provide a detailed description of methods how to combine MEA and optogenetics to control defined neuronal population in vivo. In this example, we study the role of parvalbumin-positive (PV+) GABAergic interneurons in barrel cortex signal processing. PV+ interneurons represent about 40% of all GABAergic interneurons [9] providing inhibition to local principal cells and may also inhibit neurons located several hundreds of micrometers away [10, 11]. We provide a protocol how to perform virus injection specifically in a defined barrel and how to perform MEA (8-shank 128 channels) recording in barrel cortex of lightly anesthetized mice. We introduce a custom-made two laser instrument which can provide optical stimulation through a multimode optic fiber to inhibit PV+ interneurons expressing ArchT. We then provide a straightforward method to estimate how many neurons are modulated by typical optogenetic stimulation. We take the example of our ArchT stimulation, but this method can be easily extended to other experiments. In the end, we provide an outlook on the combination of these methods.

2 Methods

2.1 Identify Barrels by Intrinsic Imaging Method

2.1.1 Surgery

1. All instruments have to be cleaned and sterilized by hot bead sterilizer or autoclave.

2. 1 to 2-month-old PV-Cre mice were initially anesthetized with 2% isoflurane. An incision was made in the middle of head skin, and the skull was exposed. The periost covering the skull was removed.

3. A custom-made stainless steel holder was glued on the cranium of the left hemisphere, 1.5% low melting agarose was then applied on top of the cortex covered with a glass coverslip. The head of the mouse was then firmly fixed in a stereotaxic frame and anesthesia was maintained with 0.75–1% isoflurane (Fig. 1a).

2.1.2 Intrinsic Optical Imaging

1. A LED lamp at 625 nm was used for illumination. A MiCAM CMOS camera was positioned on top of the craniotomy and the optical axis was kept perpendicular to the cortical surface (Fig. 1b). The optical signal was recorded using the MiCAM program in a 2.6 mm^2 area of the cortical surface.

2. We selected C1, C2 or B1, B2 as our targeted barrels. A single whisker was deflected with a protruding device consisting of a miniature solenoid actuator which was controlled by a transistor-transistor logic (TTL) pulse. The actuator was placed orthogonal to the base of the whisker approximately 2 mm from the snout. Whiskers were moved rostrocaudally with a deflection for 26 ms to reach the maximal 1 mm whisker displacement. The peak stimulus velocity is 1114 °/s.

3. The protocol for intrinsic imaging consisted of 4 s baseline, 5 s 5 Hz stimulation and then 1 s post-stimulation recording. The inter-stimulus-interval was 8 s (Fig. 1b–d).

2.2 Injecting the ArchT-GFP Vector into Identified Barrel

1. After identifying the location of the stimulated barrels by intrinsic imaging, we performed ArchT-GFP-encoding virus injection in a barrel not containing large blood vessels. (Fig. 1d).

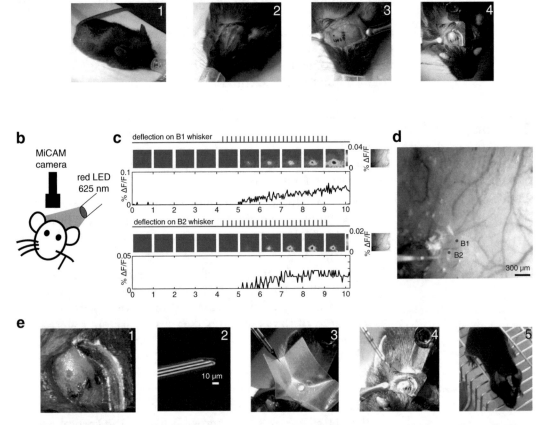

Fig. 1 Procedures for intrinsic optical imaging and virus injection. (**a**) Perform a skin incision and glue a custom-made stainless steel holder on the cranium of the left hemisphere. (**b**) Schematic illustration of the experimental setup for intrinsic optical imaging. (**c**) Two examples of evoked intrinsic imaging responses by single whisker stimulation (**B**1 and **B**2 whiskers respectively). (**d**) Virus injection was performed in **B**2 barrel according to the location of evoked **B**2 intrinsic imaging response. (**e**) The procedure of the virus injection

2. A small craniotomy of ~0.5 mm in diameter was performed on top of the selected barrel (Fig. 1e1). The dura was kept intact.

3. A microtiter pipette pulled to a tip diameter of around 5 μm (Fig. 1e2) (Ringcaps Hirschmann Laborgeräte GmbH, Eberstadt, Germany) was connected to a flexible plastic tube of a 200 ml syringe.

4. A piece of Parafilm was placed on the craniotomy, and then 1–2 μl of virus solution was placed on the Parafilm. After immersing the tip of the glass pipette into this drop of virus solution, negative pressure was applied, and about 600 nl of viral solution (rAAV2-FLEX-ArchT-GFP, 2×10^{11} viral particles per ml) was slowly aspirated into the pipette (Fig. 1e3).

5. The pipette was inserted into the selected barrel at a 60° angle at 400 μm depth. The virus solution was slowly injected into

the cortex at a rate of 300–400 nl/5 min. After injection, the pipette remained in position for at least 5 min to allow the virus solution to diffuse into the brain tissue prior to slowly retracting the pipette (Fig. 1e4).

6. The skin incision was closed with Vetbond tissue adhesive (3 M, Maplewood, Minnesota, USA), and the mice were returned to the cage for recovery (Fig. 1e5).

2.3 MEA Recording and Optical Inhibition of PV+ Interneurons Expressing ArchT

2.3.1 Inserting MEA into Targeted Barrels Exhibiting ArchT-GFP Expression

1. One month after virus injection, the mouse was anesthetized using urethane (1.5 g/kg). The mouse's head was fixed in the stereotaxic apparatus using an aluminum holder fixed on the occipital bones with dental cement.

2. If there was any sign of distress, 10–15% of the initial dose of urethane was injected (i.p.). After drying the skull, the position of the previous craniotomy for virus injection could be clearly observed (Fig. 2a1).

3. A silver wire was inserted into the cerebellum as a ground electrode. The mouse was placed on a heating blanket and kept at a constant temperature of 37 °C.

4. A 2 × 2 mm^2 craniotomy was performed over the barrel cortex of the right hemisphere (centered at previous small craniotomy for virus injection, Fig. 2a2).

5. According to the location of the virus injection, an eight-shank 128-channel electrode (NeuroNexus Technologies, Ann Arbor, MI) was labeled with DiI (1,10-dioctadecyl-3,3,30,30-tetramethyl indocarbocyanine, Molecular Probes, Eugene, OR, USA) and then inserted perpendicular into the barrel cortex (Fig. 2a3,4 and b).

6. To reduce damage of cortical tissue, the insertion speed was below 200 μm/min. 40 min after the electrode insertion, a single whisker was stimulated to functionally identify the recorded barrel-related column.

2.3.2 Optical Stimulation

1. The light for excitation of ArchT was delivered by a 50 mW solid-state laser at 552 nm wavelength (Sapphire, Coherent, Dieburg, Germany) placed in a custom-built optical setup (Fig. 3). The laser beam was coupled to a 200 μm multimode fiber with a numerical aperture of 0.39 (Thorlabs, Munich, Germany) using a fiber collimator (Schäfter + Kirchhoff, Hamburg, Germany).

2. The output power at the end of the fiber was measured with a power meter (Nova 2, Ophir, Newport, Irvine, CA) prior to each experiment.

3. The initial power density was calculated with $I_0 = i/\pi r^2$ where i is the initial power and r the radius of the fiber core. A mechanical shutter controlled the light pulse (Uniblitz, Rochester

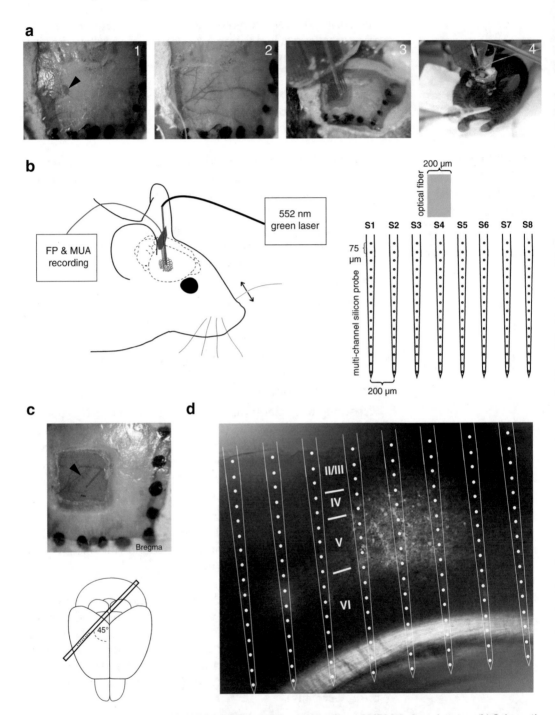

Fig. 2 Multi-electrode array (MEA). (**a**) Surgery procedure for insertion of MEA into barrel cortex. (**b**) Schematic illustration of the experimental setup for MEA recording combined with optical fiber (left) and the spacing of electrodes. (**c**) At the end of the experiment, eight red points marked by DiI could be observed. According to the position of the red points, we perform brain slicing at a 45° angle from the midline. (**d**) Fluorescent image of ArchT-GFP expression pattern (*green*) and position of DiI-covered electrode tracks (*red*). An eight-shank 128-channel electrode probe is superimposed on this image according to the position of the DiI tracks

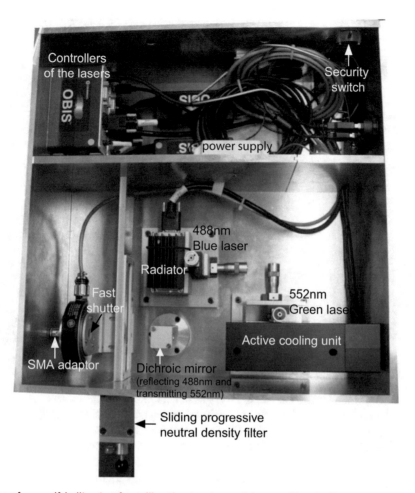

Fig. 3 Image of our self-built setup for collimating two lasers into a multimode fiber

USA), and was connected to a stimulator (Master8, A.M.P.I., Jerusalem, Israel).

4. The fiber was positioned parallel to the electrodes using a micromanipulator and barely touched the cortical surface (Fig. 2a3 and b). In order to keep the geometric shape of light illumination, the tip of the fiber must be cut precisely using a diamond pen (fiber scribe S90R Thorlabs) (Fig. 4, *see* **Notes 1** and **2**).

2.3.3 Defining the Suitable Light Intensity for the Experiment

1. Opsins display a sigmoid dose-response curve to light. For excitatory cation channels, stimulation should be applied at a frequency preventing the inactivation of the peak current (*see* Chaps. 3 and 13) (*see* **Note 3**), e.g., a brief pulse train of 10 ms pulse at up to 30 Hz for ChR2 (*see* **Note 4**), spaced by at least 10 s. For proton and chloride pumps the illumination should be continuous as their sustained activity relies on the continuous absorption of photons.

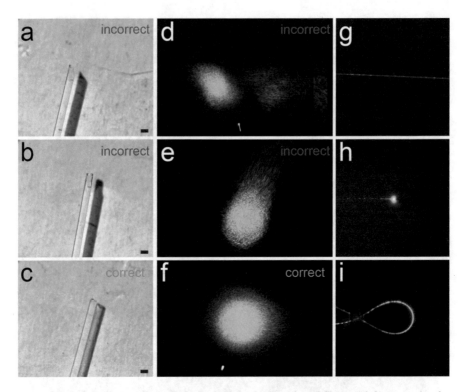

Fig. 4 Proper cutting of the optical fiber. (**a, b**) Examples of unevenly cut fibers. (**c**) An example of a properly cut fiber. (**d, e**) Examples of the deformation of the beam exiting unevenly cut fibers. (**f**) An example of the beam from a properly cut fiber. (**g**) Unbroken fiber with only little light leakage. (**h**) Broken fiber exhibiting significant leakage. (**i**) An example of leakage induced by excessive bending of the fiber

2. The first step of any recordings including optogenetic manipulation is to establish a dose-response curve (*see* **Note 5**). The optimal power should be in the linear range of the curve, use the minimal power sufficient to elicit a direct effect on the network, be aware that using highest power levels may lead to phototoxicity. Here, we found that inhibiting PV+ interneuron enhanced the evoked response by single whisker stimulation. We applied increasing intensities of 552 nm light and chose 150 mW/mm^2 as the suitable light intensity for our experiment due to its efficiency in enhancing the sensory evoked responses, and still ranging in the linear dose-response curve (Fig. 5).

2.3.4 Perfusion and Brain Slicing

1. At the end of the experiment, the mouse was perfused with 0.1 M sodium phosphate buffer and fixed with 4% paraformaldehyde.

2. The brain was sliced following the insertion angle (around 45° from the midline) of the eight-shank 128-channel electrode (200 μm thickness) (Fig. 2c).

3. Positions of individual electrodes were identified according to the fluorescence of DiI, and ArchT-PV+ interneurons were identified by fluorescence microscopy (Fig. 2d).

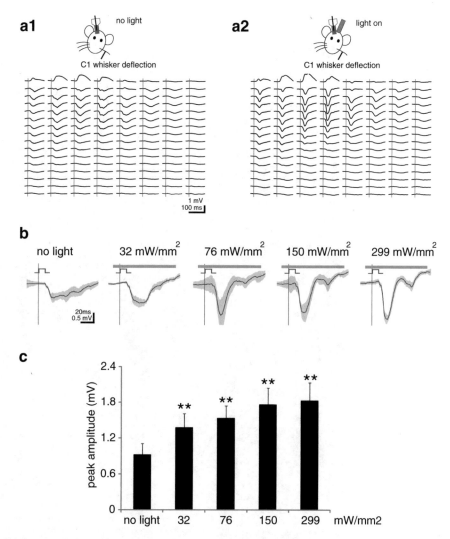

Fig. 5 Establishment of a dose-response curve by optical light illumination. (**a**) Average LFP response to 10 stimulations of the C1 whisker under control conditions (**a**1, only whisker stimulation) and during optogenetic inactivation of PV interneurons (**a**2, whisker stimulation and light illumination). (**b**) Enhancement of sensory-evoked response with increasing laser intensities (average of 70 evoked LFP responses from 7 mice). (**c**) The peak amplitude increase with increasing laser intensities ($n = 7$ mice). Paired t test, $**p < 0.01$)

2.4 Identify ArchT Expression in PV+ Interneurons by Immuno-histochemistry

2.4.1 Expression Quality

Confocal microscopy is the standard for assessing the quality of expression. The expression should be uniform in the cell membrane (Fig. 6). With common fluorophores, like GFP or YFP, the expression is smooth. Note that mcherry tends to form intracellular aggregates. Overexpression is characterized by progressive degeneration of the cell, producing highly fluorescent punctae (Fig. 6).

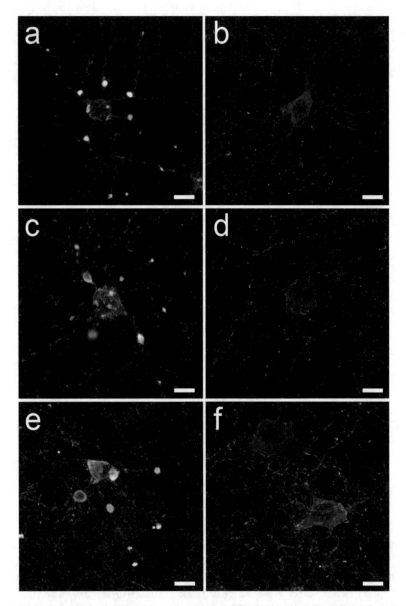

Fig. 6 Examples of opsin expression. (**a**, **c**, **e**) Typical confocal micrographs of ArchT overexpressing PV neurons. (**b**, **d**, **f**) Typical micrographs of healthy ArchT expressing PV neurons, scale bar: 10 μm

2.4.2 Assessing Cell Density and Area of Expression

Cell density has to be assessed for subsequent modeling of light distribution.

1. Measure the volume of expression. It is defined as the area containing cell bodies strongly expressing the opsin (*see* **Note 6**).

2. Two options for calculating the cell density:

(a) Counting all cells in the expressing area in every expressing slice. Resulting in: Cell number (units)/expressing volume (mm^3) equals cell density.

(b) Using a stereology software to estimate the average expression density.

2.4.3 Immunofluorescence Stainings to Assess the Cell Type Specificity of Expression

Checking for expression specific to the targeted cell type using complementary antibodies (in our case PV and CamKII antibodies) to ensure the promoter exhibits no leakage (*see* Chap. 1).

Here, individual slices around the center of expression were selected for immunohistochemistry.

1. For permeabilization, slices were incubated with 0.1% Triton X-100 and 5% normal donkey serum (Invitrogen, Life Technologies, Carlsbad, CA) in phosphate buffer solution for 90 min.

2. Slices were incubated with goat anti-PV (1:200, Swant, Marly, Switzerland) or rabbit anti-CamKII (1:200, Epitomics, Burlingame, CA) at 4 °C overnight.

3. On the next day, slices were incubated with the secondary antibodies Cy-2 donkey anti-goat (1:200, Jackson Immuno Research, West Grove, PA), or Cy-2 donkey anti-rabbit (1:200, Jackson Immuno Research, West Grove, PA). In addition a fluorescent Nissl stain (red 615 nm, Neurotrace, Molecular Probes, Life Technologies, Carlsbad, CA) was performed. Slices were mounted using antiquenching Vectashield (Vector Laboratories, Burlingame, CA).

2.4.4 Modeling of the Above-Threshold Illuminated Area

To calculate the average number of optogenetically modulated neurons, not only the relative density of opsin-expressing cells needs to be estimated (see above), but also the light distribution in tissue (*see* Chap. 13 for further details) (*see* **Note 7**).

1. Each opsin is characterized by its dose-response curve. Using this curve it is possible to determine the current imposed on the cell during the optogenetic stimulation provided the power density at a certain depth is known. It is then possible to estimate at which depth the stimulation ceases to be efficient. The dose response curves of the most common opsins can be found in the papers delineated in Table 1 (*see* **Note 8**).

2. Use the Kubelka-Munk model (*see* **Note 9**) for a geometrical estimation of light spread, but note that this model is only partially reflecting the actual light distribution (see Chap. 13). This model underestimates the later spread of light close to the fiber tip, while it is relatively accurate in determining the efficient penetration depth. To determine the effective penetration depth, defined as the depth at which the light power density falls below the activation threshold of the respective opsin [17, 18]:

Table 1
The most common opsins used for optogenetic experiments

Opsin	Wavelength	Tau off	Photocurrent references
ChR2 (H134R)	470	18 ms	[12]
ChIEF	450	10 ms	
ChETA	545	34–38 ms	[13]
C1V1	540	156 ms	[14]
ChloC			[15]
pumps			
ArchT (proton pump)	566	n/a	[16]
eNpHR3.0 (chloride pump)	590	n/a	

(a) The optical fiber is defined by a few key parameters, such as the diameter of the core and the numerical aperture. We can consider the initial light source as a circle defined by the core diameter. The numerical aperture is related to the half angle at which light will exit the fiber: $NA = ni \sin(\theta)$ where ni is the scattering coefficient of the media (1.36 for the gray matter) [19]. Thus $\theta = \arcsin\left(\frac{NA}{ni}\right)$. In our case $\theta = 16.6$

(b) The intensity decreases with depth following $I(z)/I(z = 0) = \rho^2/(z + \rho)^2$

(c) The geometrical factor ρ is $\rho = r*\text{sqrt}((ni/NA)^2 - 1)$

(d) The final equation including scattering:

$$I(z)/I(z = 0) = \rho^2/\left((Sz + 1)(z + \rho)^2\right)$$

with $Sz = 11.2$ mm^{-1} in mice and 10.3 mm^{-1} in rats.

(e) Plot the results as depth against power (see Fig. 7)

3. Once the penetration depth is known, calculate the size of the truncated cone corresponding to the stimulated volume $V = \pi r^2 \left(\frac{h}{3}\right)$ (Fig. 8).

4. Then calculate the overlapping volume between this frustum and the expressing regions (in the cortex layer II/III and V).

5. **Note 10** shows the detailed calculation of the modulated ArchT-positive neurons based on the protocols described above.

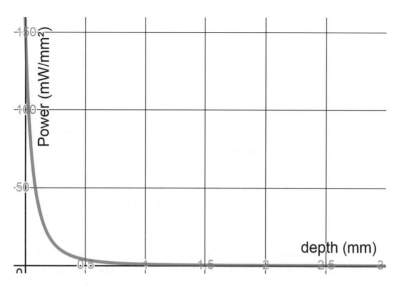

Fig. 7 Plot of the decrease of light intensity with depth using the Kubelka-Munk model

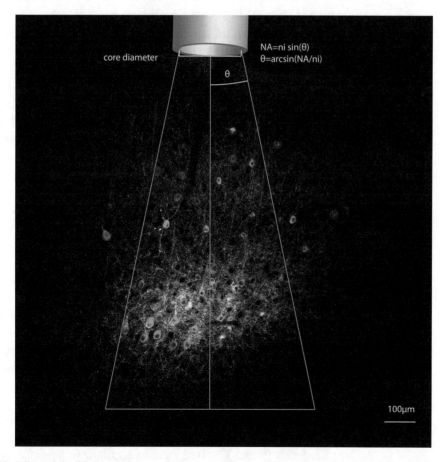

Fig. 8 Modeling of the illuminated area. *Green*, area illuminated with above-threshold light power density; *Orange*, highlighted cell bodies; Formula, relationship between numerical aperture and half angle

3 Notes

1. Optical fibers are flexible up to a certain angle, exceeding this angle leads to light emission along the length of the fiber as the total internal reflection is not maintained inside the core, resulting in a lower output power. Moreover they are relatively fragile and should be checked from time to time for damage (Fig. 4).

2. The delivering end of the fiber can be cleaned with ethanol and/or acetone after the experiment. The proper propagation of light should be checked before every experiment using either a fiber inspection scope (FS201 Newport, Thorlabs, Newton, NJ, USA) or by simply pointing it at a white wall to check that the spot of light is circular (Fig. 4). If any irregularity is detected, the end of the fiber should be re-cut to reach the proper quality.

3. The stimulation frequency should be devised in relation to the cell type, the opsin used and the research hypothesis, one would not consider a 100 Hz tetanic stimulation physiologically relevant in excitatory neurons for example.

4. ChR2 has been used at higher frequencies in the literature, but frequencies higher than 30 Hz become hard to evaluate in our hands. It is always best to check the literature beforehand and especially the original articles for each opsin.

5. In most cases the dose response curves are available in the literature. However whole cell electrophysiology after your specific viral transduction strategy in the respective brain region remains the gold standard for controls in optogenetics experiments, be it in slices or in vivo.

   The main parameters to control for patch clamp are:

   (a) The dose response curve between light density and photocurrent (obtained by ramping up the light power between stimulation and measuring the current in voltage clamp).

   (b) The power necessary for efficient silencing or excitation of the cells of interest (how much light power is necessary to inhibit action potential firing in current clamp in our case).

6. Only measure the area containing cell bodies. Neurites can extend very far and it can be difficult to obtain a precise value.

7. Here we address the case of classical multimode optical fibers as they are more commonly used. More complex systems require the establishment of specific models. Multiple tools are available online and in the literature to model the scattering of light in brain tissue. The best ratio of accuracy to accessibility is the

Kubelka-Munk model for penetration of light in brain tissue. The easiest approximation is just to consider the geometry of a truncated cone with the upper surface being the core of the fiber and the angle as calculated from the numerical aperture. When this volume is calculated, it is possible to calculate the volume overlapping with the expressing region. The last step is to multiplying this volume with the cell density calculated earlier to result in the number of affected cells. Most models give similar results regarding the axial penetration, they differ most in the degree of lateral scattering [20].

8. For excitatory opsins, the necessary power is straightforward due to the threshold property of action potential generation, a certain current depending on the cell type leads to the membrane potential exceeding the threshold for opening of voltage-dependent sodium channels, commencing an action potential. In the case of inhibitory opsins it is possible to evaluate the power at which most action potentials will be inhibited.

9. An online tool based on the Kubelka-Munk model can be found on these two websites; http://web.stanford.edu/group/dlab/cgi-bin/graph/chart.php www.open optogenetics.org

10. In our experimental conditions the initial power is about $I(z = 0) = 150)$mW/mm^2 (i.e., 4.7 mW out of a 200 μm fiber, thus $4.7/0.031 = 151$ mW/mm^2). The experiment were performed in mouse cerebral cortex, thus a refraction index of $n = 1.36$ and a scattering coefficient of $S = 11.2$ mm^{-1}. In this case $\rho = 0.2651$.
Plotting

$$I(z) = \left(\frac{\rho^2}{(Sz + 1)(z + \rho)^2} \right) \times I(z = 0)$$

Results in the graph in Fig. 7.

In case of PV interneurons we assumed a minimal inhibitory current required for inhibition ranging at 200–250 pA [21]. To reach this current by ArchT activation, a local light density of >1.5 mW/mm^2 is required [22]. If we refer to Fig. 7 we see that this power is reached around 0.75 mm. We thus efficiently illuminate a frustum with an upper radius of 0.1 mm and a height of 0.75 mm.

The volume of such a frustum is defined by:

$$V = \left(\frac{\pi h}{3} \right) \times \left(R^2 + Rr + r^2 \right)$$

Here $r = 0.1$ mm $h = 0.75$ mm and $R = h^* \tan \theta$ thus $R = 0.224$ mm and $V = 0.0648$ mm^3.

The last step consists in multiplying this value by the cell density in the expressing region in order to estimate the number of cells efficiently illuminated. Thus with a density of about 1216 neurons/mm^3 a final number about 79 neurons is efficiently illuminated. Note that this value should be interpreted as an order of magnitude estimation.

11. We observed that LFP recordings were impacted by light illumination. To assess the scope of optically induced artifacts we performed control experiments in wild-type mice without ArchT expression (Fig. 9). Under control no-light conditions, mechanical deflection of whisker C5 elicited the large LFP responses recorded at the principal shank (PS) electrode located in the C5 column (Fig. 9a). Light illumination (100 ms, 150 mW/mm^2) altered evoked LFP recordings following single whisker stimulation. The artifact was most notable in the three superficial electrodes, corresponding to layer I and upper layer II/III (Fig. 9b). Only light illumination produced obvious LFP shift (Fig. 9c). The correct evoked response under light illumination can be calculated by subtracting the

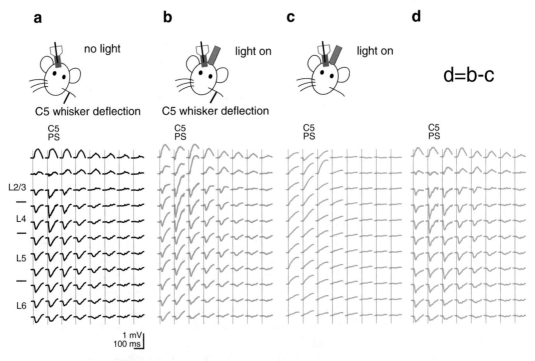

Fig. 9 Optically induced artifact. (**a**) Average LFP response to 100 stimulations of the C5 whisker under no light condition in a C57BL/6 wild type mouse at postnatal day 44. (**b**) Average LFP response to 100 stimulations of the C5 whisker combination with 150 mW/mm^2 *green light* pulse illumination (light on). (**c**) LFP response to pure 150 mW/mm^2 *green light* pulse illumination. (**d**) The corrected evoked LFP by subtracting **c** from **b**

LFP under only light-on condition from the evoked LFP under single whisker stimulation with light-on condition (Fig. 9d).

4 Outlook

1. AAV virus and opsin delivery.

   Multiple experimental parameters impact the scope of optogenetic network activation: First AAV virus injections inherently result in varying infection rates per neuron due to a multiplicity of infection, leading to heterogeneity of opsin expression [23], this in turn affecting inhibitory photocurrents. However, while this adds jitter, it might create a less artificial pattern of inhibition (see Chaps. 3 and 14). Approaches encoding regulatory elements resulting in more homogenous expression levels are currently not feasible due to the limited capacity of AAVs and furthermore require integration to a single site of the genome, in contrast to the episomal or extrachromosomal persistence of the AAVs used in this study (see Chap. 1). Second, the position of the fiber in relation to the opsin-expressing region will impact the area of above threshold illumination. Third, while light penetration can be estimated by established models, the highly heterogeneous brain tissue, in particular, the highly light scattering and absorbent microvasculature can only be modeled up to a certain extent. Finally, for optical inhibition even more so than for excitation, efficient inhibition of action potentials depends on many physiological parameters like neuromodulation, the state of the animal, membrane resistance, and synaptic inputs. This inherently limits the accuracy of any approach assessing the scope of optogenetic network modulation.

2. Recent development of MEA and opto-MEA neural interfaces.

   The development of MEA progresses rapidly from only a few recording sites to current CMOS-based high-density MEA with up to 1028 recording sites [4]. In the future, the advance of nanofabrication technology will help to generate new types of MEA with a greater number of electrodes that are smaller and less invasive [24, 25].

   Multiple teams across the world have been developing tools to make the combination of MEA recording and optical stimulation more convenient. One approach has been to confine light delivery in order to attain layer or region specific optogenetic actuation [8]. Another approach has been to integrate stimulation into recording devices [25].

3. Strategies to minimize the optically induced artifacts.

   We observed optically induced artifacts in our LFP recording independent of ArchT, brought about by photons interacting with electrons in the metallic recording probe, termed Becquerel or photoelectric effect (PE), as already reported by others [26, 27]. Several methods were proposed to reduce PE on the LFP. Here, we propose to reduce the PE by subtracting the LFP under pure light-on condition from the whisker-evoked LFP with light-on condition (Fig. 9). Another study used a band-stop filter method to reduce the light-induced artifact in MEA recording. First, they compared the difference of LFP frequency distribution with and without light illumination. Afterwards, the LFP data was filtered by specific filter settings that passed most unaltered frequencies but attenuated those frequencies dominated by light illumination to very low levels [28]. Besides using these mathematic methods to reduce the PE on LFP recording, the new designed MEA fabricated from transparent material [29, 30] or coated with light-proof material [31] can be used to diminish the optically induced artifacts.

4. Modeling of the optogenetic stimulation.

From early on the optogenetic community explored and debated the spatial specificity of optogenetics [32]. While initial studies employed the Kubelka-Munk model, assuming a conical spread, experimental data revealed that this approach largely underestimates the lateral spread [17, 20]. Applying more complex modeling approaches based on Monte Carlo simulations led to rather ellipsoid geometries, better matching experimental data [17, 33]. It has to be noted that anyhow due to the aforementioned experimental variables, any assessment can only provide order-of–magnitude estimations.

Using the Kubelka-Munk model of scattering in combination with a geometrical model and a histological assessment of cell densities represents a very straightforward way to estimate the number of cells potentially activated (or in our example inhibited) by optogenetics. In the near future, the field of optogenetic interrogation of circuits has to move from rather mechanistic connectivity studies to approaches taking into account the scope of optogenetic network modulation. Therefore a generalization of this practice in the field would be advisable, even more as network effects drastically depend on the architecture of the respective cortical circuits.

References

1. Reyes-Puerta V, Sun J-J, Kim S et al (2015) Laminar and columnar structure of sensory-evoked multineuronal spike sequences in adult rat barrel cortex in vivo. Cereb Cortex 25:2001–2021. doi:10.1093/cercor/bhu007

2. Buzsáki G, Stark E, Berényi A et al (2015) Tools for probing local circuits: high-density silicon probes combined with optogenetics. Neuron 86:92–105. doi:10.1016/j.neuron.2015.01.028

3. Berényi A, Somogyvári Z, Nagy AJ et al (2014) Large-scale, high-density (up to 512 channels) recording of local circuits in behaving animals.

J Neurophysiol 111:1132–1149. doi:10. 1152/jn.00785.2013

4. Fiáth R, Beregszászi P, Horváth D et al (2016) Large-scale recording of thalamocortical circuits: in vivo electrophysiology with the two-dimensional electronic depth control silicon probe. J Neurophysiol 116:2312–2330. doi:10.1152/jn.00318.2016

5. Fois C, Prouvot P-H, Stroh A (2014) A roadmap to applying optogenetics in neuroscience. Methods Mol Biol 1148:129–147. doi:10. 1007/978-1-4939-0470-9_9

6. Lee J, Ozden I, Song Y-K, Nurmikko AV (2015) Transparent intracortical microprobe array for simultaneous spatiotemporal optical stimulation and multichannel electrical recording. Nat Methods 12:1157–1162. doi:10. 1038/nmeth.3620

7. Wu F, Stark E, Im M et al (2013) An implantable neural probe with monolithically integrated dielectric waveguide and recording electrodes for optogenetics applications. J Neural Eng 10:56012. doi:10.1088/1741-2560/10/5/056012

8. Wu F, Stark E, Ku P-C et al (2015) Monolithically integrated μLEDs on silicon neural probes for high-resolution Optogenetic studies in behaving animals. Neuron 88:1136–1148. doi:10.1016/j.neuron.2015.10.032

9. Rudy B, Fishell G, Lee S, Hjerling-Leffler J (2011) Three groups of interneurons account for nearly 100% of neocortical GABAergic neurons. Dev Neurobiol 71:45–61. doi:10.1002/dneu.20853

10. Helmstaedter M, Sakmann B, Feldmeyer D (2009) Neuronal correlates of local, lateral, and translaminar inhibition with reference to cortical columns. Cereb Cortex 19:926–937. doi:10.1093/cercor/bhn141

11. Kätzel D, Zemelman BV, Buetfering C et al (2011) The columnar and laminar organization of inhibitory connections to neocortical excitatory cells. Nat Neurosci 14:100–107. doi:10.1038/nn.2687

12. Lin JY, Lin MZ, Steinbach P, Tsien RY (2009) Characterization of engineered channelrhodopsin variants with improved properties and kinetics. Biophys J 96:1803–1814. doi:10. 1016/j.bpj.2008.11.034

13. Gunaydin LA, Yizhar O, Berndt A et al (2010) Ultrafast optogenetic control. Nat Neurosci 13:387–392. doi:10.1038/nn.2495

14. Schneider F, Gradmann D, Hegemann P (2013) Ion selectivity and competition in channelrhodopsins. Biophys J 105:91–100. doi:10. 1016/j.bpj.2013.05.042

15. Wietek J, Wiegert JS, Adeishvili N et al (2014) Conversion of channelrhodopsin into a light-gated chloride channel. Science 344:409–412. doi:10.1126/science.1249375

16. Chow BY, Han X, Dobry AS et al (2010) High-performance genetically targetable optical neural silencing by light-driven proton pumps. Nature 463:98–102. doi:10.1038/nature08652

17. Yizhar O, Fenno LE, Davidson TJ et al (2011) Optogenetics in neural systems. Neuron 71:9–34. doi:10.1016/j.neuron.2011.06.004

18. Schmid F, Wachsmuth L, Schwalm M et al (2015) Assessing sensory versus optogenetic network activation by combining (o)fMRI with optical Ca^{2+} recordings. J Cereb Blood Flow Metab 36(11):1885–1900. doi:10. 1177/0271678X15619428

19. Sun J, Lee SJ, Wu L et al (2012) Refractive index measurement of acute rat brain tissue slices using optical coherence tomography. Opt Express 20:1084. doi:10.1364/OE.20. 001084

20. Stroh A, Adelsberger H, Groh A et al (2013) Making waves: initiation and propagation of corticothalamic Ca^{2+} waves in vivo. Neuron 77:1136–1150. doi:10.1016/j.neuron.2013. 01.031

21. Galarreta M, Hestrin S (2002) Electrical and chemical synapses among parvalbumin fast-spiking GABAergic interneurons in adult mouse neocortex. Proc Natl Acad Sci U S A 99:12438–12443. doi:10.1073/pnas. 192159599

22. Han X, Chow BY, Zhou H et al (2011) A high-light sensitivity optical neural silencer: development and application to optogenetic control of non-human primate cortex. Front Syst Neurosci 5:18. doi:10.3389/fnsys.2011.00018

23. Aschauer DF, Kreuz S, Rumpel S (2013) Analysis of transduction efficiency, tropism and axonal transport of AAV serotypes 1, 2, 5, 6, 8 and 9 in the mouse brain. PLoS One 8:e76310. doi:10.1371/journal.pone.0076310

24. Rios G, Lubenov EV, Chi D et al (2016) Nanofabricated neural probes for dense 3-D recordings of brain activity. Nano Lett 16:6857–6862. doi:10.1021/acs.nanolett. 6b02673

25. Alivisatos AP, Andrews AM, Boyden ES et al (2013) Nanotools for neuroscience and brain activity mapping. ACS Nano 7:1850–1866. doi:10.1021/nn4012847

26. Cardin JA, Carlén M, Meletis K et al (2010) Targeted optogenetic stimulation and recording of neurons in vivo using cell-type-specific

expression of Channelrhodopsin-2. Nat Protoc 5:247–254. doi:10.1038/nprot.2009.228

27. Laxpati NG, Mahmoudi B, Gutekunst C-A et al (2014) Real-time in vivo optogenetic neuromodulation and multielectrode electrophysiologic recording with NeuroRighter. Front Neuroeng 7:40. doi:10.3389/fneng.2014.00040

28. Shew WL, Bellay T, Plenz D (2010) Simultaneous multi-electrode array recording and two-photon calcium imaging of neural activity. J Neurosci Methods 192:75–82. doi:10.1016/j.jneumeth.2010.07.023

29. Park D-W, Brodnick SK, Ness JP et al (2016) Fabrication and utility of a transparent graphene neural electrode array for electrophysiology, in vivo imaging, and optogenetics. Nat Protoc 11:2201–2222. doi:10.1038/nprot.2016.127

30. Park D-W, Schendel AA, Mikael S et al (2014) Graphene-based carbon-layered electrode array

technology for neural imaging and optogenetic applications. Nat Commun 5:5258. doi:10.1038/ncomms6258

31. Kozai TDY, Vazquez AL (2015) Photoelectric artefact from optogenetics and imaging on microelectrodes and bioelectronics: new challenges and opportunities. J Mater Chem B Mater Biol Med 3:4965–4978. doi:10.1039/C5TB00108K

32. Aravanis AM, Wang L-P, Zhang F et al (2007) An optical neural interface: in vivo control of rodent motor cortex with integrated fiberoptic and optogenetic technology. J Neural Eng 4:S143–S156. doi:10.1088/1741-2560/4/3/S02

33. Yona G, Meitav N, Kahn I, Shoham S (2016) Realistic numerical and analytical modeling of light scattering in brain tissue for Optogenetic applications(1,2,3). eNeuro. doi:10.1523/ENEURO.0059-15.2015

Chapter 9

Concepts of All-Optical Physiology

Jan Doering, Ting Fu, Isabelle Arnoux, and Albrecht Stroh

Abstract

All-optical physiology represents a methodological approach using light for both readout from and manipulation of individual neurons. This approach paves the way for a true, causal analysis of neuronal activity with single-cell and single action potential resolution and is therefore highly desirable for the investigation of neural networks. The following chapter addresses the general concepts of all-optical interrogations by shedding light on all critical steps needed for these experiments: Calcium-sensitive probes for readout, next-generation two-photon-excitable optogenetic actuators for manipulation and advanced optics for efficient stimulation of and real-time readout from individual neurons. The chapter also provides a step-by-step protocol on an all-optical strategy using an optical parametric oscillator (OPO) for photo-stimulation of opsin-expressing cells alongside simultaneous two-photon calcium imaging.

Key words All-optical, Two-photon optogenetics, Optical parametric oscillator, Synthetic and genetically encoded calcium indicators, Optogenetic stimulation paradigms

1 Introduction

The key concept of all-optical physiology in the nervous system is to use light for both readout and manipulation of neuronal activity [1, 2]. The main advantage of optical methods is evident: Neurons from genetically defined cell populations can be specifically targeted in a non-invasive way with different wavelengths for recording and stimulation [3–6]. Several achievements in neuroscience brought all-optical physiology in rodents within reach: With the development of fluorescent calcium indicators in 1980 the group of the late Robert Tsien [7] devised an optical method for detection of neuronal activity, which was rapidly adopted by many laboratories all over the world. Around a decade later the establishment of two-photon (2-P) laser scanning microscopy by the group of Winfried Denk [8, 9] revolutionized brain imaging techniques and allowed imaging with single-cell resolution in highly diffracting brain tissue well beyond the typical effective imaging depth amenable with confocal microscopy of about 100 μm. In 2003 Arthur Konnerth and colleagues [10] combined the two methods, achieving in vivo

Albrecht Stroh (ed.), *Optogenetics: A Roadmap*, Neuromethods, vol. 133,
DOI 10.1007/978-1-4939-7417-7_9, © Springer Science+Business Media LLC 2018

2-P calcium imaging of neuronal microcircuits, typically in mouse cortex layer II/III at an imaging depth of 200–400 μm, paving the way to an optical detection of single action potentials in the intact circuitry of living animals. As the group of Karl Deisseroth introduced the first optogenetic actuators in 2005 and made it possible to optically manipulate the activity of genetically defined individual neurons [11], all-optical physiology seemed within reach. Still, to achieve all-optical interrogations of neurons the combination of both, in vivo 2-P calcium imaging and optogenetics, had to be accomplished.

An all-optical approach requires determining a suitable probe/sensor couple for experimentation [2]. Recently, the diversity of available optogenetic actuators has considerably expanded but yet only few of them are suitable for 2-P excitation essential for achieving spatially localized excitation with single-cell resolution deep in the brain tissue (*see* also Chap. 10).

2   General Concepts of All-Optical Physiology

All-optical approaches face unique challenges: Optogenetic actuators are driven by strong light intensities for reliable and fast activation several orders of magnitude higher than intensities for excitation of calcium indicators [12]. This allows for the use of opsin/detector pairs driven by the same (one-photon /1-P) excitation wavelength, such as Oregon Green 488 BAPTA-1 (OGB-1)/GCaMP and Channelrhodopsin-2 (ChR2), as the low intensities used for exciting the indicator will typically not suffice to activate the opsin. The high light intensities of the optogenetic pulse, ranging at >10 mW/mm^2—compared to 0.01 mW/mm^2 for excitation of calcium indicators, e.g., using optic fibers—will cause strong excitation of the calcium indicator with corresponding strong emission and thus lead to optical artifacts or even detector saturation, as the emission wavelength cannot be blocked, as it contains the signal of interest. And even in the case of an opsin/indicator pair with non-overlapping spectra, the induction of autofluorescence of the tissue during the high-intensity pulse can generate a large artifact resulting in data loss during the critical photostimulation period [13, 14]. However, 1-P based all-optical approaches can be meaningfully employed using, e.g. OGB-1/ChR2 pairs, if the neuronal response of interest is temporally delayed such as in the case of slow oscillation-associated calcium waves [15]. Also, the optic artifact can be subtracted from the data revealing underlying neuronal responses [16]. But still, the strong light pulses may lead to fast bleaching of the calcium dye [17]. Practically it also remains challenging to achieve strong expression of both sensor and actuator in the same neurons. Acute loading of synthetic calcium indicators into previously virus-injected animals can be tedious and allows for only one imaging experiment per animal (*see* Sect. 3.2), whereas co-expression of actuator and indicator requires precise titration to achieve functional expression of both genes. In a pioneering study, the group of Loren L. Looger [17] devised two fusion proteins, one red-shifted actuator with a green-shifted sensor and one blue-shifted actuator with a red-shifted sensor, allowing equimolar co-expression of both proteins. In cultured cells spectral overlap was reduced for the blue/green actuator/sensor-couple [17].

2.1 Two-Photon-Based Approaches of All-Optical Physiology

2-P-based approaches for simultaneous imaging and optogenetics have the potential to overcome most of the aforementioned problems faced by 1-P stimulation. As only a small focal volume is excited and infrared wavelengths are used, autofluorescence is low. In addition, true single-cell specificity of optogenetic stimulation can be achieved. ChR2, a light-gated ion channel, was the first opsin successfully activated by 2-P stimulation, leading to neuronal spiking [18]. In this pioneering study, the excitation of ChR2 by 2-P laser pulses was sufficient to reach firing threshold in vitro in cultured neurons. But, long pixel dwell times and the necessity of specific scan trajectories such as spiral scans prevented the wide spread use of ChR2-based all-optical approaches. In addition, the spectral overlap, while less problematic than in 1-P approaches (see above), prevented the combination with the most efficient reporters of neuronal activity such as OGB-1 and GCaMP6. However, blue-light activated channelrhodopsins with suitable properties for 2-P excitation may be used in combination with red calcium indicators such as R-CaMP2 [19] or Cal-590 [20]. Future studies may include newly developed blue-light-activated opsins such as Chronos [21] and CatCh [22]. Recently, the 2-P stimulation of Chronos and ReaChR have been performed in vitro by holographic illumination [23, 24] (*see* Chap. 10). Chronos and ReaChR exihibit fast kinetics and high photocurrents upon 2-P illumination evoking spikes with sub-millisecond precision and allowing for repeated firing up to 100 Hz (Chronos) and 35 Hz (ReaChR). The use of Chronos and ReaChR is therefore promising for truly mimicking endogenous microcircuit activity.

To allow for the use of currently most efficient blue-light-activated calcium sensors, mutations of the C1V1 family have been performed to obtain red-light-activated C1V1 variants (C1V1$_T$ and C1V1$_{T/T}$) with larger photocurrents than ChR2 [4]. The 2-P activation of C1V1 variants successfully elicited spikes in culture, in slices, and in vivo, without the need for specific trajectories such as spiral scans. Interestingly, whole-cell recordings of neurons expressing the C1V1$_{T/T}$ variant indicated that they are able to drive action potentials at high frequencies (up to 40 Hz) [4]. This development resulted in a breakthrough for 2-P-based all-optical approaches in the mouse brain in vivo, pioneered by the group of Michael Haeusser [6]. In this study, optogenetic activation of selected neuronal ensembles in mouse somatosensory cortex by using sculpted light had been combined with simultaneous resonant-based 2-P calcium imaging during different behavioral states. A spatial light modulator split the beam into multiple, individual beamlets targeting several cells simultaneously. The group of David Tank [5] used two 2-P light sources, one for photostimulation at 1064 nm to achieve fast perturbation of multiple hippocampal CA1 pyramidal neurons during spatial navigation through a virtual reality environment, while using the other one at 920 nm for simultaneous large-scale 2-P calcium imaging of network activity. Applying temporal focusing [25], fast photostimulation of multiple cells could be achieved.

On the 2-P stimulation of inhibitory opsins, at this point, only proof-of-principle studies exist for Arch3.0-expressing neurons. Raster scans yielded prolonged outward currents inhibiting neuronal spiking generated by current injection in vitro [4]. However, this 2-P excitable archaerhodopsin has not yet been applied in all-optical experiments. Other promising red-light-activated opsins, such as eNpHR3.0 [26] and Jaw [27], and the blue-light-activated chloride-conducting channelrhodopsin (ChloC) [28] have not been tested with regard to 2-P and all-optical experiments yet.

Here, we describe the basic procedures for combining in vivo high-speed 2-P calcium imaging with optogenetics by coexpressing the red-shifted opsin C1V1T/T and the genetically encoded calcium indicator (GECI) GCaMP6f in the same population of neurons in layer II/III of mouse visual cortex. Implementing an Optical Parametric Oscillator (OPO) into a custom-made 2-P microscope system allows cross-talk-free photo-activation of multiple neurons at single-cell level in the far-infrared (>1100 nm) alongside calcium imaging at 925 nm. In this method-centered chapter, we will discuss tried and tested strategies and provide troubleshooting advice in Sect. 4.

2.2 Technical Implementation

Standard 2-P systems are usually based on Ti:Sa lasers, delivering light in the spectral range between 680 and 1060 nm. This limitation poses two problems with regard to biomedical imaging research: First, most dyes and fluorescent proteins excited by red 1-P wavelengths are only 2-P excitable at far infrared wavelengths (>1000 nm) difficult to achieve with Ti:Sa lasers. Second, the implementation of an all-optical strategy requires high density light pulses for a sufficient photo-activation of the opsin and a simultaneous illumination for monitoring network activity. With a standard 2-P microscope the light pulses for optogenetic stimulation would disturb imaging due to spectral overlap with imaging wavelengths ranging from 800 nm (e.g., OGB-1) to 925 nm (e.g., GCaMP6f).

An optical parametric oscillator (OPO) (Fig. 1a) might overcome this limitations. It is a passive device pumped by a mode-locked Ti:Sa laser consisting of a laser resonator and a nonlinear optical crystal. It needs constant pumping by the Ti:Sa laser to convert the input Ti:Sa light wave ranging from 700 nm to 880 nm with defined frequency into two tunable output waves with lower frequencies between 1000 nm and 1600 nm (Fig. 1b). The output waves are termed signal and idler, which energetically add up to the original pumping wavelength [29]. All in all an OPO converts the short Ti:Sa light into light with longer wavelengths. This makes an OPO a very valuable light source in biomedical research, being used for laser spectroscopy [30] and 2-P laser scanning microscopy (TPLSM) [31]. For the latter the OPO enables the 2-P excitation of red fluorophores [32], and spectral separation when applying two fluorophores, e.g., eGFP and mCherry [33]. It increases imaging depth, improves second harmonic generation and significantly reduces phototoxicity and photobleaching compared to with conventional TPLSM [32, 34].

However, in addition to avoiding spectral overlap, another important technical aspect needs to be considered: For fast imaging of neuronal microcircuits enabling single-action potential resolution, the use of resonant scanners became standard in the field (Fig. 1c). These resonant scanners provide full-field sampling rates of 30 Hz, ideally suited for microcircuit imaging. Yet, the scanning frequency is fixed; consequently, the pixel dwell time is fixed as well and relatively short (5 μs). In an all-optical experiment using 2-P laser scanning microscopy (TPLSM) equipped only with a single resonant scanner, the two 2-P wavelengths (e.g., 925 nm for GCaMP6 and 1100 nm for C1V1TT) will be coupled to the same resonant scanning mirror (Fig. 1c 1,2). The 925 nm beam will be used for continuously scanning of the entire field of view, while the 1100 nm beam will be used for ROI-based optogenetic modulation, controlled by an electro-optical modulator (EOM). However, the short pixel dwell times might not suffice for an efficient excitation of the opsin, particularly in vivo and in deeper layers of the cortex. SLMs enable temporally decoupling of imaging and stimulation (see above and Chap. 10). Here, an alternative approach is highlighted; longer pixel dwell times and independent scanning patterns are achievable by implementing two independent scanning mirrors. In this configuration, only the 2-P beam used for imaging (ranging between 800 and 925 nm) is coupled to the resonant scanner, allowing for fast (30 Hz) full-field scanning. The second OPO wavelength (>1100 nm) is coupled to an additional galvanometric-scanner, allowing for spatiotemporal independent region of interest (ROI)-based optogenetic modulation. In addition, in cases in which single-cell specificity in not required, two 1-P lasers (488 nm, e.g. for excitation of ChR2 and 552 nm, e.g. for the

a

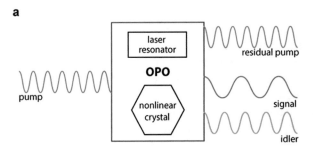

b

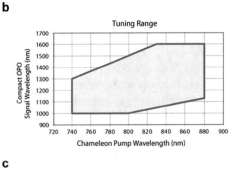

c

(1) OGB-1 Imaging   (2) GCaMP-Imaging

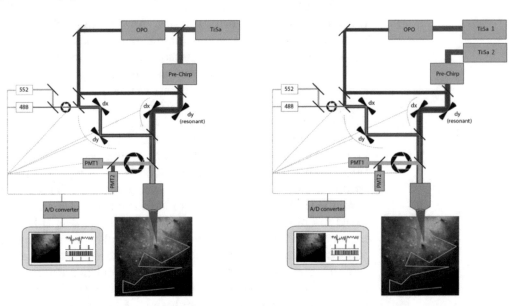

Fig. 1 Technical Implementation of all-optical concepts. (**a**) Schematic of OPO, adapted from https:/www.rp-photonics.com/optical_parametric_ oscillators.html (**b**) *Left*: Typical OPO tuning range, *right*: typical OPO output power, adapted from Chameleon compact OPO Manual by Coherent. (**c**) Schemes of 2-P microscope setup, customized system from LaVision Biotec (Bielefeld, Germany) (**1**) OGB-1-Imaging and C1V1T/T-excitation can be performed with one laser source. OPO is efficiently pumped by imaging wavelength of 800 nm, (**2**) Additional second laser is necessary to perform GCaMP6f imaging at 925 nm and optogenetic stimulation of C1V1 as the OPO cannot be efficiently pumped at 925 nm

excitation of C1V1, Arch or NpHR) are coupled to this galvanometric scanner as well. Of course, the temporal and spatial indepen-
dence is restricted to the imaging plane.

However, current Ti:Sa-OPO configurations pose restrictions
on the possible range of wavelengths (Fig. 1b). While the emission
spectrum of the OPO in principle ranges between 1000 and
1600 nm, this is only achievable by specific pumping wavelengths
provided by the Ti:Sa. The spectrum of pumping wavelengths
between 740 nm and 880 nm, does not allow emission of the entire
tuning range (Fig. 1b left panel). Also, note that the light power of
OPO emission decreases with increasing wavelength (Fig. 1b right
panel), but the OPO still provides sufficient power for optogenetic
modulation even at highest wavelengths, as the power used for
efficient optogenetic stimulation ranges at 20–40 mW [6], well
below the minimal power of the OPO, even when considering an
additional loss of power in the optical path of about 30%.

Whether the restriction of the pumping/tuning range will
result in problems for all-optical approaches depends very much
on the opsin/indicator pair used:

(a) OGB-1/C1V1$_{T/T}$

Using this indicator/opsin pair is amenable for the use of one
Ti:Sa/OPO combination for both, optogenetic stimulation and
calcium imaging, by splitting the beam path. Given the high
power of the Ti:Sa, typically ranging at 4 W, the Ti:Sa beam can
be split, using 20% for imaging at 800 nm and 80% for pumping of
the OPO. The optimal excitation wavelength of OGB-1 is 800 nm
and therefore within the range to effectively pump the OPO
(Fig. 1c 1) and tune it from 1000 to 1500 nm, covering almost
the entire emission spectrum of the OPO.

(b) GCaMP6/C1V1$_{T/T}$

The increasingly popular GECI GCaMP6 cannot be efficiently
excited below 900 nm. While baseline fluorescence is still achievable at 880 nm (Fig. 2a), the highest pumping value, no functional
transient can be recorded (Fig. 2b). The optimal excitation wavelength of 925 nm is well outside the pumping range of the OPO
(Fig. 1b, left panel). Consequently, this approach needs the implementation of a second Ti:Sa laser, resulting in two independently
tunable beams, one for imaging, allowing full Ti:Sa spectrum, and
one for optogenetic stimulation, allowing full OPO spectrum.

2.3 Functional Assessment of Opsin Expression

The first step of experimental design of an all-optical experiment
represents the choice of a suitable opsin [35] (Chaps. 1–3, 10) for
2-P optogenetics. While as mentioned above, new 2-P excitable
opsins are emerging, here, we describe the experimental procedure
for C1V1$_{T/T}$ [36], a red-shifted mutant of ChR2 [11]. Using viral

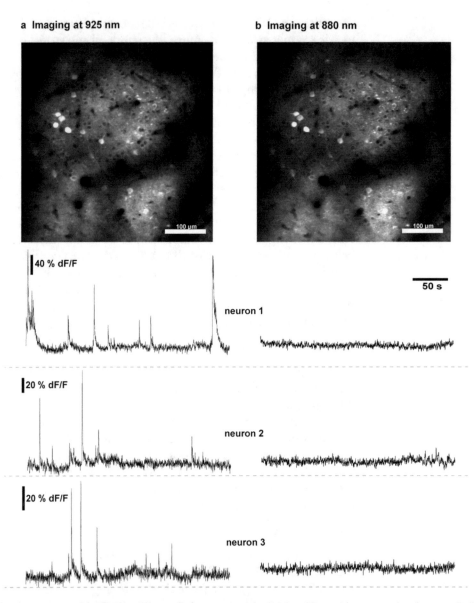

Fig. 2 2-P imaging of GCaMP6f at different Ti:Sa wavelengths (880 vs 925 nm) in mouse visual cortex. Reliable detection of spontaneous calcium transients at 925 or 920 nm, at lower wavelengths (860 or 880 nm) no clear functional transients can be identified

gene transfer, we injected the opsin-carrying adeno-associated virus (AAV) through a small craniotomy in mouse visual cortex. Two weeks after stereotactic injection confocal microscopy revealed strong membrane-bound expression in layer II/III and layer V/VI of the mouse visual cortex, whereas cells in layer IV hardly expressed $C1V1_{T/T}$ (Fig. 3a, b). Besides this morphological analysis, we tested the functionality of this opsin using an optic fiber for stimulation combined with local field potential (LFP) readout [15].

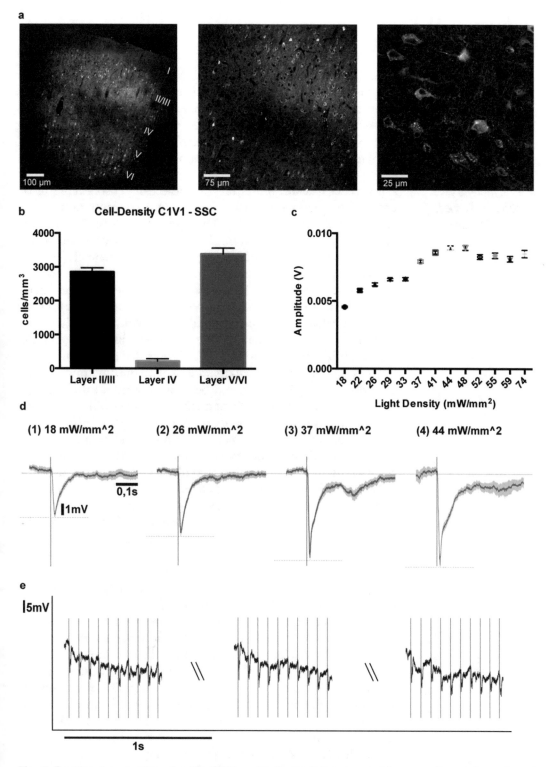

Fig. 3 Functional expression of opsin C1V1$_{T/T}$. (**a**) Confocal images revealing smooth membrane bound C1V1$_{T/T}$ expression in mouse cortex, cortical layers are indicated. (**b**) Densities of C1V1$_{T/T}$-expressing cells in cortical layers II/III, IV, and V/VI, data from five brain slices (70 μm), 416 cells. (**c**) Light power density (mW/mm^2) versus amplitude (V) of LFP response upon light stimulation of C1V1$_{T/T}$ expressing cortical neurons. (**d**) Averages of LFP responses ($n = 29$ experiments) upon optogenetic stimulation using an optic fiber (552 nm) at varying light intensities. (**e**) Individual responses upon 15 Hz stimulation trains

This easy way to implement a functional readout is highly useful for establishing light intensities effectively activating the opsin. As delineated, the light intensities applied demonstrate a roughly linear relationship between amplitude of LFP deflection and light intensity (Fig. 3c), for low light intensities. $ClV1_{T/T}$ ($\tau_{off} = 40$ ms [4]), responded to every stimulus applied up to 15 Hz (Fig. 3e).

2.4 Synthetic vs. Genetically Encoded Calcium Indicators

Fluorescent calcium indicators are molecular sensors reacting on changes of the cytoplasmatic free calcium concentration with a change in fluorescence [37]. They possess at least one calcium binding site and a fluorophore. In neurons, firing of action potentials leads to an opening of voltage gated calcium channels (VGCC) and consequently to fast and sharp elevation of cytoplasmatic calcium concentration [38]. Fluorescent calcium indicators can therefore be used to monitor suprathreshold neuronal activity. Freely diffusible calcium ions are in equilibrium with the calcium ions bound to endogenous calcium buffers like parvalbumin, calbindin-D28k and calretinin [39]. The fact that calcium indicators act as exogenous calcium buffers adding up to the buffering capacity of endogenous proteins should be kept in mind, as the cytosolic calcium equilibrium and cellular calcium dynamics will be impacted and this might have an influence on cellular function [40].

Calcium indicators are available with low and high affinities to calcium. The affinity is described by their dissociation constant K_d, corresponding to the concentration of calcium at which half of the indicator molecules have bound to calcium [41]. K_d is impacted by parameters such as pH, temperature or concentration of other ions and can differ, e.g., in vivo or in vitro conditions [41]. Low-affinity calcium indicators add only small buffering capacity and reflect particularly the decrease of calcium concentrations temporally more accurately. In noisy in vivo conditions high-affinity calcium indicators are first choice due to better sensitivity, reflecting the increase in intracellular calcium upon, e.g. action potential-mediated opening of voltage-gated calcium channels with a temporal jitter of only a few milliseconds [38]. However, the fast decrease of intracellular calcium due to the re-uptake into intracellular calcium stores is not temporally precisely reflected by the fluorescence changes of the indicator due to its high affinity. Therefore, the off-kinetics of a functional calcium transient cannot be interpreted as mirroring the kinetics of the decrease in intracellular calcium concentration, but rather governed by the off-kinetics of the indicator itself. This fact also limits the effective temporal resolution of the indicator in terms of the maximal frequency of action potentials which can be resolved. While typical sparse activity patterns of, e.g., excitatory cortical neurons can be nicely recognized as individual calcium transients by both GCaMP6 and OGB-1, fast spiking patterns of interneurons will lead to indicator saturation.

Fluorescent calcium indicators are divided into two groups: synthetic—or chemical—calcium indicators and GECIs. Synthetic calcium indicators consist of a calcium-sensitive chelator and a fluorophore. In a pioneering study *Roger Tsien* et al. [37] developed the synthetic calcium indicator *fura-2* based on the calcium selective chelator EGTA. Intramolecular conformational changes are induced by the binding of calcium and lead to changes in fluorescence emission. The synthetic calcium indicator introduced here, OGB-1, is based on the calcium chelator BAPTA. The excitation peak of OGB-1 in the blue spectrum ranges at 488 nm, the emission peak in the green spectrum at approximately 520 nm [41]. It can be very efficiently excited by femtosecond 2-P pulses at 800 nm [41] and has therefore been used in numerous studies in vivo [10, 42].

GECIs consist of one or two fluorescent proteins and a calcium binding domain [43]. The currently most widely used single fluorophore GECIs is the GCaMP-family. Particularly its newest generation, GCaMP6, has become tremendously popular in 2-P imaging, as for the first time, it exhibits similar, if not superior properties as synthetic calcium indicators such as OGB-1 in terms of sensitivity and signal to noise ratio, allowing for the identification of single action potentials in vivo. GCaMP is comprised of the circular permutated green fluorescent protein cpEGFP, the calcium binding protein calmodulin and the calmodulin-interacting M13 peptide [44]. Increased calcium binding due to elevated intracellular calcium concentrations lead to interactions between calmodulin and M13, resulting in conformational changes causing an increase in fluorescence emission [44, 45]. As red-shifted optogenetic actuators are to some extend also activated by wavelength used for GCaMP imaging due to their blue shoulder [17], red-shifted GECIs such as R-GECO1 [46] or R-CaMP [17] would be advantageous. The latter showed less spectral overlap when combined with blue light excited ChR2 in *C. elegans* [17]. Second-generation R-CaMP2 with improved action-potential detection and fast kinetics are available by now and have been successfully used for all-optical experiments in *C. elegans* [20], but they have yet to prove the same sensitivity as GCaMP6 for 2-P imaging in rodent brain. Two fluorophore GECIs are based on Förster resonance energy transfer (FRET) from an excited donor to an acceptor fluorophore. Upon calcium binding to a calcium-sensitive domain (mainly calmodulin or troponin c) conformational changes enhanced by a conformational actuator (e.g., M13) reduce spatial distance between donor and acceptor fluorescence and FRET enables the excitation of the acceptor fluorophore [47]. Due to their radiometric properties two-fluorophore GECIs more accurately detect calcium changes at least using 1-P excitation, but due to the wider, overlapping 2-P excitation spectra, and the two fluorophores covering large spectral bandwidth, application in the context of 2-P all-optical experiments seems not advisable.

In contrast to fluorescent calcium indicators, bioluminescent proteins such as *aequorin* [48] do not depend on external excitation light and thus avoid common imaging problems such as photobleaching, phototoxicity or autofluorescence [49]. Moreover in combination with optogenetic actuators, they prevent an unintended stimulation of the actuator by imaging wavelengths [50]. Until now bioluminescent proteins have been limited by different drawbacks like low quantum yield, few color variants and problems in dye delivery or protein stability [50]. Recently developed bioluminescent proteins such as multicolor Nano-Lanterns [50] overcome important drawbacks and might open up an alternative approach to all-optical physiology.

Here, we show the principles of all-optical experiments using two different calcium indicators: OGB-1 and GECI GCaMP6f.

3   Methodological Implementation

For acute OGB-1 experiments, 2 weeks prior to calcium imaging, we injected the opsin-encoding virus stereotactically in the mouse visual cortex through a small craniotomy. For imaging, we then drilled a cranial window and injected OGB-1 AM via multi-cell bolus loading technique [51] acutely into virus infected area. Using fast resonant scanning at 800 nm spontaneous suprathreshold activity of microcircuits with more than 200 neurons could be detected, with a temporal resolution of 35 ms (customized 2-P microscope, LaVision BioTec, Bielefled, Germany). However, only few neurons expressed both C1V1$_{T/T}$ and were stained with OGB-1 (Fig. 4a). Injecting OGB-1 exactly in the C1V1$_{T/T}$ infected area is challenging under in vivo conditions without hitting blood vessels during the injection of the calcium dye. In addition, the tissue had already been impacted by the injection of the opsin-carrying virus earlier and in particular the preparation of an acute window in the region previously subjected to craniotomy for virus injection often times led to bleeding: the dura mater regularly attached to the cranium, which made it almost impossible to avoid bleeding, when flipping the cranial window open. A potential solution of this would be not to inject perpendicular to the dura, but with a rather shallow angle, so that the parenchyma dorsal of the opsin-expressing region is not affected. While this approach is generally advisable for all opsin injections, here, it did not solve the aforementioned disadvantage of combining a synthetic indicator and virus injections, as still the area of injection and expression needs to be perfectly matched, and the rather large opening required for imaging still contained the location of the craniotomy.

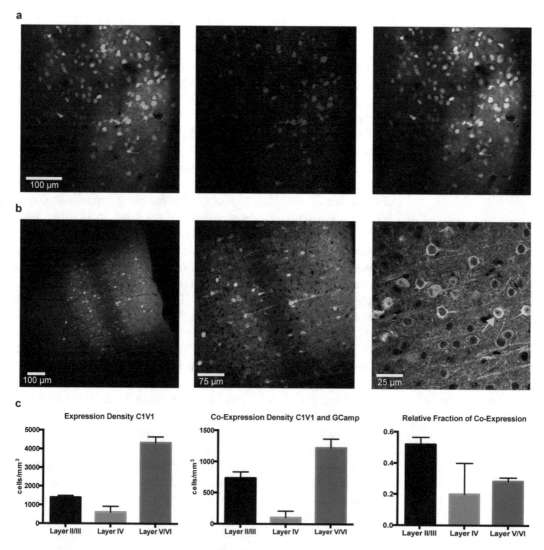

Fig. 4 2-P imaging of OGB-1/C1V1$_{T/T}$ in vivo and co-expression ratio of GCaMP6f/C1V1$_{T/T}$. (**a**) Optogenetic full-field OPO-stimulation of C1V1$_{T/T}$-mCherry expressing neurons (*red*) at 1100 nm during simultaneous 2-P calcium imaging of OGB-1 stained cells at 800 nm. A fraction of C1V1$_{T/T}$ expressing neurons are stained with the calcium indicator. (**b**) Confocal micrographs of C1V1$_{T/T}$ and GCaMP6f co-expressing neurons (*yellow arrow*), solely GCaMP6f expressing neurons are indicated with a *green*, and solely C1V1$_{T/T}$-mCherry with a *red arrow*. (**c**) *Left*: Density of C1V1$_{T/T}$ expressing cells in different cortical layers of mouse visual cortex, *Middle*: Density of co-expressing cells in different cortical layers, *Right*: Relative fraction of co-expressing cells in different cortical layers, data from five brain slices (70 μm), 175 cells

Employing a GECI in combination with a chronic window implantation allows for a single surgery, avoiding the aforementioned complications, and would in our view be advantageous. For these all-optical experiments, we prepared a virus mix containing AAVs encoding for C1V1$_{T/T}$ and GCaMP6f, ideally resulting in strong co-expression of both proteins in the same neurons. However, now, the titers of both viruses needed to be carefully titrated, keeping in mind that both proteins, the indicator and the opsin, need to be expressed in high copy numbers for an all-optical experiment to work (Fig 4b,c). Finding the right titer for an adequate level of co-expression can be tedious, as too high titers lead to cytotoxicity (*see* Chap. 8).

We employed AAVs encoding for GCaMP6f under the control of the neuronal hSYN-promoter [52] (*see* Chaps. 1 and 2), while we specifically targeted excitatory neurons using the CamKII-promoter [53] controlling ClVl$_{T/T}$ expression. We prepared a chronic window at the region of the mouse primary visual cortex (V1) and could repeatedly image a local microcircuit in layer II/III of V1 over several weeks. However, the optimal wavelength to excite GCaMP6f peaks at 925 nm (see Fig. 2), well outside of the Ti:Sa pumping range for the OPO (see Fig. 1b). As a consequence, we needed to integrate a second Ti:Sa laser to achieve to independently tunable light beams. One Ti:Sa laser was dedicated for GCaMP imaging and the other one for optogenetic ClVlT/T stimulation (Fig. 1c).

The protocol delineated below highlights the approach to yield co-expression of GECI and opsin in mouse cortex.

3.1 Stereotactic Virus-Injections and Chronic Window Implantation

First, the animal should be well fixed in a stereotaxic frame, and pre-operative analgesia should be administered. If inhalation anesthesia is used, anesthetics should start with low dose and increased very slowly until the animal is neither gasping nor shows pain reflexes (*see* also Chaps. 8 and 13). The incision should be performed at the midline sagittal suture. For chronic window implantation the entire scalp and part of the neck muscles should be removed to allow the cement to better attach to the cranium. A clean and dry surface of the skull is very important for the adhesion of the cement. After navigating to the corresponding coordinates for injection a small craniotomy (acute injection) or a window-opening (chronic injection) is performed using a dental drill. Keep the tissue under the opening moisturized with saline solution. Load the virus mix into injection pipette devoid of air bubbles and slowly inject into the brain at a rate around 0.2 μl/min. Withdraw the micropipette slowly over several minutes to minimize upward flow of viral solution.

After the injection, tissue and skull should be gently cleaned with saline solution or PBS. Suture the scalp if no chronic window is implemented. To create the chronic window place a coverslip (slightly larger than the opening) on top of the moisturized exposed tissue. Gently press the coverslip onto the skull and use surgical glue (tip-loaded in a fine-tip-micropipette) to seal it. Then carefully apply the cement with proper viscosity on the exposed dry skull carefully without covering neither the coverslip nor the skin. If any holder is needed, insert it in the cement before it dries out.

Apply post-operative analgesia and 5% glucose solution to aid recovery via subcutaneous injection and let the animal recover in the cage on a heating-plate (adjust around 37 °C) 1 h. Check the animal every day.

3.2 Virus-Titration and Coexpression

Intracranial injection of viral vectors has several technical advantages over other labeling techniques (*see* Chaps. 1 and 2). The injection parameters can be tailored to individual experiments by adjusting location, volume and rate of injection, the age of the animal, AAV serotype and the promoter driving gene expression.

Besides the injected volume, the virus titer is an important factor to tightly regulate for achieving optimized expression levels (see Figs. 4 and 5). If the virus titer is too high, toxicity and compromised cellular physiology can arise (*see* Chap. 8). In contrast, if virus titer is too low, neither efficient optogenetic induction of action potentials nor the identification of action potential-related functional calcium transients can be achieved. Therefore, virus titration is an indispensable procedure before starting an experimental series. Depending on different characteristics of individual viruses, as well as different requirements of each experiment, the proper titer of virus for the injection can vary a lot.

3.3 Coexpression

There are different approaches to achieve the coexpression of optogenetic probes and calcium indicators. One approach is to combine opsins (e.g., $C1V1_{T/T}$) expressed upon viral gene transfer and acutely injected green synthetic calcium dyes (e.g., OGB-1 AM) (Fig. 4a). In this case, the virus encoding for the opsin (e.g., $C1V1_{T/T}$) should be injected into the brain first. Two weeks post injection, the cranium will be reopened. The calcium dye will be injected acutely into the opsin-expressing region. Using this approach, titration of virus is relatively easy, as only the opsin needs to be expressed, so the entire capacity of the neuron to express exogenous proteins can be exploited, but significant challenges need to be overcome as mentioned above and below. For approaches using two genetically encoded proteins for manipulation (opsin) and monitoring (GECI), the titers of both need to be carefully balanced, as the capacity of a neuron to express these exogenous proteins without significantly impacting cellular physiology is limited (Figs. 4 and 5). However, upon obtaining this golden ratio, a co-expression of 40% of total expressing neurons in the same brain region is easy to be obtained (Fig. 4c). Note, that using current AAVs, only few neurons in layer IV can be transduced (*see* also Chap. 8).

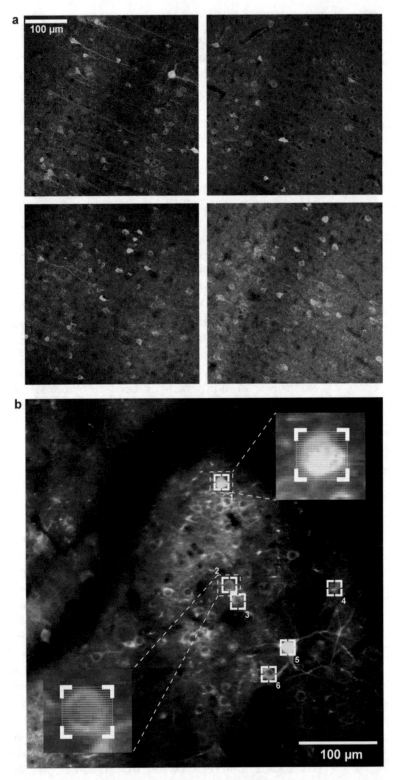

Fig. 5 Histological assessment of titers for co-expression, and in vivo single-cell optogenetic stimulation of C1V1$_{T/T}$/GCaMP6f co-expressing neurons in mouse visual cortex. (**a**) Confocal images of four combinations of different virus titers for C1V1$_{T/T}$ and GCaMP6f. (**b**) Selection of 6 regions of interests (ROIs) for OPO stimulation in a line-scanning pattern, ROIs are placed on C1V1$_{T/T}$ (*red*) and GCaMP6f (*green*) co-expressing cells

3.4 Two-Photon Stimulation Paradigm

Here, we stimulated multiple ClV1$_{\text{T/T}}$ expressing neurons individually by 2-P excitation using an OPO and recorded the response by either simultaneous fast resonant 2-P calcium imaging or electrophysiological recordings (LFP) (see Figs. 3 and 5). Up to 6 ClV1 expressing cells could be selected for sequential stimulation with OPO excitation within an acceptable time of 50 ms. For that, square ROIs (~20 × 20 µm) were placed onto the cells of interest (Fig. 5b). Line scanning of each ROI was then performed with a pixel size of 0.6 µm and a pixel-dwell time of 5 µs. These parameters were established by Prakash et al. [4]. The stimulation of a duration of 50 ms was performed every 10 s to allow for the recovery of the peak current (*see* Chaps. 3, 5, and 13). The maximum power was 29.5 mW at 1100 nm. A subset of experiments was performed with 1-P full-field stimulation (552 nm) directed through the same galvanometric scanner, using the same duration and frequency of stimulation. The maximum power delivered was 2.5 mW.

4 Notes

1. Interference of the imaging wavelength with the spectral activation-bandwidth of the actuator may lead to unintended activation of the optogenetic actuator disabling the recording of spontaneous baseline activity [54, 55]. To avoid this problem, use actuators and sensors with well-separated excitation spectra.

2. By underfilling of the back aperture of the objective the temporal resolution can be improved whilst increasing the focal spot size and decreasing the number of positions scanned by the beam [56].

3. State light intensities in mW/mm^2, just stating light powers (e.g., 7 mW laser power) does not allow for comparing different modes of stimulation and setups.

4. Titrate light intensities using a simple-to-implement functional readout, e.g., LFP. Make sure to stay in the linear section of the dose-response curve (see [35]) to avoid unnecessary heating and other light-induced artifacts such as the Bequerel effect (*see* Chap. 8).

5. The events induced by opsin activation could significantly differ from those mediated by physiological inputs (*see* also Chap. 13). Several opsins can conduct for a rather long time even though the excitation pulse is short depending on their off kinetics (*see* also Chap. 3). For example, the opsins ClV1 and VChR1 exhibit rather long off kinetics (156 and 133 ms, respectively) [57]. A prolonged opening of channels is an advantage for strong and efficient stimulation but passage of

ions and membrane depolarization/hyperpolarization for such long periods can evoke non-physiological events. Indeed, it is questionable whether prolonged activation or inhibition of neuronal activity mimics natural inputs.

5   Outlook

Depending on the respective research hypothesis, different approaches to an all-optical experiment should be implemented. Direct physical access to the brain through acute craniotomies allows calcium-sensitive dye imaging to be easily combined with different electrophysiological methods, such as electrical recording, whole cell recording, electric LFP recordings [58], as well as local pharmacological manipulation. Acute dye injections could also be combined with viral-based expression of opsins. Yet this preparation poses the aforementioned challenges: several weeks after recovery from virus injection, the cranium needs to be reopened and the scar tissue could cause a bleeding. In addition, longitudinal experiments are not possible.

Chronic cranial window implantation allows for long-term, high resolution imaging in various brain regions. A single surgical procedure allows a variety of in vivo optical imaging techniques on a single specimen over an extended period of time, potentially even longer than one year [59]. External stimuli, as well as behavioral training could also be implemented in anesthetized and awake animals. However, light scattering caused by skull regrowth occurred frequently. In addition, unlike OGB-1 yielding homogenous staining of the microcircuit and linear relation of number of action potentials and fluorescence emission, the expression of GCaMP6f varies and a dense expression is difficult to achieve, also the linearity between number of action potential and fluorescence emission is not ensured, complicating quantitative analysis of activity levels. Moreover, once a window is attached, it is difficult to regain access to the brain for cellular manipulations; however, emerging techniques are being developed to combine chronic imaging and direct cellular recordings or manipulation in larger mammals [60, 61], and rodents [59, 62].

From the imaging side, due to the long pixel dwell times needed for an efficient 2-P excitation of opsins, a combination of fast resonant imaging and optogenetic manipulation using only one resonant scanning mirror seems not to be advisable. For longer dwell times the excitation beam needs to be temporally uncoupled from the imaging beam. Here, we present an approach using two independent mirrors, while the use of spatial light modulators represents another promising approach, highlighted in Chap. 10.

In conclusion, all-optical interrogation of (cortical) microcircuits opens up the possibility of a causal manipulation of individual neurons in vivo and the simultaneous assessment of the response of the local circuit. However, while these experiments are achievable, they pose significant technical and methodological challenges, requiring dedicated hardware, elaborated co-expression schemes, and seamless surgical manipulations. Future developments of novel opsins optimized for efficient and fast 2-P excitation represent in our view the next critical step needed for the broad implementation of all-optical approaches.

References

1. Stroh A, Diester I (2012) Optogenetics: a new method for the causal analysis of neuronal networks in vivo. e-Neuroforum 3(4):81–88. doi:10.1007/s13295-012-0035-8

2. Emiliani V, Cohen AE, Deisseroth K, Hausser M (2015) All-optical interrogation of neural circuits. J Neurosci 35(41):13917–13926. doi:10.1523/JNEUROSCI.2916-15.2015

3. Akerboom J, Chen TW, Wardill TJ, Tian L, Marvin JS, Mutlu S, Calderon NC, Esposti F, Borghuis BG, Sun XR, Gordus A, Orger MB, Portugues R, Engert F, Macklin JJ, Filosa A, Aggarwal A, Kerr RA, Takagi R, Kracun S, Shigetomi E, Khakh BS, Baier H, Lagnado L, Wang SS, Bargmann CI, Kimmel BE, Jayaraman V, Svoboda K, Kim DS, Schreiter ER, Looger LL (2012) Optimization of a GCaMP calcium indicator for neural activity imaging. J Neurosci 32(40):13819–13840. doi:10.1523/JNEUROSCI.2601-12.2012

4. Prakash R, Yizhar O, Grewe B, Ramakrishnan C, Wang N, Goshen I, Packer AM, Peterka DS, Yuste R, Schnitzer MJ, Deisseroth K (2012) Two-photon optogenetic toolbox for fast inhibition, excitation and bistable modulation. Nat Methods 9(12):1171–1179. doi:10.1038/nmeth.2215

5. Rickgauer JP, Deisseroth K, Tank DW (2014) Simultaneous cellular-resolution optical perturbation and imaging of place cell firing fields. Nat Neurosci 17(12):1816–1824. doi:10.1038/nn.3866

6. Packer AM, Russell LE, Dalgleish HW, Hausser M (2015) Simultaneous all-optical manipulation and recording of neural circuit activity with cellular resolution in vivo. Nat Methods 12(2):140–146. doi:10.1038/nmeth.3217

7. Tsien RY (1981) A non-disruptive technique for loading calcium buffers and indicators into cells. Nature 290(5806):527–528

8. Denk W, Strickler JH, Webb WW (1990) Two-photon laser scanning fluorescence microscopy. Science 248(4951):73–76

9. Yuste R, Denk W (1995) Dendritic spines as basic functional units of neuronal integration. Nature 375(6533):682–684. doi:10.1038/375682a0

10. Stosiek C, Garaschuk O, Holthoff K, Konnerth A (2003) In vivo two-photon calcium imaging of neuronal networks. Proc Natl Acad Sci U S A 100(12):7319–7324. doi:10.1073/pnas.1232232100

11. Boyden ES, Zhang F, Bamberg E, Nagel G, Deisseroth K (2005) Millisecond-timescale, genetically targeted optical control of neural activity. Nat Neurosci 8(9):1263–1268

12. Adelsberger H, Grienberger C, Stroh A, Konnerth A (2014) In vivo calcium recordings and channelrhodopsin-2 activation through an optical fiber. Cold Spring Harb Protoc 2014 (10). doi:10.1101/pdb.prot084145. pdb prot084145

13. Zhang YP, Oertner TG (2007) Optical induction of synaptic plasticity using a light-sensitive channel. Nat Methods 4(2):139–141. doi:10.1038/nmeth988

14. Wilson NR, Runyan CA, Wang FL, Sur M (2012) Division and subtraction by distinct cortical inhibitory networks in vivo. Nature 488(7411):343–348. doi:10.1038/nature11347

15. Stroh A, Adelsberger H, Groh A, Ruhlmann C, Fischer S, Schierloh A, Deisseroth K, Konnerth A (2013) Making waves: initiation and propagation of corticothalamic Ca2+ waves in vivo. Neuron 77(6):1136–1150. doi:10.1016/j.neuron.2013.01.031

16. Schmid F, Wachsmuth L, Schwalm M, Prouvot PH, Jubal ER, Fois C, Pramanik G, Zimmer C, Faber C, Stroh A (2016) Assessing sensory versus optogenetic network activation by combining (o)fMRI with optical Ca2+ recordings. J Cereb Blood Flow Metab 36 (11):1885–1900. doi:10.1177/0271678X15619428

17. Akerboom J, Carreras Calderon N, Tian L, Wabnig S, Prigge M, Tolo J, Gordus A, Orger MB, Severi KE, Macklin JJ, Patel R, Pulver SR, Wardill TJ, Fischer E, Schuler C, Chen TW, Sarkisyan KS, Marvin JS, Bargmann CI, Kim

DS, Kugler S, Lagnado L, Hegemann P, Gottschalk A, Schreiter ER, Looger LL (2013) Genetically encoded calcium indicators for multi-color neural activity imaging and combination with optogenetics. Front Mol Neurosci 6:2. doi:10.3389/fnmol.2013. 00002

18. Rickgauer JP, Tank DW (2009) Two-photon excitation of channelrhodopsin-2 at saturation. Proc Natl Acad Sci U S A 106 (35):15025–15030. doi:10.1073/pnas. 0907084106

19. Inoue M, Takeuchi A, Horigane S, Ohkura M, Gengyo-Ando K, Fujii H, Kamijo S, Takemoto-Kimura S, Kano M, Nakai J, Kitamura K, Bito H (2015) Rational design of a high-affinity, fast, red calcium indicator R-CaMP2. Nat Methods 12(1):64–70. doi:10. 1038/nmeth.3185

20. Tischbirek C, Birkner A, Jia H, Sakmann B, Konnerth A (2015) Deep two-photon brain imaging with a red-shifted fluorometric Ca2+ indicator. Proc Natl Acad Sci U S A 112 (36):11377–11382. doi:10.1073/pnas. 1514209112

21. Klapoetke NC, Murata Y, Kim SS, Pulver SR, Birdsey-Benson A, Cho YK, Morimoto TK, Chuong AS, Carpenter EJ, Tian Z, Wang J, Xie Y, Yan Z, Zhang Y, Chow BY, Surek B, Melkonian M, Jayaraman V, Constantine-Paton M, Wong GK, Boyden ES (2014) Independent optical excitation of distinct neural populations. Nat Methods 11(3):338–346. doi:10.1038/nmeth.2836

22. Kleinlogel S, Feldbauer K, Dempski RE, Fotis H, Wood PG, Bamann C, Bamberg E (2011) Ultra light-sensitive and fast neuronal activation with the Ca(2)+−permeable channelrhodopsin CatCh. Nat Neurosci 14(4):513–518. doi:10.1038/nn.2776

23. Chaigneau E, Ronzitti E, Gajowa MA, Soler-Llavina GJ, Tanese D, Brureau AY, Papagiakoumou E, Zeng H, Emiliani V (2016) Two-photon holographic stimulation of ReaChR. Front Cell Neurosci 10:234. doi:10.3389/fncel.2016.00234

24. Ronzitti E, Conti R, Papagiakoumou E, Tanese D, Zampini V, Chaigneau E, Foust AJ, Klapoetke N, Boyden ES, Emiliani V (2016) Sub-millisecond optogenetic control of neuronal firing with two-photon holographic photoactivation of Chronos. bioRxiv. doi:10.1101/062182

25. Oron D, Tal E, Silberberg Y (2005) Scanningless depth-resolved microscopy. Opt Express 13(5):1468–1476

26. Gradinaru V, Zhang F, Ramakrishnan C, Mattis J, Prakash R, Diester I, Goshen I,

Thompson KR, Deisseroth K (2010) Molecular and cellular approaches for diversifying and extending optogenetics. Cell 141(1):154–165. doi:10.1016/j.cell.2010.02.037.   [doi]. S0092-8674(10)00190-X [pii]

27. Chuong AS, Miri ML, Busskamp V, Matthews GA, Acker LC, Sorensen AT, Young A, Klapoetke NC, Henninger MA, Kodandaramaiah SB, Ogawa M, Ramanlal SB, Bandler RC, Allen BD, Forest CR, Chow BY, Han X, Lin Y, Tye KM, Roska B, Cardin JA, Boyden ES (2014) Noninvasive optical inhibition with a red-shifted microbial rhodopsin. Nat Neurosci 17 (8):1123–1129. doi:10.1038/nn.3752

28. Wietek J, Wiegert JS, Adeishvili N, Schneider F, Watanabe H, Tsunoda SP, Vogt A, Elstner M, Oertner TG, Hegemann P (2014) Conversion of channelrhodopsin into a light-gated chloride channel. Science 344(6182): 409–412. doi:10.1126/science.1249375

29. Giordmaine JM, R. (1965) Tunable coherent parametric oscillation in LiNbO3 at optical frequencies. Phys Rev Lett (APS) 14:973

30. Johnson MJ, Haub JG, Orr BJ (1995) Continuously tunable narrow-band operation of an injection-seeded ring-cavity optical parametric oscillator: spectroscopic applications. Opt Lett 20(11):1277–1279

31. Masters BR, So PT, Gratton E (1997) Multiphoton excitation fluorescence microscopy and spectroscopy of in vivo human skin. Biophys J 72(6):2405–2412. doi:10.1016/S0006-3495 (97)78886-6

32. Herz J, Siffrin V, Hauser AE, Brandt AU, Leuenberger T, Radbruch H, Zipp F, Niesner RA (2010) Expanding two-photon intravital microscopy to the infrared by means of optical parametric oscillator. Biophys J 98 (4):715–723. doi:10.1016/j.bpj.2009.10.035

33. Shaner NC, Campbell RE, Steinbach PA, Giepmans BN, Palmer AE, Tsien RY (2004) Improved monomeric red, orange and yellow fluorescent proteins derived from Discosoma sp. red fluorescent protein. Nat Biotechnol 22 (12):1567–1572. doi:10.1038/nbt1037

34. Andresen V, Alexander S, Heupel WM, Hirschberg M, Hoffman RM, Friedl P (2009) Infrared multiphoton microscopy: subcellular-resolved deep tissue imaging. Curr Opin Biotechnol 20(1):54–62. doi:10.1016/j.copbio. 2009.02.008

35. Fois C, Prouvot PH, Stroh A (2014) A roadmap to applying optogenetics in neuroscience. Methods Mol Biol 1148:129–147. doi:10. 1007/978-1-4939-0470-9_9

36. Yizhar O, Fenno LE, Prigge M, Schneider F, Davidson TJ, O'Shea DJ, Sohal VS, Goshen I,

Finkelstein J, Paz JT, Stehfest K, Fudim R, Ramakrishnan C, Huguenard JR, Hegemann P, Deisseroth K (2011) Neocortical excitation/inhibition balance in information processing and social dysfunction. Nature 477 (7363):171–178. doi:10.1038/nature10360

37. Tsien RY (1980) New calcium indicators and buffers with high selectivity against magnesium and protons: design, synthesis, and properties of prototype structures. Biochemistry 19 (11):2396–2404

38. Grienberger C, Konnerth A (2012) Imaging calcium in neurons. Neuron 73(5):862–885. doi:10.1016/j.neuron.2012.02.011

39. Baimbridge KG, Celio MR, Rogers JH (1992) Calcium-binding proteins in the nervous system. Trends Neurosci 15(8):303–308

40. Neher E, Augustine GJ (1992) Calcium gradients and buffers in bovine chromaffin cells. J Physiol 450:273–301

41. Paredes RM, Etzler JC, Watts LT, Zheng W, Lechleiter JD (2008) Chemical calcium indicators. Methods 46(3):143–151. doi:10.1016/j.ymeth.2008.09.025

42. Rochefort NL, Garaschuk O, Milos RI, Narushima M, Marandi N, Pichler B, Kovalchuk Y, Konnerth A (2009) Sparsification of neuronal activity in the visual cortex at eye-opening. Proc Natl Acad Sci U S A 106 (35):15049–15054. doi:10.1073/pnas.0907660106

43. Tian L, Hires SA, Looger LL (2012) Imaging neuronal activity with genetically encoded calcium indicators. Cold Spring Harb Protoc 2012(6):647–656. doi:10.1101/pdb.top069609

44. Nakai J, Ohkura M, Imoto K (2001) A high signal-to-noise Ca(2+) probe composed of a single green fluorescent protein. Nat Biotechnol 19(2):137–141. doi:10.1038/84397

45. Tian L, Hires SA, Mao T, Huber D, Chiappe ME, Chalasani SH, Petreanu L, Akerboom J, McKinney SA, Schreiter ER, Bargmann CI, Jayaraman V, Svoboda K, Looger LL (2009) Imaging neural activity in worms, flies and mice with improved GCaMP calcium indicators. Nat Methods 6(12):875–881. doi:10.1038/nmeth.1398

46. Zhao Y, Araki S, Wu J, Teramoto T, Chang YF, Nakano M, Abdelfattah AS, Fujiwara M, Ishihara T, Nagai T, Campbell RE (2011) An expanded palette of genetically encoded Ca(2) (+) indicators. Science 333(6051):1888–1891. doi:10.1126/science.1208592

47. Truong K, Sawano A, Mizuno H, Hama H, Tong KI, Mal TK, Miyawaki A, Ikura M (2001) FRET-based in vivo Ca2+ imaging by a new calmodulin-GFP fusion molecule. Nat Struct Biol 8(12):1069–1073. doi:10.1038/nsb728

48. Shimomura O, Johnson FH, Saiga Y (1962) Extraction, purification and properties of aequorin, a bioluminescent protein from the luminous hydromedusan, Aequorea. J Cell Comp Physiol 59:223–239

49. Xu X, Soutto M, Xie Q, Servick S, Subramanian C, von Arnim AG, Johnson CH (2007) Imaging protein interactions with bioluminescence resonance energy transfer (BRET) in plant and mammalian cells and tissues. Proc Natl Acad Sci U S A 104(24):10264–10269. doi:10.1073/pnas.0701987104

50. Takai A, Nakano M, Saito K, Haruno R, Watanabe TM, Ohyanagi T, Jin T, Okada Y, Nagai T (2015) Expanded palette of Nano-lanterns for real-time multicolor luminescence imaging. Proc Natl Acad Sci U S A 112 (14):4352–4356. doi:10.1073/pnas.1418468112

51. Garaschuk O, Konnerth A (2010, 2010) In vivo two-photon calcium imaging using multicell bolus loading. Cold Spring Harb Protoc 10. doi:10.1101/pdb.prot5482. pdb prot5482

52. Shevtsova Z, Malik JM, Michel U, Bahr M, Kugler S (2005) Promoters and serotypes: targeting of adeno-associated virus vectors for gene transfer in the rat central nervous system in vitro and in vivo. Exp Physiol 90(1):53–59. doi:10.1113/expphysiol.2004.028159

53. Zhang F, Gradinaru V, Adamantidis AR, Durand R, Airan RD, de Lecea L, Deisseroth K (2010) Optogenetic interrogation of neural circuits: technology for probing mammalian brain structures. Nat Protoc 5(3):439–456. doi:10.1038/nprot.2009.226. [doi] nprot.2009.226 [pii]

54. Zhang F, Vierock J, Yizhar O, Fenno LE, Tsunoda S, Kianianmomeni A, Prigge M, Berndt A, Cushman J, Polle J, Magnuson J, Hegemann P, Deisseroth K (2011) The microbial opsin family of optogenetic tools. Cell 147 (7):1446–1457. doi:10.1016/j.cell.2011.12.004

55. Venkatachalam V, Cohen AE (2014) Imaging GFP-based reporters in neurons with multiwavelength optogenetic control. Biophys J 107(7):1554–1563. doi:10.1016/j.bpj.2014.08.020

56. Helmchen F, Denk W (2005) Deep tissue two-photon microscopy. Nat Methods 2 (12):932–940. doi:10.1038/nmeth818

57. Rein ML, Deussing JM (2012) The optogenetic (r)evolution. Mol Genet Genomics 287

(2):95–109.   doi:10.1007/s00438-011-0663-7

58. Steriade M (2006) Grouping of brain rhythms in corticothalamic systems. Neuroscience 137 (4):1087–1106. doi:10.1016/j.neuroscience. 2005.10.029. [doi] S0306-4522(05)01153-X [pii]

59. Roome CJ, Kuhn B (2014) Chronic cranial window with access port for repeated cellular manipulations, drug application, and electrophysiology. Front Cell Neurosci 8:379. doi:10.3389/fncel.2014.00379

60. Yousef T, Bonhoeffer T, Kim DS, Eysel UT, Toth E, Kisvarday ZF (1999) Orientation topography of layer 4 lateral networks revealed by optical imaging in cat visual cortex (area 18). Eur J Neurosci 11(12):4291–4308

61. Arieli A, Grinvald A, Slovin H (2002) Dural substitute for long-term imaging of cortical activity in behaving monkeys and its clinical implications. J Neurosci Methods 114 (2):119–133

62. Goldey GJ, Roumis DK, Glickfeld LL, Kerlin AM, Reid RC, Bonin V, Schafer DP, Andermann ML (2014) Removable cranial windows for long-term imaging in awake mice. Nat Protoc 9(11):2515–2538. doi:10.1038/nprot. 2014.165

Chapter 10

Two-Photon Optogenetics by Computer-Generated Holography

Eirini Papagiakoumou, Emiliano Ronzitti, I-Wen Chen, Marta Gajowa, Alexis Picot, and Valentina Emiliani

Abstract

Light patterning through spatial light modulators, whether they modulate amplitude or phase, is gaining an important place within optical methods used in neuroscience, especially for manipulating neuronal activity with optogenetics. The ability to selectively direct light in specific neurons expressing an optogenetic actuator, rather than in a large neuronal population within the microscope field of view, is now becoming attractive for studies that require high spatiotemporal precision for perturbing neuronal activity in a microcircuit. Computer-generated holography is a phase-modulation light patterning method providing significant advantages in terms of spatial and temporal resolution of photostimulation. It provides flexible three-dimensional light illumination schemes, easily reconfigurable, able to address a significant excitation field simultaneously, and applicable to both visible or infrared light excitation. Its implementation complexity depends on the level of accuracy that a certain application demands: Computer-generated holography can stand alone or be combined with temporal focusing in two-photon excitation schemes, producing depth-resolved excitation patterns robust to scattering. In this chapter, we present an overview of computer-generated holography properties regarding spatiotemporal resolution and penetration depth, and particularly focusing on its applications in optogenetics.

Key words Light patterning, Phase modulation, Spatial light modulator, Temporal focusing, Optogenetics, Opsin kinetics

1 Introduction

The coordinated activation of neuronal microcircuits is proposed to regulate brain functioning in health and disease. A common approach to investigate the mechanisms that reduce network complexity is to outline microcircuits and infer their functional role by selectively modulating them. Combined with suitable illumination approaches, optogenetics offers today the possibility to achieve such selective control with its ever-growing toolbox of reporters and actuators.

Albrecht Stroh (ed.), *Optogenetics: A Roadmap*, Neuromethods, vol. 133,
DOI 10.1007/978-1-4939-7417-7_10, © Springer Science+Business Media LLC 2018

Wide-field single-photon (1P) illumination was the first method employed to activate optogenetic actuators [1–8], and continues to be widely used for neural circuit dissection [9, 10] (*see* also Chap. 9). Using genetic tools, including viruses, Cre-dependent systems, and transgenic lines to target optogenetic actuators to neurons of interest (*see* Chaps. 1 and 2), investigators have used wide-field illumination to dissect correlation and causal interactions in neuronal subpopulations both in vitro [11–15] and in vivo [13, 16–18]. With this approach, population specificity is achieved through genetic targeting, and temporal resolution and precision are only limited by the channels' temporal kinetics and cell properties (e.g., opsin expression level and membrane potential). Suitable combinations of opsins have also enabled independent optical excitation of distinct cell populations [19]. The primary drawback of wide-field illumination is that all opsin-expressing neurons are stimulated simultaneously, and thus wide-field schemes lack the temporal flexibility and spatial precision necessary to mimic the spatiotemporal distribution of naturally occurring microcircuits activity.

Replacing 1P visible light excitation with two-photon (2P) near infrared light illumination enables improved axial resolution and penetration depth [20]. However, the small single-channel conductance of actuators such as ChR2 (40–80 fS; [21]), in combination with the low number of channels excitable within a femtoliter-two-photon focal volume, makes it difficult to generate photocurrents strong enough to bring a neuron to firing threshold. This challenge has prompted the development of 2P-stimulation approaches that increase the excitation volume.

2P-stimulation approaches for optogenetics can be grouped in two main categories: scanning and parallel excitation techniques. 2P laser scanning methods use galvanometric mirrors to quickly scan a laser beam across several positions covering a single or multiple cells [22–25] (*see* also Chap. 9). Parallel approaches enable to simultaneously cover the surface of a single cell using a low-numerical aperture (NA) beam [26], or multiple cells using computer-generated holography [27–31] and generalized phase contrast [32]. In this chapter, we will specifically focus on the description of computer-generated holography and its application to optogenetic neuronal control. A broader overview on the different approaches for 2P optogenetics can be found in Refs. [33–36] and Chap. 9.

2   Computer-Generated Holography

Originally proposed for generating multiple-trap optical tweezers [37], the experimental scheme for computer-generated holography (CGH) (Fig. 1a) consists in computing with a Fourier transform-

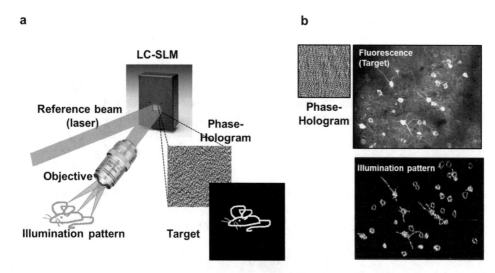

Fig. 1 Computer-generated holography. (**a**) In CGH an arbitrary light distribution (*target*) is used as the input source to a Fourier transform-based iterative algorithm to calculate the interference pattern or phase hologram, that after interfering with the reference beam (*laser*) at the diffraction plane would reproduce the target at the imaging plane. The calculated phase hologram is addressed to a liquid-crystal spatial light modulator (LC-SLM). The reference beam plane, after diffraction through the LC-SLM, will generate at the objective focal plane a light distribution (*illumination pattern*) reproducing the original target shape. (**b**) For precise light stimulation, the whole fluorescence image or a defined region of interest can be used to select the excitation targets that will be sent to the iterative algorithm for producing the corresponding phase hologram

based iterative algorithm [38] the interference pattern or phase-hologram that back-propagating light from a defined target (input image) will form with a reference beam, on a defined "diffractive" plane. The computer-generated phase-hologram is converted into a grey-scale image and then addressed to a liquid-crystal matrix spatial light modulator (LC-SLM), placed at the diffractive plane. In this way, each pixel of the phase-hologram controls, proportionally to the analogous grey-scale-level, the voltage applied across the corresponding pixel of the LC matrix such as the refractive index and thus the phase of each pixel can be precisely modulated. As a result, the calculated phase-hologram is converted into a pixelated refractive screen and illumination of the screen with the laser beam (or reference beam) will generate at the objective focal plane a light pattern reproducing the desired template. This template can be any kind of light distribution in two (2D) or three dimensions (3D), ranging from diffraction-limited spots or spots of bigger sizes (bigger surface) to arbitrary extended light patterns.

Precise manipulation of neuronal activity via holographic light patterns requires an accurate control of the spatial colocalization between the generated light pattern and the target. To do so, a few years ago we proposed to generate the template for the calculation of phase-holograms on the base of the fluorescence image [39].

Briefly, a fluorescence image of the preparation is recorded and used to draw the excitation pattern. In this way it is possible to generate a holographic laser pattern reproducing the fluorescence image or a user defined region of interest (Fig. 1b) [33, 40].

In CGH, the pixel size and number of pixels of the LC-SLM define the lateral and axial field of excitation (FOE). The maximum lateral FOE, FOE_{xy}, is given by [41–43]:

$$FOE_{xy} = \Delta X \times \Delta Y = \left(2 \cdot \frac{\lambda f_{eq}}{2dn}\right) \times \left(2 \cdot \frac{\lambda f_{eq}}{2dn}\right), \quad (1)$$

where λ is the excitation wavelength, d the LC-SLM pixel size, n is the medium refractive index, and f_{eq} is the equivalent focal lens including all the lenses located between the LC-SLM and the sample plane (*see* **Note 1**).

Within this region, the diffraction efficiency, $\delta(x, y)$, defined as the intensity ratio of the incoming to the diffracted beam, depends on the lateral spot position coordinates, x, y:

$$\delta(x, y) = b\left(\sin c^2\left(\frac{\pi d}{\lambda f_{eq}}x\right)\sin c^2\left(\frac{\pi d}{\lambda f_{eq}}y\right)\right), \quad (2)$$

with b being a proportionality factor considering the frontal window LC-SLM reflectivity. Consequently, $\delta(x, y)$ reaches its maximum value at the center of the FOE_{xy} and the minimum one at the borders of the FOE_{xy}. Existing LC-SLM devices permit nowadays to reach diffraction efficiency values of ~95% at the center and ~38% at the border, which are the values close to the limit imposed by theory. The remaining light is distributed among the higher diffraction orders and an un-diffracted component, so-called zero-order, resulting in a tightly focused spot at the center of the FOE. Depending on the applied phase profile, the intensity of the zero-order spot can reach 25% of the input light. This value can be reduced down to 2–5%, regardless of the projected hologram, by performing ad hoc pre-compensation of the LC-SLM phase pixel values [44]. The focused zero-order spot can be removed from the FOE by adding a block or diaphragm at a plane conjugated to the sample plane [45, 46]. Nevertheless, this limits the accessible FOE. Alternatively, the intensity of the zero-order component can be strongly reduced by using a cylindrical lens placed in front of the LC-SLM, which stretches the zero order into a line [47]. A phase hologram compensating the cylindrical lens effect is then addressed onto the LC-SLM in addition to the original phase hologram generating the target spot, so that the holographic-pattern-shape is restored (Fig. 2) (*see* **Note 2**). Importantly, the use of cylindrical lenses enables suppressing excitation from the zero-order component in the FOE_{xy} without using intermediate blocks, thus having access to the entire FOE.

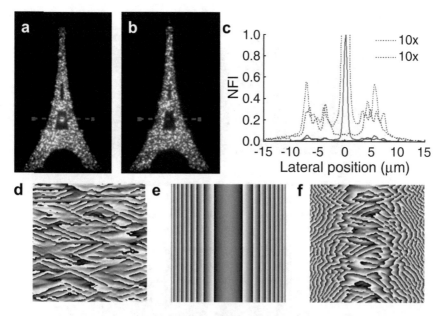

Fig. 2 Zero order removal by using cylindrical lenses. Comparison between a computer-generated holographic image of the Eiffel Tower (**a**) without and (**b**) with a single 1-m cylindrical lens, aberrating the zero order in the optical path. (**c**) 2P normalized fluorescence intensity (NFI) profiles along the lines drawn in (**a**), *red*, and (**b**), *blue*. *Dotted lines* represent the signal of solid lines multiplied by 10 for better view. (**d**) Phase mask reproducing the image of the Eiffel Tower at the focal plane of the objective, calculated with a Gerchberg-Saxton-type algorithm. (**e**) Conjugated cylindrical Fresnel lens hologram added to the one of (**d**) for aberration compensation. (**f**) Final corrected phase mask addressed to the SLM. Reproduced from Ref. [47]

Intensity inhomogeneities due to diffraction efficiency are a limiting factor for applications requiring lateral displacing of a single spot or multiple spots within the FOE. Therefore, we have proposed approaches that compensate those inhomogeneities for keeping the spot intensity constant independently on the lateral position. In the case of single spot generation, a homogenization of light distribution can be achieved by projecting one or multiple spots outside the FOE_{xy} and tuning their brightness or size to compensate the intensity loss due to the diffraction efficiency curve. Thus, a constant intensity value in the excitation spot is maintained for each position of the FOE_{xy}. The extra spots can be blocked by adding an external diaphragm placed at an intermediate imaging plane of the optical system, conjugated to the sample plane [43]. For multispot excitation, one can use graded input images in order to generate brighter spots into regions in the border of the FOE, where the diffraction efficiency is lower, and dimmer spots into the central part of the FOE, where diffraction efficiency is higher [43, 48, 49] (Fig. 3). Graded input patterns can also be used to compensate for sample inhomogeneity. For example they can be applied to equalize photocurrents from cells with different expression levels [48].

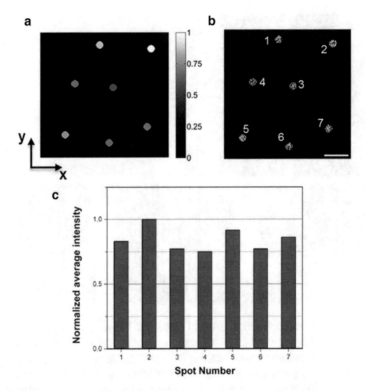

Fig. 3 Diffraction efficiency compensation. (**a**) A graded input image to the iterative algorithm is used to equalize the light distribution across multiple spots (10 μm in diameter) generated on regions of different diffraction efficiency. (**b**) The corresponding light distribution is visualized by illuminating a uniform rhodamine fluorescent layer and by collecting the fluorescence on a CCD camera. Scale bar 30 μm. (**c**) Histogram of the normalized average intensity on the spots shown in (**b**). Intensity is averaged over the area of the spots. Intensity variations among the spots are of the order of 9%

CGH pattern generation also suffers from "speckle", i.e., undesired intensity variations of high spatial frequency, within the same spot. This is an intrinsic limitation of CGH and it is due to phase discontinuities at the sample plane inherent to the Gerchberg-Saxton algorithm [38], the most commonly used Fourier transform-based iterative algorithm. Speckle fluctuations reach 20% in 1P and 50% in 2P CGH implementations. Different approaches have been proposed to reduce or eliminate speckles, each with its advantages and limitations. Temporally averaging of speckle-patterns can be achieved by mechanical rotating a diffuser [50] or by generating multiple shifted versions of a single hologram [51]. Also smoother intensity profiles can be created by ad hoc algorithms that remove phase vortices in the holographic phase mask [52]. Alternatively, the interferometric method, generalized phase contrast (GPC) (*see* **Note 3**) [53], generates speckle-free 2D extended shapes with adequate precision, e.g., to precisely reproduce the shape of a thin dendritic process [32]. Recently, researchers showed that GPC can be also extended to 3D by combining it with CGH, an approach called Holo-GPC [54]. In that case, a holographic phase mask is used to multiplex a GPC pattern in different lateral or axial positions.

2.1 Spatial Resolution

In general, the lateral spatial resolution of an optical microscope is defined on the basis of the maximum spatial frequency that can be transferred through the focusing objective. That is related to the maximum angle of convergence of the illumination rays, i.e., to the objective angular aperture. Consequently, in CGH the smallest obtainable illumination pattern is a diffraction-limited Gaussian spot whose full width at half maximum (FWHM) is equal to $\Delta x \approx \lambda/NA_{eff}$, where NA_{eff} is the effective numerical aperture (with $NA_{eff} < NA$ for an under-filled pupil). Conjointly with the concept of resolution, it is useful to introduce the notion of spatial localization accuracy, i.e. the precision to target a certain position at the sample plane [55]. This is ultimately related to the minimum displacement, $\Delta\delta_{min}$, of the illumination spot that is possible to achieve by spatially modulating the phase of the incoming light beam. In particular, an illumination spot can be laterally shifted by a certain step $\Delta\delta$ by applying at the objective back aperture a prism-like phase modulation of slope $\alpha \approx \Delta\delta/f_{obj}$, where f_{obj} is the objective focal length. The spatial localization accuracy therefore depends on the SLM capability to approximate a prism-like phase shift [55]. This ultimately is limited by the number of pixels, N, and grey levels, g, of the SLM. More precisely the theoretical upper limit for the minimum step $\Delta\delta_{min}$ is inversely proportional to $N \cdot g$ [55, 56].

In CGH, axial resolution scales linearly with the lateral spot size and inversely with the objective numerical aperture (NA). Precisely, defining as s the holographic spot radius and $\sigma \approx \lambda/(NA\sqrt{8 \cdot \ln(2)})$ the speckle size, the 1P and 2P axial resolution is twice the axial distance, Δz, at which the 2P intensity drops at 50% (FWHM), where Δz is given by:

$$\Delta z(s)_{1P} = \frac{2\sqrt{3}z_R\sigma}{\sqrt{s^2 + \sigma^2}}; \Delta z(s)_{2P} = \frac{2z_R\sigma}{\sqrt{s^2 + \sigma^2}}, \qquad (3)$$

with $z_R = \pi \cdot s^2/2\lambda$. This means that a spot size of, e.g., 10 μm in diameter will correspond to an axial resolution of 14 μm using a $NA = 0.9$ objective and 2P illumination at 900 nm [57]. The corresponding illumination volume has roughly the size of a cell soma (yellow circle, Fig. 4a), thus enabling in principle optical photostimulation with single-cell precision (Fig. 4a).

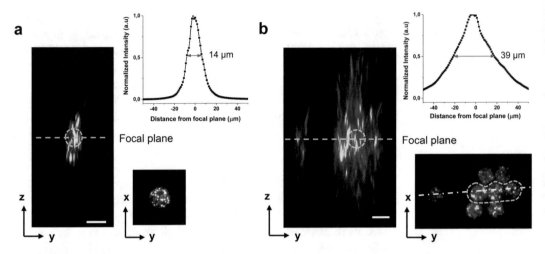

Fig. 4 Axial propagation of holographic beams. Experimental *y-z* and *x-y* intensity cross-sections for holographic beams generated to produce at the objective focal plane (**a**) a circular spot, or (**b**) multiple spots of 10-μm diameter. *y-z* cross-section in (**b**) is shown along the *white dash-dotted line*. The *yellow circle* in both panels approximates the size of a cell soma. Integrated intensity profiles of *y-z* cross-sections around the circular spot (**a**) and in an area covering three spots (*dashed yellow*) of the multispot light configuration (**b**) are shown on the top of the panels. For comparison, the full width at half maximum of the axial integrated intensity profile of the single 10-μm spot is around 14 μm. Scale bars: 10 μm

Optical stimulation with near cellular resolution was indeed achieved in freely moving mice using 1P holographic stimulation [58]. Briefly, holographic light patterning coupled to a fiber bundle with a micro-objective at the end, was used to photostimulate and monitor functional responses in cerebellar molecular layer interneurons coexpressing a calcium indicator (GCaMP5-G) and an opsin (ChR2-tdTomato) in freely behaving mice (Fig. 5). These experiments proved optical photostimulation with near cellular resolution using sparse staining and sparse distribution of excitation spots. However, a similar approach would not reach the same precision if applied for multisite photostimulation of a densely labeled neuronal population as in this case the axial resolution would quickly deteriorate both using 1P and 2P excitation (Fig. 4b).

A few years ago, we demonstrated that micrometer-size optical sectioning independent of the lateral spot dimension [50] can be achieved by combining CGH and GPC with temporal focusing. Briefly, the technique of temporal focusing (TF), originally demonstrated to perform wide-field 2P microscopy [59, 60], uses a dispersive grating to diffract the different frequencies comprising the ultra-short excitation pulse toward different directions. The various frequencies thus propagate toward the objective focal plane at different angles, such that the pulse is temporally smeared above and below the focal plane, which remains the only region irradiated at peak powers efficient for 2P excitation.

The combination of TF with CGH is achieved by adding to the conventional TF optical path a LC-SLM and a focusing lens (L1) so

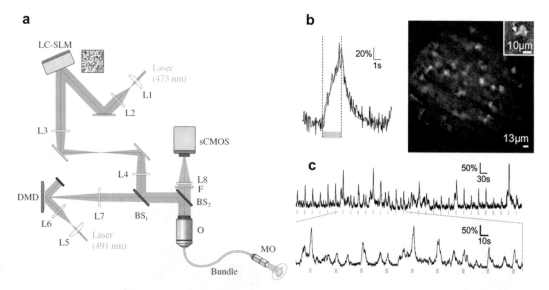

Fig. 5 Holographic photostimulation and functional imaging in freely behaving mice. (**a**) Schematic of the holographic fiberscope composed of two illumination paths: one for photoactivation with CGH including a LC-SLM, and a second for fluorescence imaging including a digital micromirror device (DMD). Backward fluorescence was detected on a scientific complementary metal oxide semiconductor (sCMOS) camera. Both paths were coupled to the sample using a fiber bundle attached to a micro-objective (MO). *L* Lens, *BS* beam splitter, *O* microscope objective. (**b**) *Left*, Calcium signal triggered by photoactivation (*blue line*; $p = 50$ mW/mm^2) with a 5-μm diameter holographic spot placed on the soma of a ChR2-expressing cell recorded in a freely behaving mouse coexpressing GCaMP5-G and ChR2 in cerebellar molecular layer interneurons (MLIs). *Right*, Structure illumination image recorded in a freely behaving mouse and showing MLI somata and a portion of a dendrite (insert). Scale bar: 10 μm. (**c**) *Top*, The same photoactivation protocol as in (**a**) was repeated every 30 s for 15 min (photostimulation power, 50 mW/mm^2; imaging power, 0.28 mW/mm^2). *Bottom*, Expansion of the top trace showing that spontaneous activity frequently occurs between evoked transients. Adapted from Ref. [58]

that the TF grating lies at the focal plane of L1 and is illuminated by the holographic pattern. A second telescope, made by a second lens (L2) and the objective, conjugates the TF plane with the sample plane, thus enabling the generation of spatiotemporally focused patterns (Fig. 6a). Notably, TF enables decoupling lateral and axial resolution so that the same axial resolution is achieved independently on the lateral extension of the excitation spot (Fig. 6b, c) [50]. TF combined with low-*NA* Gaussian beams, GPC and CGH has enabled efficient 2P optogenetic excitation with micrometer axial resolution and millisecond temporal resolution both in vitro and in vivo [23, 26, 27, 32].

Although wavefront shaping and TF enable precise sculpting of the excitation volume, the ultimate spatial precision achievable for 2P optogenetics depends also on the opsin distribution within the expressing neurons: opsins are efficiently trafficked to the membrane of cell soma, as well as to dendrites and axons. Consequently, illumination with a theoretically micrometer-sized focal volume

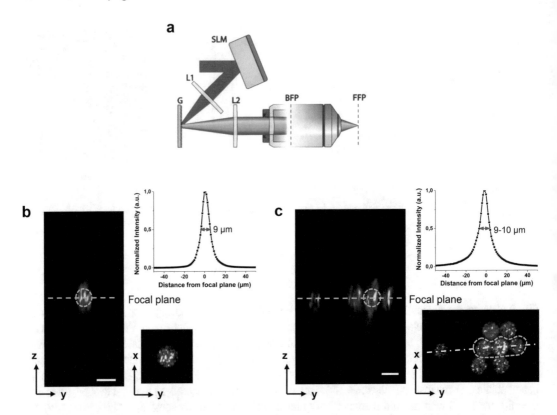

Fig. 6 Computer-generated holography and temporal focusing. (**a**) Schematic representation of an experimental setup combining CGH with TF. *G* grating, *L1 and L2* lenses, *BFP* Back focal plane, *FFP* Front focal plane. (**b–c**) Experimental *y-z* and *x-y* intensity cross-sections for temporally focused holographic beams generated to produce at the objective focal plane a circular spot (**b**), or multiple spots of 10-μm diameter (**c**). *y-z* cross-section in (**c**) is shown along the *white dash-dotted line. Yellow circles* approximate the size of a cell soma. Integrated intensity profiles of *y-z* cross-sections around the circular spot (**b**) and in an area covering three spots (*dashed yellow*) of the multispot light configuration (**c**) are shown on the top of the panels. For comparison, the full width at half maximum of the axial integrated intensity profile of the single 10-μm spot is around 9 μm. The axial confinement thanks to temporal focusing is well preserved, even when multiple spots are projected close together. Scale bars: 10 μm

could depolarize all cells whose processes (dendrites and axons) cross the target excitation volume, even if their somata are located micrometers away from the illumination spot (Fig. 7). This activation cross talk needs to be carefully considered, for example when performing connectivity experiments, as it could prevent distinguishing if a postsynaptic response, recorded while photostimulating a presynaptic cell, originates from a true connection between the two cells rather from direct stimulation of postsynaptic dendrites or axons crossing the photostimulation volume. Reaching a true cellular resolution for exciting densely labeled samples requires combining optical focusing with molecular strategies enabling confined opsin expression in restricted cell areas (soma or axonal hillock) [49, 61].

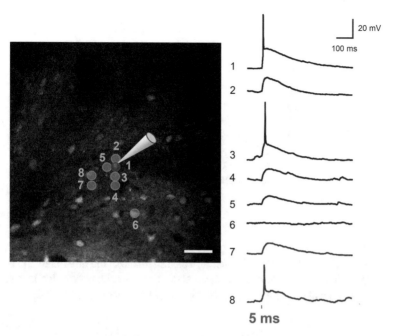

Fig. 7 Two-photon holographic photostimulation of ReaChR-expressing cells in vivo. The opsin ReaChR was expressed in neurons at layer 2/3 of mouse visual cortex via injection of the viral vector rAAV1-Ef1α-ReaChR-P2A-tdTomato. Positive neurons expressing ReaChR (fluorescent cells in the 2P fluorescence image of the left panel) in isoflurane-anesthetized mouse were photostimulated with a 12-μm diameter excitation spot and 5-ms illumination duration at 0.15 mW/μm^2 and $\lambda = 1030$ nm (*red shaded area in the right panel*). Action potentials were induced by holographic excitation of one positive cell soma (spot 1), whose membrane potential was measured using 2P-guided whole-cell recording (trace 1). Sub-threshold or supra-threshold activation was induced in the patched cell (trace 2–8) upon holographic excitation of spots targeting around its soma (respective radial distance from spot 1 for spots 2–8: 12 μm, 12 μm, 24 μm, 12 μm, 67.6 μm, 40.4 μm, 35 μm), caused by exciting opsin-channels distributed into axon, proximal, and distal dendrites of the patched neuron. Scale bar: 40 μm

2.2 Temporal Resolution

Parallel light illumination enables simultaneous excitation of all selected targets. The temporal resolution is therefore only limited by the illumination time needed to evoke, e.g., an action potential (AP), a detectable Ca^{2+} response or a defined behavioral change [33, 34]. This ultimately depends on opsin conductance, virus promoter, serotype, titer, kinetics parameters and excitation power (*see* Chaps. 1–3). In the following we will specifically focus on reviewing how the opsin kinetics parameters determine temporal resolution, temporal precision (temporal jitter) and AP spiking rate.

Light illumination of an opsin-expressing cell with a hundred-millisecond illumination pulse generates a characteristic photocurrent trace (Fig. 8a) where one can distinguish an activation, inactivation and deactivation part, characterized by a temporal decay, τ_{on}, τ_{inact} and τ_{off} respectively. The overall kinetics of the current, as well as the peak-to-plateau ratio of current can be qualitatively well

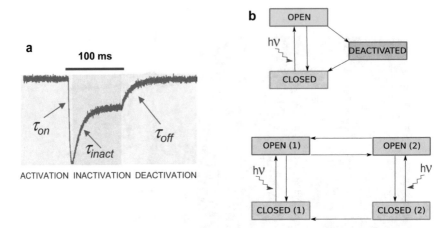

Fig. 8 ChR2 photocurrent and photocycle. (**a**) Typical photocurrent trace of a CHO cell expressing ChR2 under visible light illumination (1P excitation) for 100 ms. (**b**) *Top*, schematic of the three-state model with a closed/ground state, an open state and a closed/desensitized state. *Bottom*, schematic of the four-state model with two closed, and two open states. For a detailed description of the models see Ref. [66]

Table 1
Kinetics parameters for different opsins

	Chronos	CoChR	ReaChR
τ_{on} (ms)	0.73	2.4	8
τ_{inact} (ms)	9.3	200	443
τ_{off} (ms)	4.2	31	94

Chronos, CoChR, and ReaChR were expressed in Chinese Hamster Ovary (CHO) cells following the protocol described in Refs. [27, 28]. Electrophysiological data were recorded 24—60 h after transfection using 2P ($\lambda = 950$ nm, pulse duration 4 s for ReaChR and CoChR and 1 s for Chronos; three trials at 1-min intervals holographic illumination at variable power (from 0.05 to 1.1 mW/μm^2). The current curves at saturation (defined as the power at which the peak current reaches 90% of its maximum) where fitted using a mono-exponential decay for the three transitions: activation, inactivation, and deactivation. The corresponding decay times, τ_{on} (ms), τ_{off} (ms), τ_{inact} (ms), are reported in the table and correspond to power close to saturation (0.86 mW/μm^2, 0.54 mW/μm^2, and 0.28 mW/μm^2 for Chronos, CoChR, and ReaChR, respectively (Emiliani Laboratory, unpublished data).

reproduced by using a 3- or 4-state model [62–66] (Fig. 8b), the latter being more accurate to reproduce the bi-exponential decay of the light-off current and the photocurrent voltage dependence [63]. A qualitative value for the characteristic temporal decay, τ_{on}, τ_{inact} and τ_{off} can be directly extracted by assuming a mono-exponential process for the three transitions. In Table 1 we report the values of τ_{on} (at saturation), τ_{inact} and τ_{off} measured under 2P holographic illumination of CHO (Chinese Hamster Ovary) cells

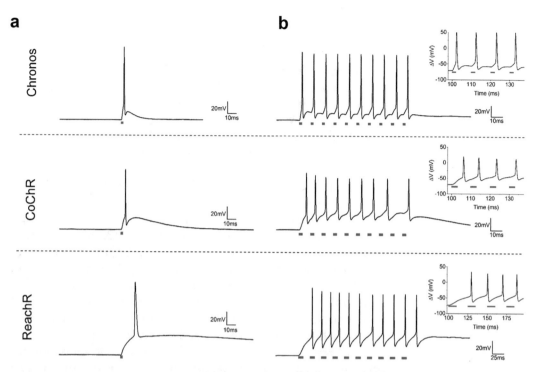

Fig. 9 Temporal resolution and spiking rate. (**a**) Light-elicited single-spike by 2P holographic illumination (spot diameter: 15 μm; $\lambda = 1030$ nm) with short light illumination pulses ($t = 2$ ms) of Chronos ($P = 0.09$ mW/μm^2), CoChR ($P = 0.1$ mW/μm^2; Emiliani Laboratory, unpublished data) and ReaChR ($P = 0.07$ mW/μm^2) expressing interneurons from layer 2/3 of the mouse visual cortex. (**b**) Light-driven firing fidelity in opsin-positive interneurons by illuminating with a train of 10 light pulses, Chronos ($t = 2$ ms; $f = 100$ Hz; $P = 0.12$ mW/μm^2), CoChR ($t = 3$ ms; $f = 100$ Hz; $P = 0.1$ mW/μm^2) and ReaChR ($t = 10$ ms; $f = 40$ Hz; $P = 0.04$ mW/μm^2). The first 4 pulses are zoomed in, in the insets. Chronos data are adapted from Ref. [29] and ReaChR data are adapted from Ref. [28]

expressing a fast (Chronos) [19], an intermediate (CoChR) [19] and a slow (ReaChR) [67] opsin.

In practice, the efficient current integration obtained under parallel photostimulation enables using photostimulation pulses much shorter than the channel rise time therefore enabling in vitro AP generation with millisecond temporal resolution and sub-millisecond temporal jitter, independently on τ_{on} [28, 29, 49] (Fig. 9a). Conversely, the value of τ_{off} has a key role in limiting the achievable spiking rate, as shown in Fig. 9b, where the in vitro spike generation under 2P holographic illumination of interneurons (layer 2/3 of visual cortex) expressing different opsins (Chronos, ReaChR, CoChR) is reported. Fast opsins, such as Chronos, enabled generation of light-evoked AP trains up to 100 Hz spiking rate [29], while for ReaChR, with a ~50 times slower τ_{off}, the light-evoked firing rate was limited at ~35 Hz (~15 Hz for pyramidal cells) [28]. Photostimulation of interneurons expressing CoChR, which has an intermediate value of τ_{off} (~30 ms), could still generate light-evoked trains at 100 Hz but the temporal precision and fidelity were progressively lost across the train.

Interestingly the efficient current integration achievable with parallel holographic illumination enables reliable AP generation with millisecond temporal resolution and sub-millisecond precision using, at depths of ~100 μm, excitation densities less than 1 mW/μm^2 or 100 μW/μm^2, with a conventional mode-locked high-repetition rate (80 MHz) laser oscillator or a low-repetition rate (500 kHz) amplifier, respectively [28, 29, 49].

2.3 Penetration Depth

As previously described, temporal focusing coupled with CGH or GPC enables micrometer-range axial confinement. A further advantage of this approach is that it also enables robust light propagation through scattering media [27, 43, 68]. Scattering deviates the original photon trajectory, thus deforming the excitation spot shape at the focal plane. For light illumination with diffraction-limited spots, this mainly translates into occurrence of aberrations and loss of axial and lateral resolution. Moreover, scattered photons do not contribute to signal arising from the focal volume, which translates in loss of light intensity [69].

For large illumination areas, the presence of scattering also generates speckle in the excitation spot due to the random interference between ballistic and scattered photons (Fig. 10). A few years ago, we demonstrated that TF combined with both CGH [27] and GPC [68] enables to reduce this effect as scattered photons have less probability to interfere with the ballistic ones (Fig. 11). Light propagation of patterns generated using CGH [27] and GPC [68] through cortical brain slices or zebrafish larvae [43] have revealed robust conservation of lateral shape and axial resolution up to depths twice the scattering length (Fig. 10), enabling in-depth optogenetic stimulation [68] (*see* also Chap. 13).

2.4 Multiplane Light Activation

One important feature of CGH is the ability to generate spots in different axial planes, a feature used to generate multi-diffraction-limited traps for optical tweezers [37, 70], 3D glutamate uncaging [42, 71–73], and, combined with spiral scanning, to achieve multi-plane 2P optogenetic stimulation [74].

Adding lens-phase modulations to 2D-phase holograms also enables remote axial displacement and 3D positioning of laterally shaped targets [75]. This configuration, combined with optogenetics, enables simultaneous control of neurons and substructures in different planes, as well as provides a flexible means to stimulate locations lying above or below the imaging plane [30]. Conventional 3D-CGH optical designs are not, however, compatible with TF because axially shifted excitation planes cannot be simultaneously imaged onto the TF grating. To overcome this limitation, we have recently developed a new optical system including two LC-SLMs that enable simultaneous TF at multiple planes (Fig. 12a)

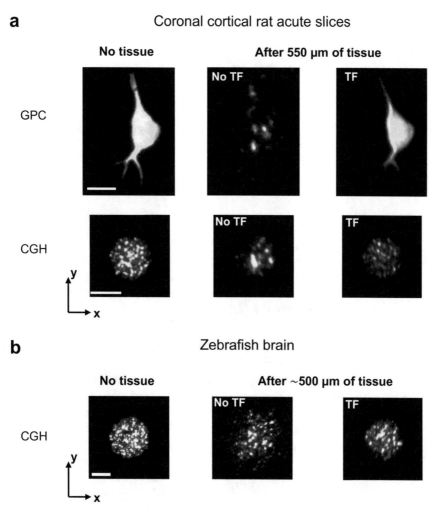

a   Coronal cortical rat acute slices

Fig. 10 Temporal focusing and penetration depth. (**a**) 2P fluorescence *x–y* cross-sections of GPC-generated excitation patterns mimicking a neuron with small processes (*top*) and a 15-μm-diameter CGH spot (*bottom*) after propagation through 550 μm of acute coronal cortical brain slices without (*middle panel*) and with (*right panel*) temporal focusing ($\lambda = 950$ nm). Non-temporally focused beams are transformed to speckled patterns after traveling through the tissue. (**b**) 2P fluorescence *x–y* cross-sections of a 20-μm-diameter CGH spot after propagation through the whole brain of a zebrafish larva (~500 μm thick) without (*middle panel*) and with (*right panel*) temporal focusing ($\lambda = 920$ nm). Left panels in all cases are obtained from unscattered beams. Scale bars: 10 μm. Adapted from Refs [27, 43, 68]

[43]. The system achieves remote axial displacement of temporally focused holographic beams, as well as multiple temporally focused planes, by shaping the incoming wave-front in two steps, using two LC-SLMs. The first LC-SLM (SLM1 in Fig. 12a), laterally shapes the target light distribution that is focused onto the TF grating, which disperses the spectral components of the illumination pattern onto the second LC-SLM (SLM2 in Fig. 12a). SLM2, conjugated to the objective back focal plane, is addressed with a single or

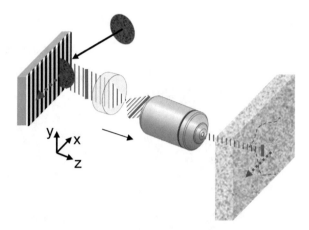

Fig. 11 Temporal focusing in the temporal domain. The propagation of a large beam (a holographic spot in this case) diffracted by the grating produces an ultrafast line scanning of the sample. Scattering events off the scanning line at a single moment in time cannot interfere with the ballistic photons in the line. Adapted from Ref. [27]

multiple Fresnel-lens phase-profiles to control the target's axial position in the sample volume. In this way, the spatial and temporal focal planes coincide at the grating, and are jointly translated by SLM2 across the sample axial extent (Fig. 12b).

Similarly, to what has been discussed for the generation of 2D multiple spots, the pixel size, d, and pixels number of the LC-SLM limit the axial FOE, FOEz, to:

$$\mathrm{FOE}_z = \Delta Z = \left(\frac{\lambda f_{eq} \sqrt{n^2 - NA^2}}{2 d NA} \right), \qquad (4)$$

where NA is the objective numerical aperture, and n is the refractive index of the immersion medium, giving rise to an axial-position-dependent diffraction efficiency [43] (Fig. 12c). Similarly to the 2D case, homogeneous light distribution can be obtained by using graded input images (Fig. 12d).

3   Notes

1. When CGH is used for microscopy, in principle it is sufficient to address the proper phase modulation at the back focal plane of the objective for having the desired illumination pattern at the sample plane. However, practically it is difficult to place a spatial light modulator there. The common practice is to image the back focal plane of the objective at another conjugated plane where the LC-SLM is placed, through a telescope of lenses. Typically, this includes a first lens (L1 of focal length

a

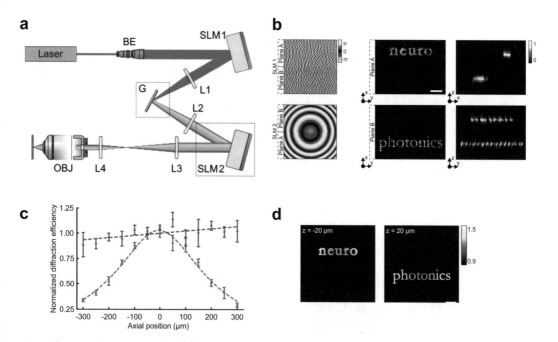

Fig. 12 Multiplane generation of spatiotemporal focused holographic patterns. (**a**) Schematic of the experimental setup for 3D-CGH-TF. *BE* Beam expander, *G* diffraction grating, *L1, L2, L3, L4* Lenses, *OBJ1* Microscope objective. (**b**) *Left panel, top*, tiled phase profiles addressed to SLM1 for encoding the words "neuro" (plane A) and "photonics" (plane B). *Bottom*, Fresnel lens-phase profiles addressed to SLM2 to axially displace each holographic pattern generated by SLM1 on separated planes at +20 µm (plane A) and −20 µm (plane B). *Middle panel*, *x–y* 2P fluorescence intensity cross-sections at planes A and B generated by phase-holograms in the *left*. *Right panel*, orthogonal maximum 2P intensity projection along *x* (*top*) and *y* (*bottom*). (**c**) Axial diffraction efficiency curve, where the intensity of each holographic spot projected at each axial plane is normalized to the one of a spot projected at the center of the nominal focal plane. Experimental data (*red points*) represent the average of four realizations and follow a Lorentzian distribution (*red dashed line*) with $\Delta z_{FWHM} = 360$ µm. Weighting the axially displaced holographic spot intensity according to the calculated diffraction efficiency enables intensity equalization between holographic patterns in separated axial positions (*blue points* and *dashed line* represent the corrected intensity ratio and fitting, respectively). Vertical error bars show the standard deviation for the different realizations. (**d**) Input patterns to the Gerchberg-Saxton algorithm used to calculate the holograms that generate the holographic patterns shown in (**b**): low-diffraction efficiency regions appear brighter over those closer to the center of the excitation field. To improve the observation of differences between conditions, the amplitude scale of the images was chosen from 0.9 to 1.5. Adapted from Ref. [43]

f_1), placed at a distance equal to f_1 after the LC-SLM, which generates a first holographic intensity pattern at its focal plane, and a second lens (L2 of focal length f_2), placed at a distance $f_1 + f_2$ from L1 and $f_2 + f_{obj}$ from the objective (focal length f_{obj}), which conjugates the LC-SLM plane to the objective back focal plane. In this case $f_{eq} = (f_1 \cdot f_{obj})/f_2$. The focal lengths f_1 and f_2 are chosen to match the LC-SLM short axis to the objective back aperture size (usually 6–8 mm), in order to achieve the best possible axial resolution for the holographic patterns. As a numerical example, considering a standard commercial LC-SLM array of 12 mm, with a pixel size of 20 μm, and a typical water-immersion 40× objective ($f_{obj} = 4.5$ mm) often used in neurophysiology experiments, the ratio f_1/f_2 is chosen to be ~2, which would give a FOE$_{xy}$ ~ 230 × 230 μm^2 or ~420 × 420 μm^2 according to Eq. (1), using $\lambda = 520$ nm or $\lambda = 950$ nm, respectively.

2. In particular, for 2P–CGH combined with temporal focusing (see Sect. 2.1), two cylindrical lenses with opposite focal lengths, orthogonally oriented, with preference at 45° relatively to the direction of the axes of the LC-SLM array, enable generating two axially displaced zero-order lines. Then, thanks to the out-of-focus pulse dispersion of temporal focusing, the zero order contribution to the 2P signal can be reduced by 4–6 orders of magnitude [47]. The choice of 45° direction is related to the orientation of the grating grooves that is used for temporal focusing and the dispersion direction of spectral frequencies. Usually, the diffraction grating is placed with its grooves perpendicularly oriented in respect to the plane of incidence (i.e., the plane which contains the grating's surface normal and the propagation vector of the incoming radiation), or simply in respect to the lab's floor. This means that dispersion occurs parallel to the incidence plane. If the cylindrical lenses are oriented with their focusing axes one perpendicular and one parallel to the grating's grooves, then at the focal plane only fluorescence that originates from the line that lies along the direction of temporal focusing, i.e. the direction where the spectral frequencies are dispersed, will be efficiently suppressed. Fluorescence from the other line will be reduced, but still present at another plane. If the cylindrical lenses are oriented with their focusing axes at 45° in respect to the grating's grooves then both zero-order lines at the sample volume undergo temporal focusing and thus out-of-focus fluorescence suppression.

3. The generalized phase contrast method (GPC) is an alternative approach to CGH for generating speckle-free arbitrarily shaped illumination patterns. It is a common-path interferometry visualization technique, i.e., the output image is obtained by the interference between a signal and a reference wave travelling along the same optical axis, based in Zernike's phase contrast [76]. The basic principle of GPC involves separating a light beam into its Fourier components by using a lens. The on-axis, low spatial frequency components are shifted in phase (usually by π or λ/2) using a small wave retarder or phase contrast filter (PCF). A second lens then recombines the high and low spatial frequencies. The introduced phase shift by the PCF causes the two different components to interfere and produces an

intensity distribution according to the phase information carried by the higher spatial frequencies [77]. By controlling the value of the input phase function at the LC-SLM plane and by choosing appropriate phase retardation at the PCF, a pure phase to intensity imaging is accomplished. A simple binary phase modulation is sufficient in this case (e.g., addressing with 0 phase dark areas and π phase bright areas of the desired pattern, when a π phase retardation is introduced by the PCF). This means that a phase modulation is finally turned to an intensity modulation, without using any iterative algorithm that provides approximate solutions for the phase hologram, resulting to speckle patterns, like in CGH. The rapid phase variations of the holographic phase profile do not longer exist when GPC is used and patterns generated are speckle-free. The phase wavefront of the output beam is smooth, similar to the one of a Gaussian beam. This results in a long axial propagation, revealing the interferometric character of the beam [32], meaning that GPC patterns lack of axial confinement. For 2P excitation, the remedy to this issue is combination of GPC with temporal focusing. Temporally focused 2P-GPC patterns result with an axial resolution similar to that obtained by 2P line-scan microscopy [32].

4 Outlook

Computer-generated holography combined with 2P excitation enables in depth optical stimulation with millisecond temporal resolution and sub-millisecond temporal precision. The combination of CGH with temporal focusing enables generation of excitation volumes with micrometer axial resolution and robust propagation through scattering media. For neuronal activation the efficient current integration achievable with parallel holographic illumination combined with laser amplifiers at low-repetition rate (500 kHz) and laser pulses of moderate excitation intensities ($100~\mu W/\mu m^2$) enables reliable AP generation with millisecond temporal resolution and sub-millisecond precision. These findings, together with high average power available nowadays in commercially available laser systems (more than 10 W at laser output), indicate that laser power is not the limiting factor for the maximum achievable number of targets using CGH. More likely, this will be limited by other factors, such as sample heating and deterioration of the photostimulation spatial resolution. Indeed, for multiple-cell stimulation, photostimulating neurites crossing the illumination volume affects cellular resolution, thus limiting the maximum number of targets that can be stimulated with single-cell precision. Recent progresses in engineering of somatic opsins should enable to solve these limitations in the close future [61].

Acknowledgments

We thank Marco Pascucci and Benoît C. Forget for helpful discussions on the fitting procedure to extract opsins kinetics parameters; Valeria Zampini for recording data on CoChR-expressing neurons (Fig. 9); Dimitrii Tanese and Nicolò Accanto for help with preparation of Figs. 3, 4 and 6. We thank the National Institutes of Health (NIH 1-U01-NS090501-01), the "Agence Nationale de la Recherche" (grants ANR-10-INBS-04-01; France-BioImaging Infrastructure network, ANR-14-CE13-0016; HOLOHUB, ANR-15-CE19-0001-01; 3DHoloPAc) and the Human Frontiers Science Program (Grant RGP0015/2016) for financial support.

References

1. Boyden ES, Zhang F, Bamberg E et al (2005) Millisecond-timescale, genetically targeted optical control of neural activity. Nat Neurosci 8:1263–1268

2. Nagel G, Brauner M, Liewald JF et al (2005) Light activation of Channelrhodopsin-2 in excitable cells of caenorhabditis elegans triggers rapid behavioral responses. Curr Biol 15:2279–2284. doi:10.1016/j.cub.2005.11.032

3. Zhang F, Wang LP, Brauner M et al (2007) Multimodal fast optical interrogation of neural circuitry. Nature 446:633–639

4. Adamantidis AR, Zhang F, Aravanis AM et al (2007) Neural substrates of awakening probed with optogenetic control of hypocretin neurons. Nature 450:420–424. doi:10.1038/nature06310

5. Gradinaru V, Thompson KR, Zhang F et al (2007) Targeting and readout strategies for fast optical neural control in vitro and in vivo. J Neurosci 27:14231–14238

6. Aravanis AM, Wang LP, Zhang F et al (2007) An optical neural interface: in vivo control of rodent motor cortex with integrated fiberoptic and optogenetic technology. J Neural Eng 4: S143–S156

7. Huber D, Petreanu L, Ghitani N et al (2008) Sparse optical microstimulation in barrel cortex drives learned behaviour in freely moving mice. Nature 451:61–64

8. Anikeeva P, Andalman AS, Witten I et al (2012) Optetrode: a multichannel readout for optogenetic control in freely moving mice. Nat Neurosci 15:163–170. doi:10.1038/nn.2992

9. Weible AP, Piscopo DM, Rothbart MK et al (2017) Rhythmic brain stimulation reduces anxiety-related behavior in a mouse model based on meditation training. Proc Natl Acad Sci U S A 114(10):2532–2537. doi:10.1073/pnas.1700756114

10. Makinson CD, Tanaka BS, Sorokin JM et al (2017) Regulation of thalamic and cortical network synchrony by Scn8a. Neuron 93:1–15. doi:10.1016/j.neuron.2017.01.031

11. Petreanu L, Huber D, Sobczyk A, Svoboda K (2007) Channelrhodopsin-2-assisted circuit mapping of long-range callosal projections. Nat Neurosci 10:663–668. doi:10.1038/nn1891. nn1891 [pii]

12. Petreanu L, Mao T, Sternson SM, Svoboda K (2009) The subcellular organization of neocortical excitatory connections. Nature 457:1142–1145. doi:10.1038/nature07709

13. Tovote P, Esposito MS, Botta P et al (2016) Midbrain circuits for defensive behaviour. Nature 534:206–212. doi:10.1038/nature17996

14. Joshi A, Kalappa BI, Anderson CT, Tzounopoulos T (2016) Cell-specific cholinergic modulation of excitability of layer 5B principal neurons in mouse auditory cortex. J Neurosci 36:8487–8499. doi:10.1523/JNEUROSCI.0780-16.2016

15. Morgenstern NA, Bourg J, Petreanu L (2016) Multilaminar networks of cortical neurons integrate common inputs from sensory thalamus. Nat Neurosci 19:1034–1040. doi:10.1038/nn.4339

16. Lee S-H, Kwan AC, Zhang S et al (2012) Activation of specific interneurons improves V1 feature selectivity and visual perception. Nature 488:379–383. doi:10.1038/nature11312

17. Adesnik H, Bruns W, Taniguchi H et al (2012) A neural circuit for spatial summation in visual cortex. Nature 490:226–231. doi:10.1038/nature11526

18. Atallah BV, Bruns W, Carandini M, Scanziani M (2012) Parvalbumin-expressing interneurons linearly transform cortical responses to visual stimuli. Neuron 73:159–170. doi:10.1016/j.neuron.2011.12.013

19. Klapoetke NC, Murata Y, Kim SS et al (2014) Independent optical excitation of distinct neural populations. Nat Methods 11:338–346. doi:10.1038/nmeth.2836

20. Denk W, Strickler JH, Webb WW (1990) Two-photon laser scanning fluorescence microscopy. Science 248:73–76

21. Feldbauer K, Zimmermann D, Pintschovius V et al (2009) Channelrhodopsin-2 is a leaky proton pump. Proc Natl Acad Sci U S A 106:12317–12322

22. Rickgauer JP, Tank DW (2009) Two-photon excitation of channelrhodopsin-2 at saturation. Proc Natl Acad Sci U S A 106:15025–15030. doi:10.1073/pnas.0907084106

23. Andrasfalvy BK, Zemelman BV, Tang J, Vaziri A (2010) Two-photon single-cell optogenetic control of neuronal activity by sculpted light. Proc Natl Acad Sci U S A 107:11981–11986

24. Prakash R, Yizhar O, Grewe B et al (2012) Two-photon optogenetic toolbox for fast inhibition, excitation and bistable modulation. Nat Methods 9:1171–1179. doi:10.1038/nmeth.2215

25. Packer AM, Peterka DS, Hirtz JJ et al (2012) Two-photon optogenetics of dendritic spines and neural circuits. Nat Methods 9:1171–1179. doi:10.1038/nmeth.2249

26. Rickgauer JP, Deisseroth K, Tank DW (2014) Simultaneous cellular-resolution optical perturbation and imaging of place cell firing fields. Nat Neurosci 17:1816–1824. doi:10.1038/nn.3866

27. Bègue A, Papagiakoumou E, Leshem B et al (2013) Two-photon excitation in scattering media by spatiotemporally shaped beams and their application in optogenetic stimulation. Biomed Opt Express 4:2869–2879

28. Chaigneau E, Ronzitti E, Gajowa AM et al (2016) Two-photon holographic stimulation of ReaChR. Front Cell Neurosci 10:234

29. Ronzitti E, Conti R, Zampini V et al (2017) Sub-millisecond optogenetic control of neuronal firing with two-photon holographic photoactivation of Chronos. J Neurosci:1246–1217. https://doi.org/10.1523/JNEUROSCI.1246-17.2017

30. dal Maschio M, Donovan JC, Helmbrecht TO, Baier H (2017) Linking neurons to network function and behavior by two-photon holographic optogenetics and volumetric imaging. Neuron 94:774–789.e5. doi:10.1016/j.neuron.2017.04.034

31. Förster D, Maschio MD, Laurell E, Baier H (2017) An optogenetic toolbox for unbiased discovery of functionally connected cells in neural circuits. Nat Commun 8(1):116

32. Papagiakoumou E, Anselmi F, Bègue A et al (2010) Scanless two-photon excitation of channelrhodopsin-2. Nat Methods 7:848–854. doi:10.1038/nmeth.1505

33. Oron D, Papagiakoumou E, Anselmi F, Emiliani V (2012) Two-photon optogenetics. Prog Brain Res 196:119–143. doi:10.1016/B978-0-444-59426-6.00007-0

34. Vaziri A, Emiliani V (2012) Reshaping the optical dimension in optogenetics. Curr Opin Neurobiol 22:128–137. doi:10.1016/j.conb.2011.11.011

35. Bovetti S, Fellin T (2015) Optical dissection of brain circuits with patterned illumination through the phase modulation of light. J Neurosci Methods 241:66–77. doi:10.1016/j.jneumeth.2014.12.002

36. Packer AM, Roska B, Häusser M (2013) Targeting neurons and photons for optogenetics. Nat Neurosci 16:805–815. doi:10.1038/nn.3427

37. Curtis JE, Koss BA, Grier DG (2002) Dynamic holographic optical tweezers. Opt Commun 207:169–175

38. Gerchberg RW, Saxton WO (1972) A pratical algorithm for the determination of the phase from image and diffraction pictures. Optik (Stuttg) 35:237–246

39. Lutz C, Otis TS, DeSars V et al (2008) Holographic photolysis of caged neurotransmitters. Nat Methods 5:821–827. doi:10.1038/nmeth.1241

40. Papagiakoumou E (2013) Optical developments for optogenetics. Biol Cell 105:443–464. doi:10.1111/boc.201200087

41. Golan L, Reutsky I, Farah N, Shoham S (2009) Design and characteristics of holographic neural photo-stimulation systems. J Neural Eng 6:66004

42. Yang S, Papagiakoumou E, Guillon M et al (2011) Three-dimensional holographic photostimulation of the dendritic arbor. J Neural Eng 8:46002. doi:10.1088/1741-2560/8/4/046002. S1741-2560(11)87640-8 [pii]

43. Hernandez O, Papagiakoumou E, Tanese D et al (2016) Three-dimensional spatiotemporal focusing of holographic patterns. Nat Commun 7:11928. doi:10.1038/ncomms11928

44. Ronzitti E, Guillon M, de Sars V, Emiliani V (2012) LCoS nematic SLM characterization and modeling for diffraction efficiency optimization, zero and ghost orders suppression. Opt Express 20:17843–17855

45. Polin M, Ladavac K, Lee S-H et al (2005) Optimized holographic optical traps. Opt Express 13:5831–5845

46. Zahid M, Velez-Fort M, Papagiakoumou E et al (2010) Holographic photolysis for multiple cell stimulation in mouse hippocampal slices. PLoS One 5:e9431

47. Hernandez O, Guillon M, Papagiakoumou E, Emiliani V (2014) Zero-order suppression for two-photon holographic excitation. Opt Lett 39:5953–5956

48. Conti R, Assayag O, De Sars V et al (2016) Computer generated holography with intensity-graded patterns. Front Cell Neurosci 10:236

49. Shemesh OA, Tanese D, Zampini V, Linghu C, Piatkevich K, Ronzitti E, Papagiakoumou E, Boyden ES, Emiliani V (2017) Temporally precise single-cell resolution optogenetics. Nat Neurosci (in press)

50. Papagiakoumou E, de Sars V, Oron D, Emiliani V (2008) Patterned two-photon illumination by spatiotemporal shaping of ultrashort pulses. Opt Express 16:22039–22047

51. Golan L, Shoham S (2009) Speckle elimination using shift-averaging in high-rate holographic projection. Opt Express 17:1330–1339. doi: 176064 [pii]

52. Guillon M, Forget BC, Foust AJ et al (2017) Vortex-free phase profiles for uniform patterning with computer-generated holography. Opt Express 25:12640. doi:10.1364/OE.25. 012640

53. Glückstad J (1996) Phase contrast image synthesis. Opt Commun 130:225–230

54. Bañas A, Glückstad J (2017) Holo-GPC: holographic generalized phase contrast. Opt Commun 392:190–195. doi:10.1016/j.optcom. 2017.01.036

55. Schmitz CHJ, Spatz JP, Curtis JE (2005) High-precision steering of multiple holographic optical traps. Opt Express 13:8678–8685

56. Engström D, Bengtsson J, Eriksson E, Goksör M (2008) Improved beam steering accuracy of a single beam with a 1D phase-only spatial light modulator. Opt Express 16:18275–18287. doi:10.1364/OE.16.018275

57. Hernandez Cubero OR (2016) Advanced optical methods for fast and three-dimensional control of neural activity. Paris Descartes University, France

58. Szabo V, Ventalon C, De Sars V et al (2014) Spatially selective holographic photoactivation and functional fluorescence imaging in freely behaving mice with a fiberscope. Neuron 84:1157–1169. doi:10.1016/j.neuron.2014. 11.005

59. Oron D, Tal E, Silberberg Y (2005) Scanningless depth-resolved microscopy. Opt Express 13:1468–1476

60. Zhu G, van Howe J, Durst M et al (2005) Simultaneous spatial and temporal focusing of femtosecond pulses. Opt Express 13:2153–2159. doi: 83023 [pii]

61. Baker CA, Elyada YM, Parra-Martin A, Bolton M (2016) Cellular resolution circuit mapping in mouse brain with temporal-focused excitation of soma-targeted channelrhodopsin. eLife 5:1–15. doi:10.7554/eLife.14193

62. Nagel G, Szellas T, Huhn W et al (2003) Channelrhodopsin-2, a directly light-gated cation-selective membrane channel. Proc Natl Acad Sci U S A 100:13940–13945

63. Williams JC, Xu J, Lu Z et al (2013) Computational optogenetics: empirically-derived voltage- and light-sensitive channelrhodopsin-2 model. PLoS Comput Biol 9:e1003220. doi:10.1371/journal.pcbi.1003220

64. Hegemann P, Ehlenbeck S, Gradmann D (2005) Multiple photocycles of channelrhodopsin. Biophys J 89:3911–3918. doi:10. 1529/biophysj.105.069716

65. Nikolic K, Degenaar P, Toumazou C (2006) Modeling and engineering aspects of ChannelRhodopsin2 system for neural photostimulation. Annu Int Conf IEEE Eng Med Biol Soc 1:1626–1629

66. Nikolic K, Grossman N, Grubb MS et al (2009) Photocycles of Channelrhodopsin-2. Photochem Photobiol 85:400–411. doi:10. 1111/j.1751-1097.2008.00460.x

67. Lin JY, Knutsen PM, Muller A et al (2013) ReaChR: a red-shifted variant of channelrhodopsin enables deep transcranial optogenetic excitation. Nat Neurosci 16:1499–1508. doi:10.1038/nn.3502

68. Papagiakoumou E, Bègue A, Leshem B et al (2013) Functional patterned multiphoton excitation deep inside scattering tissue. Nat Photonics 7:274–278

69. Booth MJ, Débarre D, Jesacher A (2012) Adaptive optics for biomedical microscopy. Opt Photonics News 23:22. doi:10.1364/ OPN.23.1.000022

70. Leach J, Sinclair G, Jordan P et al (2004) 3D manipulation of particles into crystal structures using holographic optical tweezers. Opt Express 12:220

71. Anselmi F, Ventalon C, Bègue A et al (2011) Three-dimensional imaging and photostimulation by remote-focusing and holographic light patterning. Proc Natl Acad Sci U S A 108:19504–19509. doi:10.1073/pnas. 1109111108

72. Daria VR, Stricker C, Bowman R et al (2009) Arbitrary multisite two-photon excitation in four dimensions. Appl Phys Lett 95:93701

73. Yang S, Emiliani V, Tang C-M (2014) The kinetics of multibranch integration on the dendritic arbor of CA1 pyramidal neurons. Front Cell Neurosci 8:127. doi:10.3389/fncel.2014.00127

74. Packer AM, Russell LE, Dalgleish HWP, Häusser M (2014) Simultaneous all-optical manipulation and recording of neural circuit activity with cellular resolution in vivo. Nat Methods 12:140–146. doi:10.1038/nmeth.3217

75. Haist T, Schönleber M, Tiziani H (1997) Computer-generated holograms from 3D-objects written on twisted-nematic liquid crystal displays. Opt Commun 140:299–308. doi:10.1016/S0030-4018(97)00192-2

76. Zernike F (1942) Phase contrast, a new method for the microscopic observation of transparent objects. Physica 9:686–698. doi:10.1016/S0031-8914(42)80035-X

77. Glückstad J, Mogensen PC (2001) Optimal phase contrast in common-path interferometry. Appl Optics 40:268–282

Chapter 11

Optophysiology and Behavior in Rodents and Nonhuman Primates

Golan Karvat and Ilka Diester

Abstract

The combination of optogenetic interventions and behavioral assays allows investigating causal effects of neural activity with unprecedented spatial and temporal resolution. However, utilizing the technique also requires special considerations in designing the experimental procedures and assigning controls. In this chapter, we give an overview of the requirements for behavioral-optogenetic experiments in rodents and nonhuman primates. Special emphasis is given to correct assignments of controls and how to avoid artifacts.

Key words Optogenetics, Behavioral neuroscience, Electrophysiology, Behavior, Nonhuman primate, Rodent, Rat, Mouse, Procedure, Protocol

1 Introduction

Behavior is the ultimate output of the central nervous system. Therefore, it is not surprising that shortly after the first demonstration of genetically modified neurons which were responsive to light [1], many groups started to use optogenetics during experiments in behaving animals [2]. Optogenetics offers time- and cell type-specific excitation or inhibition, and by that improves tremendously the research of neuronal circuits. Together with rapidly developing technologies to convey the light, and to measure behavioral and electrophysiological output, it has paved the way for a better understanding of behavior. However, caution must be exercised in planning an optogenetic behavioral experiment, since this attractive method adds requirements to the experimental design. Care must be taken of the covert features of the experimental system, as well as the appropriate controls, to allow reproducibility and avoid false conclusions. Hence, the aim of this chapter is to provide the milestones, and highlight the advantages and pitfalls, of designing and implementing behavioral studies utilizing optogenetics in mammals (rodents and nonhuman primates [NHP]).

Albrecht Stroh (ed.), *Optogenetics: A Roadmap*, Neuromethods, vol. 133,
DOI 10.1007/978-1-4939-7417-7_11, © Springer Science+Business Media LLC 2018

2   Equipment, Materials, and Setup

2.1   Animals

The first step in designing an experiment is to choose the appropriate experimental system. In behavioral studies, the first question is which species to use. As a rule of thumb, a trade-off exists between the ability of the animal to perform cognitively demanding tasks and the available genetic manipulation toolbox. While a vast selection of genetically modified mice exists, some expressing different opsins under specific promoters [3], the variety of genetically modified rats is much smaller [4]. At the moment, transgenic NHP remain an exciting prospect [5, 6].

The species and line of choice have a direct influence on the experimental design. When using mouse lines expressing opsins in specific cell types, no virus injection is necessary and the behavioral testing can start right after establishing light delivery to the brain area of interest (by an optical fiber or thinning of the skull and an external light source) (see Chaps. 1, 2 and 13). However, in mouse lines with brain wide opsin expression, light directed into a certain area will influence not only the cells with local somas but also cells projecting into and through that area (antidromic or back-propagating action potentials). Although this effect can be useful, for example, for circuit targeting [7], it may limit conclusions regarding the influence of a specific region on behavior. To avoid this effect, a one-injection strategy of an opsin in a recombinase-dependent cassette ("floxed") into the brain of a recombinase (e.g., Cre) expressing transgenic line is a commonly applied strategy [8]. This approach capitalizes on the vast library of already existing cre-expressing transgenic mouse lines allowing targeting of a diverse set of cell types (see Chap. 2).

With recently developed rat Cre lines [4], cre-dependent strategies are feasible in rats, too. However, the more common strategy in behavioral assessment of rats is delivering the opsins by a viral construct into wild-type animals. If projection-target specificity is desired, a dual viral vector strategy has been suggested; one viral vector carries a recombinase (delivered by a retrogradely traveling virus) at the target and another vector conveys a floxed opsin at the source [9]. Further, Herpes simplex virus, canine adenovirus 2, and pseudorabies virus can be used as a retrogradely travelling option [10, 11]. An alternative method for acute gene delivery is electroporation, in which plasmids are forced into neurons using an electric field in-utero [12]. Both viral injection and electroporation yield a spatially restricted expression of the opsin, with higher expression levels than in transgenic animals [13]. However, since electroporation is technically more demanding than viral injection, it is utilized mainly in developmental studies.

Finally, for efficient expression of opsins in NHP, larger amounts of high-titer virus for covering larger brain areas are needed, which implies multiple injections. This has a strong effect on the operational procedure (see below), and the possible length of the experiment, as expression of the virus accumulates over time and increases toxicity [14].

2.2 Behavioral Setup

A large variety of behavioral tests can be combined with optogenetic manipulation and electrophysiological measurements, depending on the relevant research question. While commonly used and well established methods can be utilized, the special requirements of optogenetics should be taken into account.

First, for experiments with freely moving animals, care should be taken of the cables and fibers not to interfere with the animal's ability to move. To this aim, sufficiently long cables and a commutator to avoid wrapping should be used. A counterbalance takes additional weight from the animal's head (see Chap. 12 for details). A different approach would be to utilize the experiment while the animal is head-restrained. When choosing this approach, special care should be taken to minimize the stress of the subject, and when working with rats and NHP, habituation to the restraining devise is necessary [15]. Although adding complexity to the experimental procedure, head-restrained experiments allow better control of different aspects of behavior, and easier access to light emission and recording devices.

Second, some behavioral tasks have to be modified in order to allow "safe" optogenetic working procedures. For example, special care should be taken as to waterproof the headstage for tests containing water (such as the Morris water maze and the forced swim test [16]). Common sense dictates that when cables are connected to the subject's head, especially cables as fragile as a fiber-optic, the area above the animal should be free of any object.

2.3 Opsin of Choice

Broadly speaking, there are four kinds of optogenetic manipulations useful for behavioral studies: excitation (gain of function), inhibition (loss of function), cell type-specific phototagging, and circuit-targeting. Optogenetic activation can be millisecond precise (e.g., ChR2(H134R)) or rather long-lasting (e.g., "step-function" opsins like ChR2-SFO). Inhibition can be achieved by ion-pumps (e.g., NpHR3.0/Arch), anion-specific channels (e.g., iChloc), or by selectively activating inhibitory neurons. Cell type-specific phototagging is done by expressing opsins selectively in a certain class of cells, most commonly by using cell type-specific promoters, e.g., CaMKII or parvalbumin (see Chap. 2 for more details). Light is used to excite the cells and responsive units are "tagged" as belonging to the targeted class. This allows characterizing their response properties during e.g., a specific behavior. Circuit-targeting is a strategy to study neuronal population activity based on its projection target. When utilizing this approach it is important to ensure neuronal response and to check whether the connection to the target is direct or indirect, by the so-called collision test: optically evoked action potentials (AP) should not occur during the cell's refractory period. Thus, if an AP triggered by axonal photostimulation collides with a somatic AP in a defined (short) timewindow, there is at least one synapse between the axon and that soma [7].

The type of opsin has an influence on the stimulation protocol, and thus on the experimental design. Therefore, when planning the experiment it is important to assure that the planned stimulation fits the biophysical properties of the opsin. Also, it is imperative that the species and strain expresses the opsin as desired and that the neuronal response to the planned stimulation protocol is as expected. For example, AAV2/5 expresses well in rodents and NHP but not in cats [17] (see Chaps. 1 and 2 for more details). Hence, it is recommended to test all optogenetic tools which are new to the laboratory for expression and functionality before launching a full behavioral assay.

2.4 Light Delivery

There are several special concerns regarding the mean of light delivery when performing behavioral experiments. First, it must not interfere with the examined behavior, nor be a visual cue for the subject. Therefore, it is important to minimize the amount of light visible to the animal (e.g., by coating the optical fiber patch cords with an opaque cover) (see Chap. 11). Second, since in most cases behavioral examinations require chronic implantation of the light emitting device or fiber, the implant has to be light and robust to damage caused by the animal (especially when applied to rats and NHP). We find it useful to cover the device with a stable recording chamber when working with rats and NHP. Finally, it has to have enough power to induce a reliable stimulation, yet not too much to avoid damage to the tissue, especially for studies lasting more than a few days (a range 100–200 mW/mm^2 is advisable to influence behavior).

2.5 The Room

Like for any behavioral study, the experimental room should be kept as quiet as possible, dimly lit (unless studying stress) and free of odors of the subjects' predators (e.g., rats when studying mice). When working with lasers, the room should also meet standards of safety, such as minimally reflecting floor and no mirrors. The height of the laser output and of the behavioral setup should be higher or lower than eye-level. It is advisable that only staff trained in working with lasers will be allowed in the room. Finally, it should be marked clearly when lasers are on (ideally in an automatic manner).

3 Procedures

3.1 Surgeries

All experiments using animals should be carried out under institutional and national guidelines. It is recommended to separate the behavioral examination room from the surgery room, in order to minimize stress and improve training. If in accordance with the goals of the study, it is better to habituate the animals to the behavioral room and do some basic training before the surgery. This is especially important for head fixed experiments with rats and NHP. Since some differences exist in the procedures between rodents and NHP, they will be described separately below.

3.2 Rodents

Surgical procedures for optophysiological experiments are similar to other injection and implantation operations carried out on rodents [18] (see also Chap. 9). When injection of viruses is needed, allow enough time for expression (e.g., 4–6 weeks for AAV2/5). It is recommended to evaluate the minimal volume needed for sufficient expression with the viral vector of choice and in the brain area of interest. For high titer AAV2/5 (i.e., in the order of 10^{12-14}), we usually limit the injected volume to 1 μl, as it roughly corresponds to 1 mm^3 of transfected area. Since the brain tissue absorbs and scatters the transmitted light, less than 40% of light energy passes 0.2–0.6 mm from the light source [19] (see also Chap. 13). In our hands, neurons at a distance of more than 1.6 mm become nonresponsive under maximal illumination of 382 mW/mm^2 [20]. Injection should be slow (~100 nl/min) to avoid tissue damage and 5–10 min waiting time should be added before withdrawing the needle to allow diffusion and prevent back flow of the viral vector along the injection track. After recovery time of 2–5 days, training of the animals in the behavior task can commerce.

When implanting electrodes and optical devices ("optrodes"), a hole is drilled above the desired area down to the level of the dura, which is then removed using fine forceps to avoid damage to the cortex. It is important to allow some flexibility around the inserted optrode and separate the brain from the dental cement, thus it is recommended to fill the craniotomy with a lubricant (e.g., Vaseline). The tip of the light source should be located above the transfected area in order to be effective. For stability, special care should be taken that the skull is completely dry and clean before applying dental cement. After surgeries, apply analgesics for 2–3 days and antibiotics for 5 days. Experimental procedures can continue when the animal is vital and back to its pre-surgery weight, which is usually the case within 2–5 days post-surgery. It is essential to keep track of the animals' body-weight throughout the whole experiment, especially when depriving of food or water.

3.3   NHP

Most standard electrophysiological approaches can be adapted to include optogenetic manipulations. In case of an already implanted recording chamber, injections can be done through burr holes or larger craniotomies within the chamber. For this a light weight injection system, similar to setups which are used for pharmacological interference experiments can be adapted [21]. This approach has the advantage of allowing injection in the awake head fixed animal [22]. This is beneficial as injection times quickly add up if a larger brain area needs to be covered with viral vectors. As a rule of thumb, 10 min injection time plus 10 additional minutes of waiting time to prevent viral vector back flow along the injection track has to be taken into account. With an expression volume of roughly 1 mm^3 the time needed to cover a large enough area quickly adds up while the time of the surgery should be kept as short as possible to reduce risks. However, injections in the anesthetized animal are an equally valid approach. For this, the same procedure and injection equipment as in rodents can be applied.

After the injections are done the choice of stimulation and recording probe comes up. Either, a classic optrode (i.e., an optical fiber glued next to an electrode) can be inserted on a daily basis via microdrives (Diester et al. [22]). As the approach is very similar to regular electrode recordings almost no new equipment is needed. On the downside the tissue damage is substantial due to the dual-tip geometry of the probe. A better design might be a single tip coaxial optrode which consists of an optical fiber core coated with conductive material for electrical recordings [23]. This coaxial optrode can also be integrated into chronic electrode arrays with the advantage of covering a larger cortical area and allowing the write in of spatiotemporal patterns [24].

3.4   Controls

The validity and power of a behavioral study depends greatly on the correct choice of controls. Optogenetics allows combining gain- and loss-of-function controls in the same experiment, thus increasing its yield. However, the technique raises the need to control for some elements.

3.4.1   Control for Expression

It is imperative to correlate the behavioral outcome with the expression of opsins, and thus neuronal activity manipulation, by a "light with no opsin" between-groups control. If transgenic animals are used, wild-type (preferably littermates) should be the control group. If a virus is injected with the opsin (usually in conjunction with a fluorescent reporter gene), the control group should be injected with a virus containing the same reporter gene, without the opsin (e.g., "YFP-controls"). If the opsin gene is injected as an inverse expressing cassette, a Cre-negative group should serve as control.

3.4.2 Control for Involvement

Next, it is important to show that the opsins' reaction to light is causing the observed effect. For that, several controls may be effective, all possible as a within-animal control: (a) no-light trials; (b) illuminating with a nonstimulating wavelength. It is important to change the illumination power to compensate for the different heat absorbance of different wavelengths [25]; here the visual spectrum of the animal species has to be taken into account (e.g., since rodents are dichromatic they see far better within the blue-green range compared to red light [26]; (c) conducting the experiment with the fiber out of the brain. The last version can be useful also for controlling for the possible effect of light as a behavioral cue for the animal. This control is important since sometimes it is not possible to completely prevent light from reaching the eyes of the animal, and since some light might reach light-sensitive organs such as retina from within the brain. Another method to tackle this problem is to "mask" the animal with light flashes in the same wavelength as the stimulating wavelength near the eyes in all trials, regardless of the actual stimulation.

3.4.3 Control for Necessity

In studies aiming to investigate the role of specific systems for behavior, as is often the case with optophysiological studies, it is important to show that the system in question is necessary for the behavior, and not just involved in it. This kind of control is somewhat harder to achieve than others, as it requires manipulation of other brain areas and showing that the observed effect is limited to the system under investigation. The difficulty thus is twofold: first, since neural networks are heavily interconnected, sometimes remote areas may have an unexpected effect; and second, practical constrains limit the plausible locations which can be manipulated. Therefore, it is a common practice to manipulate 1–2 nearby areas hypothesized to have distinct effects on behavior [16], or in unilateral studies, by stimulating the corresponding location in the hemisphere not involved in the task.

3.4.4 Light Induced Artifacts

When combining optogenetics with electrophysiology, it is important to control for currents created on the electrode directly due to illumination by the Becquerel effect [27] (see also Chap. 8). The Becquerel effect, also termed the photovoltaic or photoelectric effect, is the induction of electric current in a material, usually a metal, upon exposure to light. Hence, if we illuminate directly on the recording metal electrode, an optical artifact will appear. Several steps can be taken in order to deal with this effect: (a) recordings from animals not expressing opsins (wild-type or only reporter gene expressing) can be used as the template of the artifact; (b) reducing the metal area of the electrode exposed to the light by adjusting the illumination angle; (c) using electrodes made of materials less reactive to light (such as specially treated silicon or indium tin oxide [13]); (d) use glass electrodes; (e) covering the electrode or the

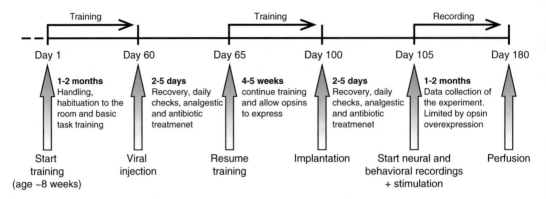

Fig. 1 Time-course of a representative behavioral optogenetics experiment with rats. Initial training duration can vary, depending on the behavioral task. The length of recovery time from surgery may change from animal to animal

glass with a dark or opaque painting (such as acrylic or lacquer, [28]); (f) reducing the light pulse width, thus allowing less time for the current induction by light, and (g) using a frequency control protocol; since opsins stop responding reliably at a specific stimulus frequency (e.g., [29]), "units" responding to frequencies exceeding this maximum could be sorted out during the spike-sorting analysis.

3.5 Finalizing the Experiment

The behavioral optogenetic experiment usually ends when the goals of the behavioral task have been reached. However, too high expression levels of the delivered proteins may hamper the interpretability of the study. A way to assess this time point is by chronically measuring the level of fluorescence signals in the infected area [22]. After completion of the experiment and sacrificing the animal, it is important to assess the location of the optical fiber and verify the opsin expression (see also Chap. 8). This step is important also as a control before interpreting the behavioral results. In long-term studies, it is important also to check for morphological abnormalities that might be caused by the opsin expression [14]. The anticipated time-course of a representative behavioral optogenetic experiment with rats is shown in Fig. 1. A flowchart with an overview of the main concerns of the experimental design is given in Fig. 2.

4   Typical/Anticipated Results

Optogenetics can be used to investigate almost any kind of behavior, hence the possible range of anticipated results of behavioral optogenetic studies is vast. Therefore, in this section we will focus on studies dealing with movement, and show how the applications of optophysiology in behavioral studies have developed in the last decade.

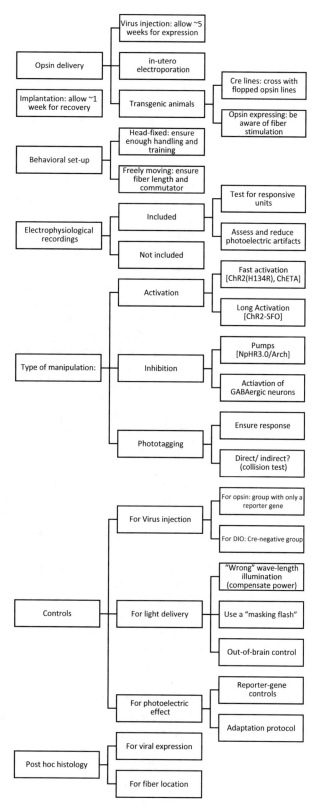

Fig. 2 Overview of the main concerns for designing behavioral/optogenetic experiment

In the first demonstration of optogenetics in a behavioral assay, researchers from the group of Karl Deisseroth [2] showed that activation of excitatory neurons in the motor cortex can control motor output. In this study, lentiviruses delivering ChR2 under the control of a CaMKIIα promoter were injected into the vibrissal motor cortex via an implanted guiding cannula. The same cannula served to guide the insertion of the optic fiber, when the rats and mice were sedated and restrained to a stereotactic frame. When illuminating with quite high power (20 mW out of a 200 μm fiber resulting in 380 mW/mm^2) and long pulses (20 s) blue light, the whiskers were deflected (by a maximum of 10 and 20° for rats and mice, respectively). Electrophysiological validation of photostimulation was done separately in acute brain slices. A group of animals injected with vehicles served as a between group control, and the behavior before and after the light-pulse as the within group control.

In the first optogenetic study of locomotion in freely moving animals [30], unilateral activation of the secondary motor cortex (M2) biased for moving to the contra-lateral side. Similar to the first movement related study from the Deisseroth lab [2], the optic fiber was guided into the brain by a cannula. However, the subjects were transgenic mice expressing ChR2 under the Thy1 promoter (in cortex expressed mainly in layer 5). The other end of the fiber was connected to an optical commutator to release tension. Blue light pulses (30 Hz/15 ms) into the right M2 induced time-locked repetitive rotations to the left (1.96 ± 0.24 rotations/10 s, compared to light-off). As this study was done on a previously tested opto-transgenic line, no electrophysiological verification was done (even though the use of optrodes was suggested). The authors also demonstrated the utility of extra-cranial light delivery through a thinned skull in activating superficial layers, in causing whisker deflection by exciting the vibrissal motor cortex with light-emitting diodes (LED). Wild-type animals not expressing ChR2 were the control group and behavior during before and after photostimulation was the within group control.

These two studies paved the way for behavioral studies by showing that photostimulation can alter behavior. Later studies showed the utility of optogenetics in the study of locomotion by manipulating down- and upstream brain areas, and thus disentangling the role of different pathways. The first empirical testing of the effect of direct and indirect pathways of the basal-ganglia on locomotion [31] showed opposite outcomes of activation of the different pathways. To this aim, the authors injected an adeno-associated virus (AAV2/1) containing a double-floxed inverted open reading frame encoding ChR2, into the dorsal striatum of mice encoding Cre under regulatory elements of the dopaminergic receptor D1 or D2, thus effecting the direct or indirect pathways, respectively. Classically, the direct pathway is associated with facilitating movements, and bilateral excitation of D1-positive neurons indeed increased locomotion and reduced freezing. Conversely, the indirect pathway is thought to inhibit movement, and excitation of D2-positive cells increased freezing and bradykinesia and decreased locomotor initiations, mimicking Parkinsonian symptoms. Here, the authors used a commercially available optrode (NeuroNexus Technologies, Ann Arbor, MI) to conduct an opto-pyhsiological measure, thus providing in vivo evidence for the effect of photostimulation in each circuit in their setup. Illumination was limited to 1 mW at the 200 μm fiber tip (~30 mW/mm^2) with constant illumination for 30 s on average. The within group control was light on/off, and the experimental design allowed for a direct comparison between the two pathways as a between-group analysis.

Following the above mentioned work, the lab of Rui Costa disentangled further the pathways of the basal ganglia by using the same transgenic mouse lines in order to identify the cell types (D1 or D2) by photostimulation [32]. They recorded the activity of these subpopulations during learning and execution of a motor sequence, and found that similar proportions of D1 and D2 positive cells responded to the initiation or termination of a sequence. This finding supports the model in which a coordinated action of the basal ganglia's direct and indirect pathways facilitate the appropriate movement while inhibiting competing movements. In addition, they found that D1 positive cells fire throughout the performance of the sequence while D2 cells are inhibited, providing the first direct evidence of differential activation of the direct and indirect pathways during motor programs. Their results refine the findings gathered from inhibiting all motor programs (by stimulating D2 neurons) or facilitating all motor programs (by stimulating D1 neurons). Thus, they exemplify that choosing the appropriate behavioral paradigm and stimulation and recording approach may contribute to finer conclusions regarding the role of a circuit.

With optogenetics, it is also possible to study the causative role of fine anatomical structures on specific aspects of behavior. The group of Karel Svoboda investigated the role of different cortical regions on decision related tongue movement by photo-induced inhibition [33]. They used VGAT-ChR2-EYFP mice (Jackson Laboratory, Bar Harbor, ME), which express ChR2 in inhibitory interneurons, and illuminated different cortical areas by projecting the 1.5 mW blue light via galvo-mirrors and through a clear-skull cap, fabricated from a thin layer of dental cement and a headpost. In this way, they showed that the anterior lateral motor cortex (ALM) is important for movement planning. Wild-type mice served as between-groups control, and no-light trials as within-group control.

Next, it was possible to disentangle heterogeneous circuits in the same area. A puzzling finding of this study was that although the ALM contains neurons with both ipsi- and contra-lateral preference, inactivating it causes a directional movement bias. To solve this puzzle, they utilized projection defined phototagging, and showed that the population of neurons projecting to the pyramidal tract are biased towards contralateral preference, and the population of cells projecting to collateral or to other cortical areas is more heterogeneous. Thus, the overt behavioral phenotype (movement) is contra-laterally biased [7].

Optogenetic control of movement in NHP is somewhat more challenging. Early studies aiming at this goal successfully modulated neuronal activity in motor cortex of NHP, yet did not evoke movements [22, 27]. One suggested explanation is that the optical stimulation did not reach the threshold for overt movement [34], or that the movements were too subtle to detect. Indeed, later studies successfully demonstrated that premotor cortex optogenetic inactivation on a reaching task leads to small, yet systematic, changes in behavior [35]. In addition, optogenetics was found useful in controlling subtle movements in NHP, as activation of ventral premotor (F5) and prefrontal cortex (frontal eye fields, FEF) reduced latency for saccade initiation [36], and inhibition of the superior-colliculus led to saccade deficits manifested in shift in saccadic end point, reduced peak velocity, and increased latency [37].

To conclude, the development of devices and technologies tailored for behavioral optogenetic studies together with refined methods allow us to investigate finer aspects of behavior and circuits in an unprecedented manner.

5   Advantages

The main advantage of adding optogenetics to behavioral experiments is the ability to investigate the causative impact of different neuronal subpopulations on behavior, with temporal, spatial and genetic accuracies superior to traditional approaches. To show causation, the experimental system has to provide evidence that excitation of the subpopulation in question evokes the behavior (gain of function), inhibition blocks it (loss of function), and that these results are not obtained by other subpopulations or projections (specificity).

Before optogenetics became available, gain of function was studied by electrical stimulation, agonist chemical agents or genetic over-expression. The main advantage of optogenetics is that unlike chemical and genetic interventions, it allows temporal resolution relevant to any behavior in question. Unlike electrical stimulation, optogenetics can be cell type-specific, and with a careful design of the stimulation neural recordings remain unaffected. This issue of simultaneous recording and stimulation is particularly relevant for brain-machine-brain interface devices, designed to convey information to the brain while reading its activity [38].

Loss of function was obtained by lesions, genetic knock-outs and pharmacological antagonists. None of these techniques allows the temporal resolution of optogenetics, and the first two also affect the animal through long time effects, which sometimes cause changes in neuronal connections and behavior in unexpected manners due to neural plasticity. Lesions in particular have the additional caveat of affecting projections of other brain areas as fibers of passage get destroyed in addition to the local cell bodies. Although some drugs can be cell type-specific, their time-course and spatial control fall short of that of optogenetics. Importantly, optogenetics allows for a study design that combines specificity, gain of function and loss of function in the same experiment, thus providing more controlled experiments with a higher yield.

An important advantage of optogenetics is the ability to control the activity in a pathway defined manner, by phototagging or retrograde travelling proteins and viruses (see Chaps. 1 and 2). As described above, this strategy provides us with the tool to disentangle fine neuronal circuits and their impact on behavior.

Since optogenetics allows both excitation and inhibition on a precise timescale with no electrical artifact, it is now possible to design real-time and closed-loop neuronal modulations in behaving animals [39]. A closed-loop design uses the output of the system as an error-signal to determine the manipulating input, in order to reach a defined target (as opposed to an open-loop, in which the input is predefined and the output is only measured). Closing the loop allows to account for the nonlinear and nonstationary dynamic nature of neural networks, such as changes in environmental context, behavioral and attentive states, neuromodulation, plasticity and more. These features can improve our basic understanding of the circuits' impact on behavior and help determining efficient photostimulation protocols with minimized undesired effects. Importantly, closing the loop can aid in adding sensory feedback to neuroprosthetic devices [38, 40].

In the case of behavioral optogenetic studies, the output can be both behavioral (e.g., location in a maze) and electrophysiological (e.g., population firing rate). The input photostimulation can be switched on or off until reaching the target [41, 42] or the illumination power can be proportional to the error and its integral over time [43]. The on/off switching is easier to implement and was demonstrated to be effective in a behavioral context [44]. The PI method is potentially more accurate and can allow actual "clamping" of neural networks to the target, but was tested to date only in vitro [43]. Future technological advances should allow for PI closed-loop in behaving animals, as well as for faster feedback to reach optimal response times of less than 2 ms [45, 46]. For a summary of the features of the different stimulation methods, see Table 1.

Table 1
Comparison between different stimulation methods

	Optogenetics	Electrical stimulation	Pharmacology	Lesion	Genetical engineering
Gain of function	✓	✓	✓	✗	✓
Loss of funciton	✓	✓ (1)	✓	✓	✓
Gain and loss in the same experiment	✓	✗	✗	✗	✗
Spatial resolution	<mm	<mm	mm-cm	mm-cm	cm-body
Temporal resolution	ms	ms	min	days	min(2)-lifetime
Longest effect time course	min(3)	sec(4)	min-hours	lifetime	lifetime
Cell-type targeting	✓	✗	✓	✓ (5)	✓
Pathway targeting	✓	✓ (6)	✗	✓ (7)	✗
Closed-loop manipulation	✓	✓ (8)	✓ (9)	✗	✗
Simultaneous recording and manipulation	✓	✓ (10)	✓	✗	✓

[1]Behavioral loss of function can be achieved by stimulating areas which inhibit the behavior
[2]Several minutes' resolution can be achieved by using an inducible transgenic line in conjunction with diet changing, substances injection, etc.
[3]Bistable step-function opsins allow activation for >30 min (e.g., [58])
[4]In some cases, such as deep-brain-stimulation (DBS) treatment to Parkinson patients, the effect of stimulation can last up to minutes.
[5]Cell type-specific ablation can be achieved by cell type-specific toxins (e.g., oxidopamine targeting dopaminergic and noradrenergic cells, [59]
[6]Electrical stimulation is thought to affect mainly fibers. However, source-target specificity is not achievable
[7]It is possible to cut fiber bundles; however, this method is possible in targeting only crude connections (e.g., corpus callosum, optic tract, spinal cord.)
[8]Electrical excitation is possible in a relevant timescale, yet in order to achieve a real closed-loop system inhibition is also needed
[9]A closed-loop system can control the administration of agonists and antagonists in a behavior-dependent manner, yet the activation timescale will limit its usefulness
[10]Simultaneous electrical stimulation and recordings can be achieved when efficient artifact removal strategies are applied

6   Notes

Even though the utility of optogenetics opens exciting venues of understanding behavior, it doesn't come without side effects. Here we list the main side effects which are important for behavioral studies as well as means to avoid artifacts and misinterpretations.

1. The first pitfall when conducting behavioral optogenetic studies deals with scales and interpretation, as we investigate the outcome of a big system (behavior of the animal) by modulating only small parts of it (a few cells up to a local network). It is very attractive to characterize the behavioral outcome in response to photostimulation, but if it is done without electrophysiology it is not possible to sort out confounding side effects. For example, long activation can lead to desensitization of the opsins, and inhibition can be followed by excitation as a "rebound" [47]. To avoid misinterpretations, it is important to monitor the neuronal response to the photostimulation by electrophysiology throughout the experiment.

2. The second side effect to take into account is the undesired damage caused to the brain tissue. The best way to avoid lesion is to deliver the light through thinned skull or cranial windows. However, this method is plausible only for investigation of superficial cortical layers. If the study deals with deeper structures, the tip of the fiber should be implanted slightly above the target, leaving the brain area of interest intact. Another type of damage to the tissue might occur from heating [19], and slight temperature increase can also lead to undesired increase in neuronal activity [48, 49]. To avoid heating, limit the power and duration of illumination. Long-lasting optogenetic activation, especially of pumps, is better be avoided also because it may change ion concentration and pH. This change might influence behavior in a manner which will be hard to interpret. If long-lasting effects are needed, consider using step-function opsins [50].

3. False-positive misinterpretations might occur if the controls are not properly assigned. Suggested controls are detailed under Sect. 3.4. False-negative errors might occur due to the focused nature of the optogenetic modulation, which may lead to subtle evoked responses. Another phenomenon that might lead to false-negative results is the resistance of the neural network to robust changes, especially during endogenously evoked activation [51]. This challenge is most probable to be met with refined design of the behavioral setup and stimulation protocol. The setup should be sufficiently sensitive to detect subtle differences in behavior, and the stimulation should be physiologically verified to the minimum necessary to excite or inhibit the targeted neurons.

7   Outlook

Opsin engineering remains an active research area providing a rapidly increasing library of opsins. Due to the better tissue penetration, red-shifted opsins are favorable for reaching deep brain structures noninvasively. Additionally, new wavelengths specificities will allow the combination of opsins for independent control of cell sub-populations. With all these new developments, intensive in vivo testing should be conducted before using these new opsins in behavioral studies to avoid negative results due to differences between in vitro and in vivo settings. For example, chloride-conducting ChRs (ChloCs) were shown to inhibit synaptically evoked action potentials in patch-clamped neurons. However, the inhibitory effect was absent in vivo, thus additional engineering was needed to achieve improved ChloCs which inhibitory effect holds in an in vivo setting [52]. Finally it should always be confirmed with a well-established behavioral paradigm (e.g., rotations induced by unilateral stimulation [30]) that the opsin is sufficiently strong to influence behavior.

Combined tools for light delivery and neural recordings require continual improvement. For behavioral studies which involve measurements over many weeks or even months, moveable drives are required. Ideally they would allow multi-site recordings and stimulation. So far these probes are either quite bulky or commercially not yet available. Wireless systems are another exciting prospect. Recently, two rather futuristic approaches have been suggested [53–55], in which light is delivered by an implantable miniaturized LED, and power is supplied wirelessly using radio-frequency (RF) technology. To be fully compatible with optophysiological behavioral studies, this exciting technology should be combined with neuronal recordings. Experiments in larger animals (e.g., NHP) call for broader volumes of illumination to facilitate achieving behavioral effects. For some applications a device with multiple outlets along the vertical axis could be helpful [56]. Alternatively, if a large horizontal area needs to get illuminated, an LED array may be preferable (e.g., [57].

The development of improved targeting strategies is of particular importance for circuit dissections. This approach promises to establish a connection between neural circuit dysfunction and psychiatric or motor diseases. For establishing this causal link, it is crucial to include a behavioral part in the set of experiments. Therefore, the combination of optogenetic manipulation, neural recordings, and behavioral analysis is one of the most exciting and promising avenues in current neurosciences.

References

1. Boyden ES, Zhang F, Bamberg E, Nagel G, Deisseroth K (2005) Millisecond-timescale, genetically targeted optical control of neural activity. Nat Neurosci 8:1263–1268. doi:10.1038/nn1525

2. Aravanis AM, Wang L-P, Zhang F, Meltzer LA, Mogri MZ, Schneider MB, Deisseroth K (2007) An optical neural interface: in vivo control of rodent motor cortex with integrated fiberoptic and optogenetic technology. J Neural Eng 4:S143. doi:10.1088/1741-2560/4/3/S02

3. Zeng H, Madisen L (2012) Mouse transgenic approaches in optogenetics. Prog Brain Res 196:193–213. doi:10.1016/B978-0-444-59426-6.00010-0

4. Witten IB, Steinberg EE, Lee SY, Davidson TJ, Zalocusky KA, Brodsky M, Yizhar O, Cho SL, Gong S, Ramakrishnan C, Stuber GD, Tye KM, Janak PH, Deisseroth K (2011) Recombinase-driver rat lines: tools, techniques, and optogenetic application to dopamine-mediated reinforcement. Neuron 72:721–733. doi:10.1016/j.neuron.2011.10.028

5. Niu Y, Shen B, Cui Y, Chen Y, Wang J, Wang L, Kang Y, Zhao X, Si W, Li W, Xiang AP, Zhou J, Guo X, Bi Y, Si C, Hu B, Dong G, Wang H, Zhou Z, Li T, Tan T, Pu X, Wang F, Ji S, Zhou Q, Huang X, Ji W, Sha J (2014) Generation of gene-modified cynomolgus monkey via Cas9/RNA-mediated gene targeting in one-cell embryos. Cell 156:836–843. doi:10.1016/j.cell.2014.01.027

6. Sasaki E, Suemizu H, Shimada A, Hanazawa K, Oiwa R, Kamioka M, Tomioka I, Sotomaru Y, Hirakawa R, Eto T, Shiozawa S, Maeda T, Ito M, Ito R, Kito C, Yagihashi C, Kawai K, Miyoshi H, Tanioka Y, Tamaoki N, Habu S, Okano H, Nomura T (2009) Generation of transgenic non-human primates with germline transmission. Nature 459:523–527. doi:10.1038/nature08090

7. Li N, Chen T-W, Guo ZV, Gerfen CR, Svoboda K (2015) A motor cortex circuit for motor planning and movement. Nature 519:51–56. doi:10.1038/nature14178

8. Kuhlman SJ, Huang ZJ (2008) High-resolution labeling and functional manipulation of specific neuron types in mouse brain by cre-activated viral gene expression. PLoS One 3:e2005. doi:10.1371/journal.pone.0002005

9. Gradinaru V, Zhang F, Ramakrishnan C, Mattis J, Prakash R, Diester I, Goshen I, Thompson KR, Deisseroth K (2010) Molecular and cellular approaches for diversifying and extending optogenetics. Cell 141:154–165. doi:10.1016/j.cell.2010.02.037

10. Callaway EM (2008) Transneuronal circuit tracing with neurotropic viruses. Curr Opin Neurobiol 18:617–623. doi:10.1016/j.conb.2009.03.007

11. Hnasko TS, Perez FA, Scouras AD, Stoll EA, Gale SD, Luquet S, Phillips PEM, Kremer EJ, Palmiter RD (2006) Cre recombinase-mediated restoration of nigrostriatal dopamine in dopamine-deficient mice reverses hypophagia and bradykinesia. Proc Natl Acad Sci 103:8858–8863. doi:10.1073/pnas.0603081103

12. Judkewitz B, Rizzi M, Kitamura K, Häusser M (2009) Targeted single-cell electroporation of mammalian neurons in vivo. Nat Protoc 4:862–869. doi:10.1038/nprot.2009.56

13. Dugué GP, Akemann W, Knöpfel T (2012) A comprehensive concept of optogenetics. Prog Brain Res 196:1–28. doi:10.1016/B978-0-444-59426-6.00001-X

14. Miyashita T, Shao YR, Chung J, Pourzia O, Feldman DE (2013) Long-term channelrhodopsin-2 (ChR2) expression can induce abnormal axonal morphology and targeting in cerebral cortex. Front Neural Circuits 7(8). doi:10.3389/fncir.2013.00008

15. Schwarz C, Hentschke H, Butovas S, Haiss F, Stüttgen MC, Gerdjikov TV, Bergner CG, Waiblinger C (2010) The head-fixed behaving rat—procedures and pitfalls. Somatosens Mot Res 27:131–148. doi:10.3109/08990220.2010.513111

16. Warden MR, Selimbeyoglu A, Mirzabekov JJ, Lo M, Thompson KR, Kim S-Y, Adhikari A, Tye KM, Frank LM, Deisseroth K (2012) A prefrontal cortex-brainstem neuronal projection that controls response to behavioural challenge. Nature 492:428–432. doi:10.1038/nature11617

17. Vite CH, Passini MA, Haskins ME, Wolfe JH (2003) Adeno-associated virus vector-mediated transduction in the cat brain. Gene Ther 10:1874–1881. doi:10.1038/sj.gt.3302087

18. Zhang F, Gradinaru V, Adamantidis AR, Durand R, Airan RD, de Lecea L, Deisseroth K (2010) Optogenetic interrogation of neural circuits: technology for probing mammalian brain structures. Nat Protoc 5:439–456. doi:10.1038/nprot.2009.226

19. Yizhar O, Fenno LE, Davidson TJ, Mogri M, Deisseroth K (2011) Optogenetics in neural systems. Neuron 71:9–34. doi:10.1016/j.neuron.2011.06.004

20. Hardung S, Epple R, Jäckel Z, Eriksson D, Uran C, Senn V, Gibor L, Yizhar O, Diester I (2017) A functional gradient in the rodent prefrontal cortex supports behavioral inhibition. Curr Biol 27:549–555. doi:10.1016/j.cub.2016.12.052

21. Noudoost B, Moore T (2011) A reliable microinjectrode system for use in behaving monkeys. J Neurosci Methods 194:218–223. doi:10.1016/j.jneumeth.2010.10.009

22. Diester I, Kaufman MT, Mogri M, Pashaie R, Goo W, Yizhar O, Ramakrishnan C, Deisseroth K, Shenoy KV (2011) An optogenetic toolbox designed for primates. Nat Neurosci 14:387–397. doi:10.1038/nn.2749

23. Ozden I, Wang J, Lu Y, May T, Lee J, Goo W, O'Shea DJ, Kalanithi P, Diester I, Diagne M, Deisseroth K, Shenoy KV, Nurmikko AV (2013) A coaxial optrode as multifunction write-read probe for optogenetic studies in non-human primates. J Neurosci Methods 219:142–154. doi:10.1016/j.jneumeth.2013.06.011

24. Lee J, Ozden I, Song Y-K, Nurmikko AV (2015) Transparent intracortical microprobe array for simultaneous spatiotemporal optical stimulation and multichannel electrical recording. Nat Methods 12:1157–1162. doi:10.1038/nmeth.3620

25. Yaroslavsky AN, Schulze PC, Yaroslavsky IV, Schober R, Ulrich F, Schwarzmaier H-J (2002) Optical properties of selected native and coagulated human brain tissues in vitro in the visible and near infrared spectral range. Phys Med Biol 47:2059. doi:10.1088/0031-9155/47/12/305

26. Szél Á, Röhlich P (1992) Two cone types of rat retina detected by anti-visual pigment antibodies. Exp Eye Res 55:47–52. doi:10.1016/0014-4835(92)90090-F

27. Han X, Qian X, Bernstein JG, Zhou H, Franzesi GT, Stern P, Bronson RT, Graybiel AM, Desimone R, Boyden ES (2009) Millisecond-timescale optical control of neural dynamics in the nonhuman primate brain. Neuron 62:191–198. doi:10.1016/j.neuron.2009.03.011

28. Cardin J (2012) Integrated optogenetic and electrophysiological dissection of local cortical circuits in vivo. In: Fellin T, Halassa M (eds) Neuronal Netw. Anal. Humana Press, Totowa, NJ, pp 339–355

29. Berndt A, Schoenenberger P, Mattis J, Tye KM, Deisseroth K, Hegemann P, Oertner TG (2011) High-efficiency channelrhodopsins for fast neuronal stimulation at low light levels. Proc Natl Acad Sci 108:7595–7600. doi:10.1073/pnas.1017210108

30. Gradinaru V, Thompson KR, Zhang F, Mogri M, Kay K, Schneider MB, Deisseroth K (2007) Targeting and readout strategies for fast optical neural control in vitro and in vivo. J Neurosci 27:14231–14238. doi:10.1523/JNEUROSCI.3578-07.2007

31. Kravitz AV, Freeze BS, Parker PRL, Kay K, Thwin MT, Deisseroth K, Kreitzer AC (2010) Regulation of parkinsonian motor behaviours by optogenetic control of basal ganglia circuitry. Nature 466:622–626. doi:10.1038/nature09159

32. Jin X, Tecuapetla F, Costa RM (2014) Basal ganglia subcircuits distinctively encode the parsing and concatenation of action sequences. Nat Neurosci 17:423–430. doi:10.1038/nn.3632

33. Guo ZV, Li N, Huber D, Ophir E, Gutnisky D, Ting JT, Feng G, Svoboda K (2014) Flow of cortical activity underlying a tactile decision in mice. Neuron 81:179–194. doi:10.1016/j.neuron.2013.10.020

34. Ohayon S, Grimaldi P, Schweers N, Tsao DY (2013) Saccade modulation by optical and electrical stimulation in the macaque frontal eye field. J Neurosci 33:16684–16697. doi:10.1523/JNEUROSCI.2675-13.2013

35. O'Shea DJ, Goo W, Diester I, Kalanithi P, Yizhar O, Ramakrishnan C, Deisseroth K, Shenoy KV (2011) Optogenetic control of excitatory neurons via a red-shifted opsin in primate premotor cortex. Program No. 306.11. Neuroscience meeting planner. Society for Neuroscience, Washington, DC (online)

36. Gerits A, Farivar R, Rosen BR, Wald LL, Boyden ES, Vanduffel W (2012) Optogenetically induced behavioral and functional network changes in primates. Curr Biol 22:1722–1726. doi:10.1016/j.cub.2012.07.023

37. Cavanaugh J, Monosov IE, McAlonan K, Berman R, Smith MK, Cao V, Wang KH, Boyden ES, Wurtz RH (2012) Optogenetic inactivation modifies monkey visuomotor behavior. Neuron 76:901–907. doi:10.1016/j.neuron.2012.10.016

38. Gilja V, Chestek CA, Diester I, Henderson JM, Deisseroth K, Shenoy KV (2011) Challenges and opportunities for next-generation intracortically based neural prostheses. IEEE Trans Biomed Eng 58:1891–1899. doi:10.1109/TBME.2011.2107553

39. Grosenick L, Marshel JH, Deisseroth K (2015) Closed-loop and activity-guided optogenetic control. Neuron 86:106–139. doi:10.1016/j.neuron.2015.03.034

40. Shenoy KV, Carmena JM (2014) Combining decoder design and neural adaptation in brain-

machine interfaces. Neuron 84:665–680. doi:10.1016/j.neuron.2014.08.038

41. Paz JT, Davidson TJ, Frechette ES, Delord B, Parada I, Peng K, Deisseroth K, Huguenard JR (2013) Closed-loop optogenetic control of thalamus as a tool for interrupting seizures after cortical injury. Nat Neurosci 16:64–70. doi:10.1038/nn.3269

42. Siegle JH, Wilson MA (2014) Enhancement of encoding and retrieval functions through theta phase-specific manipulation of hippocampus. eLife 3:e03061. doi:10.7554/eLife.03061

43. Newman JP, Fong M, Millard DC, Whitmire CJ, Stanley GB, Potter SM (2015) Optogenetic feedback control of neural activity. eLife 4:e07192. doi:10.7554/eLife.07192

44. O'Connor DH, Hires SA, Guo ZV, Li N, Yu J, Sun Q-Q, Huber D, Svoboda K (2013) Neural coding during active somatosensation revealed using illusory touch. Nat Neurosci 16:958–965. doi:10.1038/nn.3419

45. Pouille F, Scanziani M (2001) Enforcement of temporal fidelity in pyramidal cells by somatic feed-forward inhibition. Science 293: 1159–1163. doi:10.1126/science.1060342

46. Wehr M, Zador AM (2003) Balanced inhibition underlies tuning and sharpens spike timing in auditory cortex. Nature 426:442–446. doi:10.1038/nature02116

47. Allen BD, Singer AC, Boyden ES (2015) Principles of designing interpretable optogenetic behavior experiments. Learn Mem 22:232–238. doi:10.1101/lm.038026.114

48. Lee JH, Durand R, Gradinaru V, Zhang F, Goshen I, Kim D-S, Fenno LE, Ramakrishnan C, Deisseroth K (2010) Global and local fMRI signals driven by neurons defined optogenetically by type and wiring. Nature 465:788–792. doi:10.1038/nature09108

49. Schmid F, Wachsmuth L, Albers F, Schwalm M, Stroh A, Faber C (2017) True and apparent optogenetic BOLDfMRI signals. Magn Reson Med 77:126–136. doi:10.1002/mrm.26095

50. Berndt A, Yizhar O, Gunaydin LA, Hegemann P, Deisseroth K (2009) Bi-stable neural state switches. Nat Neurosci 12:229–234. doi:10.1038/nn.2247

51. Lu Y, Truccolo W, Wagner FBP, Vargas-Irwin CE, Ozden I, Zimmermann JB, May T, Agha N, Wang J, Nurmikko AV (2015) Optogenetically-induced spatiotemporal gamma oscillations and neuronal spiking activity in primate motor cortex. J Neurophysiol 113 (10):3574–3587. doi:10.1152/jn.00792.2014

52. Wietek J, Beltramo R, Scanziani M, Hegemann P, Oertner TG, Simon Wiegert J (2015) An improved chloride-conducting channelrhodopsin for light-induced inhibition of neuronal activity in vivo. Sci Rep 5:14807. doi:10.1038/srep14807

53. Kim T, McCall JG, Jung YH, Huang X, Siuda ER, Li Y, Song J, Song YM, Pao HA, Kim R-H, Lu C, Lee SD, Song I-S, Shin G, Al-Hasani R, Kim S, Tan MP, Huang Y, Omenetto FG, Rogers JA, Bruchas MR (2013) Injectable, cellular-scale optoelectronics with applications for wireless optogenetics. Science 340:211–216. doi:10.1126/science.1232437

54. McCall JG, Kim T, Shin G, Huang X, Jung YH, Al-Hasani R, Omenetto FG, Bruchas MR, Rogers JA (2013) Fabrication and application of flexible, multimodal light-emitting devices for wireless optogenetics. Nat Protoc 8:2413–2428. doi:10.1038/nprot.2013.158

55. Park SI, Brenner DS, Shin G, Morgan CD, Copits BA, Chung HU, Pullen MY, Noh KN, Davidson S, Oh SJ, Yoon J, Jang K-I, Samineni VK, Norman M, Grajales-Reyes JG, Vogt SK, Sundaram SS, Wilson KM, Ha JS, Xu R, Pan T, Kim T, Huang Y, Montana MC, Golden JP, Bruchas MR, Gereau Iv RW, Rogers JA (2015) Soft, stretchable, fully implantable miniaturized optoelectronic systems for wireless optogenetics. Nat Biotechnol 33:1280–1286. doi:10.1038/nbt.3415

56. Pisanello F, Sileo L, Oldenburg IA, Pisanello M, Martiradonna L, Assad JA, Sabatini BL, De Vittorio M (2014) Multipoint-emitting optical fibers for spatially addressable in vivo optogenetics. Neuron 82:1245–1254. doi:10.1016/j.neuron.2014.04.041

57. Grossman N, Poher V, Grubb MS, Kennedy GT, Nikolic K, McGovern B, Palmini RB, Gong Z, Drakakis EM, Neil MAA, Dawson MD, Burrone J, Degenaar P (2010) Multi-site optical excitation using ChR2 and micro-LED array. J Neural Eng 7:016004. doi:10.1088/1741-2560/7/1/016004

58. Yizhar O, Fenno LE, Prigge M, Schneider F, Davidson TJ, O'Shea DJ, Sohal VS, Goshen I, Finkelstein J, Paz JT, Stehfest K, Fudim R, Ramakrishnan C, Huguenard JR, Hegemann P, Deisseroth K (2011) Neocortical excitation/inhibition balance in information processing and social dysfunction. Nature 477:171–178. doi:10.1038/nature10360

59. Breese GR, Knapp DJ, Criswell HE, Moy SS, Papadeas ST, Blake BL (2005) The neonate-6-hydroxydopamine-lesioned rat: a model for clinical neuroscience and neurobiological principles. Brain Res Rev 48:57–73. doi:10.1016/j.brainresrev.2004.08.004

Chapter 12

Employing Optogenetics in Memory Research

Limor Regev and Inbal Goshen

Abstract

Optogenetics presents many opportunities for memory research, and was indeed warmly embraced by the field, and already employed to probe memory mechanisms in over a hundred published projects. The incorporation of optogenetics enabled scientists to causally pinpoint the real-time roles of specific neuronal populations within the brain structures underlying memory, the functional connections between them, and the dynamics of memory representation over time.

This chapter presents an elaborate point-by-point plan of designing and executing any optogenetic memory experiment, recognizes possible pitfalls, and offers solutions. The technical aspects will be discussed in light of multiple examples of the ways in which optogenetics has been used in memory research, and the exciting insight it provided.

Key words Memory fear conditioning, Spatial navigation, Recognition, Operant conditioning, Freely moving, Control group

1 Introduction: Challenges in Traditional Memory Research

Memory acquisition, consolidation, and recall have fascinated neuroscientists from the early days of the discipline. In recent decades, a variety of techniques, like physical, pharmacological, and genetic lesions of specific brain areas, combined with electrical or pharmacological stimulation of these areas, were employed by the pioneers of memory research to identify the major brain areas that are involved in various cognitive tasks. This huge body of research led to many groundbreaking findings, some of them contradictory, and raised many interesting questions regarding the precise functional connections between the different regions involved in memory processes, and the contribution of other areas that were hitherto considered irrelevant. Furthermore, none of the techniques mentioned above allows a causal implication of specific neuronal populations within these structures in real-time cognitive function.

The introduction of optogenetics, allowing real-time control of genetically defined neuronal populations with millisecond precision, provided ample opportunities for memory research, and was

Albrecht Stroh (ed.), *Optogenetics: A Roadmap*, Neuromethods, vol. 133,
DOI 10.1007/978-1-4939-7417-7_12, © Springer Science+Business Media LLC 2018

indeed warmly embraced by the field, and already employed to probe memory mechanisms in over a hundred published projects. The incorporation of optogenetics enabled scientist to causally pinpoint the real-time roles of specific neuronal populations within the brain structures underlying memory, and the functional connections between them. Specifically, new insights were gained regarding the distinct roles that different brain regions, like hippocampal subregions (CA1, CA2, DG...), dorsal vs. ventral hippocampus, amygdala, thalamus, and cortex play in specific memory forms. In addition, the insight into memory dynamics over time and the ability to artificially create memory traces were explored [1–4].

This chapter presents the principles of designing and executing optogenetic memory experiments, points to possible pitfalls, and offers solutions. The technical aspects will be discussed in light of examples of the way in which optogenetics has been used to substantiate long-standing hypotheses, modify others, and provide a basis for new memory theories.

2   Design and Execution of Optogenetic Memory Studies

This section of the chapter discusses the specific practical steps one has to follow when designing and performing a memory study combined with an optogenetic manipulation. As previous chapters discussed opsin choice and in vivo light delivery in depth (*see* Chaps. 1, 2, 9, and 10), this chapter only stresses the aspects specifically relevant to memory research regarding these topics, and refers to the relevant detailed chapters when necessary.

Figure 1 illustrates the different steps in optogenetic memory research, the timeframe in which they are performed, and the different considerations one has to take into account when planning such experiments. This section elaborates on memory testing techniques and the necessary adaptations required to make them "optogenetics friendly", the surgical and light delivery options, the appropriate control groups, and the post-experiment verifications.

2.1 Opsin Expression and Light Delivery

To induce opsin expression in discrete cell populations, a foreign DNA sequence encoding for the opsin has to be delivered into the brain. In most cases, this DNA is delivered using viral vectors injected into the area of interest (*see* Chap. 1). Within days to weeks, the opsins are expressed throughout the cell membrane, around the soma and in distant projections. Illumination is achieved by implanting an optical fiber above the region of interest, and connecting it to a light source (*see* Chap. 13). In experiments requiring projection targeting, cell bodies are infected in one brain region and their axonal projections are illuminated in another.

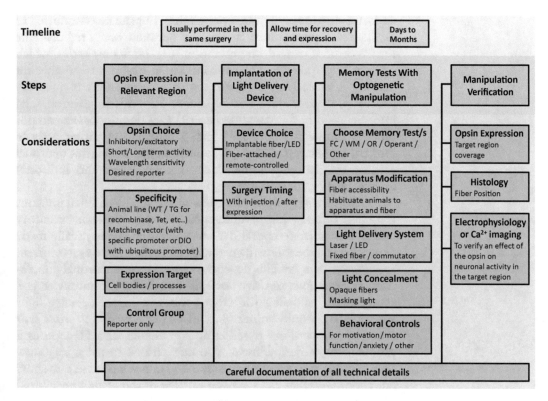

Fig. 1 Technical consideration when designing and executing an optogenetic memory study

The various choices researchers are faced with and the considerations that have to be taken into account are detailed in this section.

2.1.1 Opsin Choice

As reviewed in length in previous (Chaps. 2 and 3), a broad collection of opsins is available for researchers these days, and the variety keeps growing continuously. Beyond the clear advantages of this variety, for many investigators this endless range poses a problem in selecting the right opsin for their experiment. Here, without naming specific opsins (as new ones are published very often), we will provide the basic guidelines for choosing a suitable opsin, by the nature of the modulation it induces, its speed, its wavelength sensitivity, and the attached fluorophore, by answering the following questions:

What do you want your opsin to do? That is the basic question one has to start with when considering an opsin choice. Broadly speaking, opsins can be divided into three major functional families: (1) *Excitatory opsins*, which depolarize the neurons upon illumination. This is the largest family, offering the largest variety of choices. These opsins are suitable for proving the *sufficiency* of activating a certain neuronal population to generate a certain memory process, or even by itself serve as a memory trace. However, as excitation does not stop in the activated population, but can be transmitted to downstream populations, within or between brain regions, the results require careful interpretation. (2) *Inhibitory opsins*, which actively hyperpolarize the neuron. These opsins are suitable for proving a *necessity* of a certain neuronal population to memory processes, as they allow one to silence a specific population. The interpretation in this case is easier. (3) *Cell signaling opsins* do not directly modulate the membrane potential, but rather activate internal signaling pathways [5], which may in turn affect neuronal activity by changing intracellular calcium levels and indirectly induce spiking, for example.

How fast do you want your opsin to act? As discussed in previous chapters, one significant difference between available excitatory opsins is in their temporal dynamics. This choice is usually made depending on the following parameters: (1) *how fast do you want the manipulation to be?* This depends on your target neuronal population. For example, to drive fast-spiking inhibitory neurons, a fast-acting opsin, allowing the cell to follow a high-frequency light stimulus [6], would be necessary. (2) *How long do you want your manipulation to last?* If tethering the animal to a fiber poses a problem to your experiment, you may choose to use a very slow opsin (stable step function opsin, SSFO), in which case one short light stimulus (that can be given outside of the testing apparatus) will mildly activate the target population for over 30 min [7]. In this case, you do not directly control the spiking pattern of the cell, but rather increase its responsiveness to the naturally occurring inputs it receives. This variety in temporal dynamics is currently available only for excitatory opsins.

What wavelength do you want your opsin to respond to? Your opsin choice will determine the wavelength of the light stimulus. Beyond the peak response to their optimal wavelength, opsins are activated by a rather wide spectral range at lower efficiency, which may still be strong enough to effectively change neuronal activity [8]. As discussed in previous (Chaps. 9 and 10), different wavelengths penetrate the brain tissue at different efficiency: As a rule of thumb, the lower the wavelength, the more temporally confined your stimulation, and the higher the wavelength (towards red and infra-red) the deeper and wider it will penetrate. Initially, excitatory opsins were optimally activated by blue light, whereas inhibitory opsins were most responsive to yellow/green illumination [8]. Today, red-shifted excitatory opsins, and blue shifted inhibitory opsins are also available [7, 9–12]. It should be stressed again that each opsin is activated by a wide spectrum of wavelengths, so if attempting to combine two opsins activated by different peak wavelength, one should go to the opposite extremes of each opsin's sensitivity distribution to avoid dual activation [7].

How do you want to visualize your opsin? Opsins are usually accompanied by a fluorophore, allowing their visualization during or after the experiment. When choosing the fluorophore consider these two questions: (1) *What color do you want your fluorophore to emit?* This choice should be based on the following parameters: First if there is any previously existing fluorescence in the brain (for example, in transgenic lines expressing a fluorophore in a specific neuronal population), choose an opsin attached to a fluorophore emitting a wavelength as far as possible from the native one, to facilitate the differentiation between the two. Second, if you wish to visualize your opsin expression during the experiment, choose a fluorophore for which the excitation light is as far as possible from the excitation light of the opsin, in order not to affect neuronal activity during imaging. (2) *Where do you want your fluorophore to be?* The first opsins, and to date most opsins are fused to the fluorophore, which means that the cells produce one big protein—an opsin-fluorophore chimera. The advantages of this configuration are: First, imaging the spatial distribution of the fluorophore reports directly the position of the opsin and even the smallest neuronal processes can be imaged when the opsin-fluorophore chimera is expressed in them. Second, the intensity of the fluorophore is a good indicator to the potency of the current induced by opsin illumination [7]. The disadvantages of the chimeric configuration are: First, trafficking of the opsin to the membrane is more demanding—as a bigger protein (including a fluorophore) has to be shipped into place. Second, imaging the cell bodies expressing the opsin is difficult, as the fluorophore does not fill the cytoplasm, but is membrane-bound, thus it is hard to determine where one cell ends and the other begins, especially in highly dense regions. To overcome this problem and facilitate easier cellular identification, one can use opsins that are expressed separately from the reporter fluorophore, by inserting a P2A self-cleaving protein sequence between the two [13]. Consequently, the opsin and fluorophore are transcribed and translated together, and then cleaved, providing a one to one expression ratio, in which the opsin is expressed in the membrane, whereas the fluorophore remains in the cytoplasm, expediting imaging [14]. Alternately, an internal ribosome entry site (IRES) sequence can be inserted between the opsin and the fluorophore, in which case they will be translated separately, with a lower expression rate for the fluorophore [15].

2.1.2 Opsin Expression Specificity

Genetically defined cell-type specificity can be achieved by three main strategies: If the known unique promoter of a cellular population is specific, strong and small enough it can be packed into the viral vector together with the opsin gene, and directly drive expression in the relevant cells [16] (*see* Chaps. 1 and 2). Such a strategy is commonly used to target neurons (without differentiating between subpopulations, using the human synapsin promoter), excitatory glutamatergic neurons (using the CamKIIα promoter), Hypocretin neurons (using the Hcrt promoter), and astrocytes (using the GFAP promoter). Alternatively, if the relevant promoter is too weak to drive sufficient expression, or too large to be packed into a virus, transgenic animals can be used. Several mouse lines expressing opsins in the brain were developed [16, 17], but this approach limits the use to the transgenically expressed opsin, and does not allow the integration of newly generated opsins. A more versatile strategy is to use transgenic mouse lines that express Cre recombinase in a specific cellular population, and then introduce a virus containing any opsin DNA in an inverted orientation, flanked by two sets of incompatible Cre recombinase recognition sequences. Only in those cells that transgenically express Cre will the DNA sequence be flipped into the correct orientation, allowing transcription of the coding region under a strong ubiquitous promoter [18]. This approach was widely used to target specific populations in memory research, such as dopaminergic neurons (using TH-Cre mice and rats), cholinergic neurons (using ChAT-Cre mice and rats), parvalbumin interneurons (using PV-Cre mice), somatostatin interneurons (using Som-Cre mice), and more.

Connectivity-defined specificity: Several virus-based techniques now allow researchers to explore the role of specific cell populations based on their connectivity to other cells. Some tools use viruses that can cross synapses, whereas other tools that are becoming popular directly target terminals.

To express an opsin in the cells that project to a specific population, one can infect the target population with a pseudotyped rabies virus, engineered for monosynaptic retrograde spread. The cells forming synapses with the infected population will then express whatever transgene is carries by the virus, for example Cre recombinase [19, 20]. One can then inject a virus carrying a Cre-dependent opsin to the somatic region of these presynaptic cells, resulting in opsin expression in the neuronal population/s innervating the specific target that was initially infected with the pseudo-typed rabies virus [19, 20]. Herpes viruses can also spread between the infected neurons, both retrogradely and anterogradely. However, a specific herpes strain (the Bartha PRV strain) has a retrograde-restricted spread, and can be used for neuroscience research [19, 21] when inducing Cre expression, as explained above for rabies.

In the techniques described so far viral vectors infect the cell bodies and then travel retrogradely through synapses. Canine adenovirus (CAV) vectors, on the other hand are internalized directly by terminals and then transported through axon to the soma [22]. Thus CAV injection to the target region (rather than a specific target population) will result in Cre (for example) expression in the cell bodies of the infected projections, which can be subsequently transfected by a Cre-dependent opsin. This strategy was already used in memory research to study ACC projections to the hippocampus [23] and prefrontal projections to the amygdala [24].

Activity-defined specificity: New genetic tools now allow opsin expression based on cellular activity, providing an opportunity to specifically control a population of neurons that participated in a certain behavioral task using optogenetics. Opsin targeting to *activity*-defined populations can be achieved by using the cFos-tTA mouse line, expressing a tetracycline transactivator (tTA) under the promoter for the neuronal activity-dependent gene cFos (*see* also Chap. 2). tTA is expressed in cells that were recently active following a specific task. The cFos-tTA mice can be injected with a virus encoding for an opsin, whose expression is controlled by the tetracycline responsive element (TRE), which depends on tTA binding to drive the opsin. Thus, neuronal activity will drive the expression of tTA, which in turn will enable the expression of the opsin in the active cells only. To restrict this expression to a specific time window mice are kept on a diet containing doxycycline (Dox), which prevents tTA from binding to TRE. During the activity of interest Dox can be removed, and consequently the opsin will be expressed [25–31].

2.1.3 Illumination Strategy

Early optogenetic studies have implanted cannulas above the target region, and then inserted the optical fibers through them when illumination was required [32]. Later, implantable fibers became available, and replaced the cannulas. Briefly, a bare fiber is implanted above the target region, with its polished top part encased in a ferrule, and attached to the skull with dental cement (*see* Chap. 13 for details). When illumination is required, this ferrule is attached using a plastic sleeve to another ferrule, at the end of the fiber coming from the light source [33–36]. Implantable ferrules are commercially available, and can also be fabricated in house [36, 37].

The vast majority of experiments are performed using implantable fibers, connected to long optic fibers coming from the light source. Wireless options are also available [38, 39], but as they are yet to be used in memory research, we will not elaborate on them here. Note that when using long-acting opsins, one can activate the opsin before the memory testing or training, and have the animal behave in the apparatus with no fiber attached.

2.1.4 Opsin Expression and Light Targeting

After the choices of animal, opsin, and light delivery have been made, one has to deliver the opsin and the light to the correct location/s. Targeting the opsins surgically to the correct place is performed in a standard stereotaxic injection surgery [40], as is the fiber implantation. The first and simplest optogenetic studies in vivo were performed by expressing an opsin in a desired neuronal population and then stimulating the cell bodies with light [18, 41, 42]. In such cases, the inputs from these cell bodies to all their downstream target regions are modified. More recent studies are targeting specific projections. This can be done by molecular methods (see above), or simply by injecting a virus to the cell bodies and illuminating in the target region [23, 43]. This approach requires careful control studies and cautious interpretation, as it is possible for the spikes generated in the activated projections to antidromically activate the soma, and subsequently activate alternative projections from the same cell bodies, if such exist.

2.1.5 *Control Groups*

As the surgical procedure, the light delivery and the expression of a foreign protein are all significant processes that may affect behavior, carefully designed controls are required to make sure that the observed effects are due to the neuronal activity modulation itself. The most straightforward and commonly used control, covering all of these aspects, is to use a fluorophore-only group. In such cases, the control animals are injected with a virus of the same type and the same serotype, inducing the expression of the same fluorophore that is attached to the opsin in the experimental group, but unaccompanied by an opsin. This group is then exposed to the same illumination protocol and behavioral procedure as the experimental group. As the surgical procedure, the virus infection, the expression of a foreign protein, and the light exposure are almost identical, whatever differences are observed from the experimental group can be attributed to the activity of the opsin. Other control groups that are sometimes used are: (1) Animals expressing the same opsin but not receiving any light stimulation. This is not an appropriate control, as the light can heat the brain mildly and thus directly affect neuronal activity even when no opsin is present [44]. (2) Animals expressing the same opsin, but exposed to light in a wavelength outside the active range of the opsin. This is an acceptable, but not commonly used control. (3) Animals receiving no virus injection, and exposed to the same illumination protocol. This is not an acceptable control, especially as fluorophore-only vectors are now widely available, because the surgery and the expression of a foreign protein may by themselves change the activity of the transfected neurons. (4) When using Cre driver lines, Cre-negative animals injected with the same vector as the experimental group, which should result in no expression, and exposed to the same illumination protocol. This is not an optimal control, as the expression of two foreign proteins (Cre recombinase and the injected vector) may by itself change the activity of the transfected neurons, so it is preferable to inject Cre expressing animals with a control fluorophore-only vector.

2.1.6   Timing Considerations

When planning an optogenetic memory study, one should allow enough time for the opsin to express and for the animal to recover properly from surgery before commencing the behavioral experiment. When targeting cell bodies, sufficient levels of expression can be achieved in a few days to a few weeks (normally 3 weeks), depending on the type of the virus and its serotype. In such cases, the optical fiber can be transplanted in the same surgery as the injection, thus allowing the animal to recover while opsin expression levels are established. When projections are targeted, expression takes a longer time to accumulate in sufficient levels in distant processes, which could take 2–4 months. In such cases, it is advisable to implant the fiber in a separate surgical procedure ~2 weeks before commencing the behavioral experiment. This way, one makes sure that the implant will not be damaged before the experiment on the one hand, and that the animal has enough time to recover from surgery on the other.

2.2   Common Memory Testing Paradigms

2.2.1   Fear Conditioning and Extinction

Classical fear conditioning (FC; Fig. 2a) is a common powerful experimental model for studying the neural basis of associative learning and memory formation in mammals [45–47]. The simultaneous presentation of a neutral conditioned stimulus (CS) of any sensory modality (auditory, visual, olfactory...) and an aversive unconditioned stimulus (US) renders the formerly neutral stimulus a frightful quality, so that even when it appears by itself, without the aversive stimulus, it will elicit a fearful conditioned response.

Fear conditioning can be rapidly formed in humans and animals, even following a single conditioning trial, and is usually maintained for long periods. In rodents, the dominant behavioral fear response is freezing (complete immobility), and the most commonly used aversive stimulus is the delivery of a weak short electrical shock [48]. Other versions of the paradigm can be performed in head-fixed animals, in which case the conditioned response will be inhibition of an appetitive behavior like liquid or food consumption [49].

The conditioning process itself (i.e., the association between the neutral and aversive stimuli) is mediated primarily by the amygdala [45–47]. By using different types of conditioned stimuli the fear-conditioning paradigm enables differentiation between hippocampal dependent and independent functions: When a simple perceptual conditioned stimulus (such as a light or an auditory cue) is used, the hippocampus is not required. However, when the conditioned stimulus is a new environment, a mental representation of this new context has to be created, so that the amygdala can associate this representation with the aversive stimulus; this function depends on hippocampal functioning [46, 48, 50]. In the most commonly used version of the fear conditioning paradigm, animals are placed in a novel conditioning cage, in which they hear an auditory tone, followed by a short foot-shock. Thus, the animal

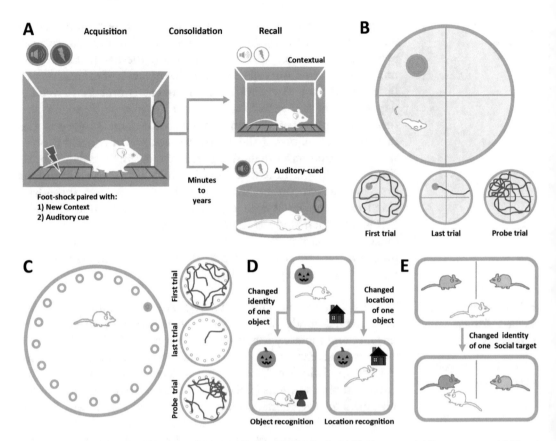

Fig. 2 Common memory testing paradigms. (**a**) Fear-conditioning*. (**b**) Water maze. (**c**) Barnes maze. (**d**) Novel object recognition. (**e**) Social recognition. *Replicated, with permission, from Goshen, Trends in Neuroscience, 2014

can associate the aversive stimulus with the new context as well as with the tone. To test contextual fear conditioning, animals are placed again in the original conditioning cage, and freezing is measured. This task is hippocampal dependent, as it cannot be performed following hippocampal lesions [48] or following opto-genetic hippocampal inhibition [51, 52]. To test the hippocampal-independent auditory-cued fear conditioning, freezing is measured when the tone is sounded in a differently shaped context. In this case, hippocampal lesions, or optogenetic hippocampal silencing have no detrimental effect on performance level [47, 48, 50, 51, 53]. Lesions to the amygdala, or optogenetic amygdalar silencing, on the other hand, impair both auditory-cued, and contextual FC [51].

Following the acquisition and recall of the conditioned fear, the FC paradigm then offers access to an additional kind of learning—extinction. In such experiments, the conditioned response to the CS is gradually extinguished by repeated exposure to the CS without co-incidence with the US. Extinction does not represent forgetting, but rather learning that a previously fearful stimulus is actually safe, at least in the extinction context. This extinction process depends on a complex neuronal circuit, involving the pre-frontal cortex (PFC), and optogenetic studies are now testing the functional connection to/from the PFC in extinction learning.

The major strengths of the FC model are the simplicity of the task, the defined anatomy of the basic neuronal circuit, and the well-defined temporal separation between different stages of memory: acquisition, consolidation, and recall. Due to its simplicity, this paradigm has been the most extensively used in combination with optogenetics both in the regular [7, 17, 24–31, 51, 52, 54–79] and the appetitive behavior inhibition [23, 80–82] versions.

2.2.2 Spatial Navigation and Memory	*The water maze* paradigm (Fig. 2b), developed by Morris et al. [83] provides insight to spatial learning and memory, procedural learning, and learning flexibility. In this task, animals are placed in a circular water pool and trained to find an escape platform, located at a particular spatial location by using extra-maze visual cues. The latency to find the hidden platform serves as an indicator for the learning process—the stronger the memory, the faster the animal will reach the platform, improving from trial to trial. In the hippocampal-dependent task, the platform remains in a fixed position, but the entry point to the maze is randomly changed from trial to trial. Hippocampal lesions abolish the ability to navigate in the maze [84]. In the non-spatial (hippocampal independent) versions either the platform is visible above water surface, or the entry point does not vary between trials, thus spatial learning is not required, and only procedural learning is required (climbing the platform in order to stop swimming and be removed from the maze), and rodents acquire this learning rapidly.

Following the full acquisition of the spatial task, a probe test (also known as transfer test) can be performed to examine the memory of the platform location: The platform is removed from the maze, and the animal is placed in the maze for a short trial, in which the time it swims in the quadrant in which the platform was located, or the number of times it crosses the former location of the platform, are measured [83]. Additionally, reversal learning trials can be performed, in which the hidden platform is moved to a different location, and the animal's ability to update the former learning is tested. The water maze is a commonly used, and relatively easy to perform and interpret, but involves high stress levels. On the one hand, the stress motivates the animals to find the platform, which is why performance relies on water temperature, and animals learn slower when the temperature is pleasant [85]. On the other hand, optogenetic effects on stress rather than on cognitive performance must be controlled for. The coming paragraphs will describe less stressful spatial memory tasks. Another challenge when using the water maze with optogenetics is that mice (but not rats) may find it difficult to swim when carrying the weight of an implant and of the light deliver fiber connected to the light source. For these reasons, this paradigm was not extensively used in combination with optogenetics [86, 87].

The Oasis Maze is a dry version of the water maze, in which thirsty animals have to find a well containing a small water reward in a circular arena containing hundreds of other empty wells, by using extra-maze visual cues [88]. The latency to find the correct hole or the number of errors (nose pokes in wrong holes) serves as an indicator for the learning process. The oasis maze task is suitable for use with optogenetic manipulation, but requires extensive pre-training trials (in order for the animals to learn to find water in the maze, regardless of location), and then a longer training protocol than used in the water maze.

The Barnes Maze [89] (Fig. 2c) is a dry elevated brightly lit circular arena, with ~20 holes around its circumference, one of them leading to a dark escape box, which the animal has to find by using extra-maze visual cues. As rodents prefer narrow dark places, and find open bright spaces aversive, they are motivated to find the location of the escape box. This paradigm involves only mild stress, and does not require any previous deprivation, which makes it cleaner in terms of stress-related effects on the one hand, but also decreases the motivation of the animals to acquire the task on the other. The latency to find the correct hole or the number of errors (nose pokes in wrong holes) serves as an indicator for the learning process. In a probe trial, entry to the escape box is blocked, and the number of nose pokes into holes in the target quadrant is quantified. Whereas the learning trials of this paradigm would be very challenging to combine with optogenetics (the fiber may be damaged when the mouse enters the escape box), the probe trial can be performed under optogenetic manipulation [86].

The radial arm maze [90] is another spatial memory task, in which the animal does not freely explore an open arena (as in the water or Barnes mazes), but its movement is restricted to closed arms (usually 8 of them). This makes performance easier to quantify manually in a nonbiased way even without an automated tracking system, by simply counting arm entries.

2.2.3 Recognition Memory

Novel object recognition (NOR; Fig. 2d) is a memory task for objects or their location [91]. In the training trial, the animal is exposed to two unfamiliar objects (chosen based on similar average exploration levels in calibrations trials). In the testing trial, which can be performed minutes to hours later, one of the objects is either replaced or moved to a different location. In the object recognition version, one object is replaced with a different object, and this newly presented object attracts more exploratory behavior than the previously encountered object, because of its novelty. In the novel object location version, the two objects presented in the training trial remain the same, but one of them is moved to a new location. This novel spatial position attracts more exploratory behavior from the animal. NOR is easily quantified either manually or automatically, and is very simple to execute. However, it involved minimal motivation levels, and in some mouse strains there is not enough exploration in the training to support recognition in the test. Rats normally show more robust exploration. This method was already successfully combined with optogenetics [81].

Social recognition (SR; Fig. 2e) is a similar task, in which the exploration motivation is increased by using social targets as the recognition objects [92]. As social memory has higher significance than neutral object recognition (as it may be a source of mating or threat, for example), it can last for much longer periods of time. Social recognition is easily quantified manually with no special equipment, and was already successfully combined with optogenetics [93].

2.2.4 Operant Conditioning

In operant conditioning, animals gradually learn to perform a certain action (lever press, direction choice in a maze...) in order to receive a reward or avoid punishment. This learning is achieved by introducing the desired outcome when the correct behavior is performed. As opposed to classical conditioning (like FC) in which the measured behavior is natural and reflexive, and thus can be acquired in a single trial, the operant acquisition process is longer, requiring tens to thousands of exposures, depending on the complexity of the task. Operant conditioning tasks are usually employed in attention and reward studies, rather than memory research per se, and thus will be less elaborated in this chapter.

In simple *Operant conditioning* the animal has to learn to perform a certain activity (e.g., lever press) in order to receive a reward. Originally, the rewards were "real" (food, drink..), then it was shown that animals can be trained to perform an activity in order to receive electrical or chemical brain stimulation [94], and recently optogenetic activation of dopaminergic neurons was also shown to induce operant conditioning [33]. This technique is commonly used to study reward circuitry, and was extensively employed together with optogenetic manipulation [33, 95–99].

In *working memory* tasks, another degree of complexity is introduced. Now, to be rewarded, the animal does not only have to perform the correct action, but also to preform it only when an additional cue is presented. For example, in delayed nonmatch to place tasks, the animal has to take the correct turn in a maze based on a stimulus presented at a different location than the turning point, or even based on the location of the reward in the previous trial (like in the T maze alternation task). Such tasks were repeatedly used in combination with optogenetics [100–104], and in one interesting study, the optogenetic manipulation itself served as the cue, based on which the animal has to choose the correct response [77].

2.2.5 Other Memory Tasks

Various other memory tests, like the Y, H and crossword mazes exist, and were used together with optogenetic manipulations [105–107].

Of special note is the emerging option of performing many of the above-mentioned paradigms using a touchscreen system, in which visual objects are presented to rats or mice, which in turn have to react by touching the screen in a correct location. Such a system can be used for operant conditioning, object recognition, and even spatial learning [108, 109]. This system is not yet in wide use, but it is clearly very suitable to combine with optogenetics as it requires less movement than "real" tasks. To date, only a single published research employed the two technologies (touchscreen and optogenetics) together, to perform an object recognition task [110].

2.3 Apparatus Modifications for Optogenetic Experiments

Most optogenetic studies are performed with the animal connected to the light source via an optical fiber. Thus, the testing apparatus has to be modified to allow a smooth movement of the fiber together with the animal, and protect both from unnecessary strain. In addition, the fiber adds a significant visual noise that can affect automatic tracking. These technical issues and ways to limit their effects are discussed in this section.

2.3.1 Free Movement and Protection for Both Fiber and Subject

Clear a fiber path: The first step in making your behavioral setup optogenetics-compatible is cutting a hole (or preferably a groove) in the top of the apparatus, to allow the connection of the free fiber coming from the light source to the fiber implanted in the brain of the behaving animal inside. Note that in tasks that have an auditory component (like fear conditioning, and some operant tasks) this will render the behavior box not sound-proof.

Use an optical commutator: In long-duration tasks that last more than a few minutes, and require a lot of movement or extended training (e.g., operant training) or in tasks that are motorically demanding (e.g., water maze), it is advisable to use a commutator between the light source and the mouse. The commutator will allow a smooth rotation of the fiber, which has two benefits: First, it puts less mechanical strain on the animal's head, and when mounted on a counterbalanced arm it carries a significant portion of the fiber's weight. Second, the commutator reduces the strain on the fiber, thus extending its life span. Using a commutator also has its prices, though: First, financial. A good commutator will cost a few hundred dollars (and combined with an electrical commutator—a few thousands). Second, adding another fiber coupling point to the optical system will reduce the light intensity by

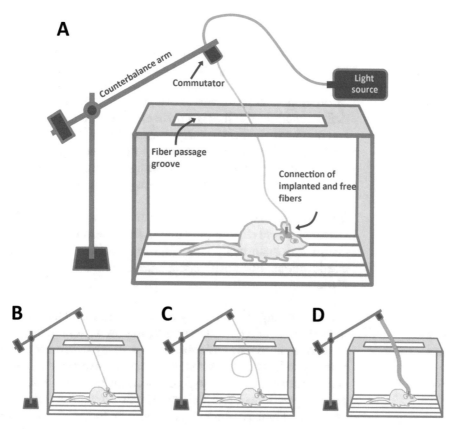

Fig. 3 Apparatus modification for optogenetic memory research. Panel A is modified, with permission, from Witten et al., Neuron, 2011

tens of percent. When using a laser as a light source this will not be a problem, as most lasers emit light intensities that are orders of magnitude higher than needed at the tip of the fiber. When using an LED device, especially for wavelengths in the green-yellow range that tends to be of lower power, this may pose a serious problem. Third, if the optic fiber is additionally used for fluorometric Ca^{2+} recordings, loss of emission light due to the commutator may be critical (*see* Chap. 13).

Use the right fiber length: Beyond the optically relevant properties of the fiber (like diameter and numerical aperture) it is important to consider simple physical parameters. First, the fiber length should allow easy free movement of the animal, and enable it to cover the whole apparatus (see Fig. 3a). The fiber should not be too short, and thus limit exploration (Fig. 3b). On the other hand, it should not be too long, such that it tangles and puts extra weight and strain on the animal as the trial progresses and more loops are twisted (Fig. 3c). Another disadvantage of excess length fibers is that it makes them accessible to the animals to chew on.

Use the right fiber coating: choosing the coating will affect both the animal's performance and the duration of the fiber's life. In order to conceal the light from the animal subject, it is advisable to use a fiber with an opaque, dark colored coating. However, this coating should be of minimal diameter (and thus minimal weight), in order not to put unnecessary strain on the animal. Often the default coating of a fiber is a thick protective sleeve (Fig. 3d) that is too heavy for a mouse to carry. Such a fiber will not pose a problem for a rat, and in fact, it is advisable to use even bigger fiber protection (like a protective metal spring) when performing long-duration experiments in rats [33], to prevent chewing.

2.3.2 Tracking System Adjustments

The presence of a visible fiber can pose a serious problem for tracking systems in the quantification of various parameters, especially those quantifying lack of movement (freezing), but also precise locations. For possible strategies to minimize these effects, *see* **Note 1**.

2.4 Post-Experimental Verification

After collecting the necessary behavioral data one has to verify that indeed the observed differences represent a causal effect of opsin-induced neuronal modulation on cognitive performance. In simple terms, you have to show that your opsin is expressed in the correct cell population and in the right location, that your fiber is appropriately positioned, and that the effects on neuronal activity are as expected. These requirements are met by following these steps:

2.4.1 Verify that Your Opsin Is Expressed as Planned

This step is performed at the very end of the experiment, after all behavioral and physiological exams are done. For histological analysis perfuse the animal first with PBS to wash auto-fluorescing blood cells, and then with paraformaldehyde to fix the tissue. Then slice the brain and image the spread of the fluorophore to assess the location, strength and expression pattern of the opsin. This verification is important for several reasons. For example: (1) Partial expression that does not cover your entire target region may distort your results. (2) Extensive expression that spreads to neighboring regions adjacent to the target one can introduce a bias. (3) High expression levels of a foreign protein can sometimes result in cell death, which may alter your results. Note that this step should be performed in a few animals before attempting the behavioral experiment, to confirm that the virus spreads correctly, and combined with immunohistochemistry to assure high specificity and penetrance (*see* also Chap. 8). These parameters often vary between regions, so good penetrance and specificity in one region does *not* assure the same in a different region, even when using exactly the same animal model, the same opsin and the same viral vector.

2.4.2 Verify that Your Fiber Was in the Correct Location	The fiber track is easy to observe in fixated brain slices. Make sure that your fiber was indeed placed above the target region; close enough to provide ample illumination, but far enough not to physically damage the area.
2.4.3 Verify that Your Opsin Is Producing the Expected Effect on Neuronal Activity	It is crucial to show that the opsin is producing the expected effect. This point is not trivial, as the same opsin can have different effects in different brain regions. Effects on activity can be shown by electrophysiology or calcium imaging, either in-vivo or in brain slices. These verifications can be performed before or after the behavioral experiment.
2.5 Good Citizenship in the Scientific Community	As the complexity of experiments continuously increases, researchers should be careful not to neglect to carefully document and report even the smallest technical details in order to make it easier for their fellow scientists to accurately understand and successfully replicate their results. For example, one should estimate as precisely as possible the size of the modulated area. The spread of opsin expression can be easily quantified, and the parameters governing light spread through the brain tissue are generally known, allowing a good estimation of the upper boundaries of the modulated area size. However, this data is often missing from papers, making it difficult to precisely interpret and compare them (*see* also Chap. 8 for concepts estimating the scope of optogenetic network activation).

3 Challenges and Limitations

Optogenetics offers a variety of advantages, described throughout this review. However, one should keep in mind that this technique is not free of limitations. This section details the major conceptual and technical challenges, possible ways to overcome them, where such exist, and suggestions for careful interpretation in light of the above.

3.1 Your Manipulation Is Now a Part of the Memory *3.1.1 Illumination as a Learning Cue*	Several memory tasks, especially those involving conditioning, rely on association between sensory cues. One should always bear in mind that the illumination used for activating the opsin can, by itself, serve as a learning cue. The solution to this obstacle is to mask the opsin excitation light. This can be done in different ways: First, use a fiber with a dark coating. This will block the light coming from the fiber itself, but will sometimes leave some residual visible illumination through the coupling sleeve between the implanted fiber and the fiber coming from your light source. Some light from the lower part of the implanted fiber can also leak through dental cement cap holding the implantable fiber into place—which can be solved by painting it black with a marker during surgery, while it is drying. These solutions are usually sufficient. However, in tasks that attempt to compare behavior between light ON and light OFF

trials *within* subjects, additional masking is required. In such cases, opsin illumination is masked by a bright light that is presented at all trials (with and without opsin illumination), thus hiding the lower intensity light leak.

3.1.2 Neuronal Activity Modulation as a Cue

Many studies had shown that just as animals can sense external physical stimuli in different modalities and use them as learning cues, they are also capable of sensing neuronal activity in the absence of real external stimuli, and change their behavior in response. Thus, not only the visible light itself, but the modulation of neuronal activity per-se can serve as a learning cue. Indeed, as will be detailed in the following section, several optogenetic studies had demonstrated behavioral consequences to neuronal manipulation in the absence of other cues. For example, amygdalar and cortical stimulation can directly induce fear [54, 55, 62], and thus serve as an independent US [62, 111]. Light stimulation can also serve as a "sensory" CS [112], and mice can even sense reduction in neuronal activity, and use it as a learning cue [77].

3.2 General Challenges in Using Optogenetics

3.2.1 Noninvasive? Very Invasive!

Compared to physical lesions, and big-diameter cannulation for pharmacology, optogenetics is considered less invasive. However, optogenetics requires fiber insertion into the brain, usually combined with intracranial virus injection. These technical issues are the simplest to control for. First, the effect that a strong expression of a foreign protein may have on cellular function should always be controlled for by expressing a fluorophore alone under the same promoter in the control group. Second, as new red-shifted opsins are developed, the fiber position can move farther away from the area of interest (even above the dura, e.g., [7]), because longer wavelengths better penetrate the tissue. Hopefully, within a few years fiber implantation will become redundant.

3.2.2 Cellular Specificity in Light of Inherent Complexity

A major strength of optogenetics is the ability to perturb the function of a specific neuronal population in real time. However, one should keep in mind that what is considered to be specific at this point in time will not necessarily stay so, and that specificity may sometimes lead to the wrong (or at least limited) conclusions.

Population specificity is a dynamic concept that changes throughout scientific history. In the past, neurons were considered a discrete population (as opposed to glia), then came the differentiation between excitatory, inhibitory, and neuromodulatory neurons, and their subpopulations. However, even if the tools to genetically distinguish between subpopulations even further do not yet exist, it is clear that there are many more sub-classifications to be made within "specific" populations, based on location, anatomy, and function (*see* also Chap. 2). For example, whereas dopaminergic neurons are considered a "specific" population (the shared genetic marker with noradrenaline aside), they were recently demonstrated to secrete not only dopamine, but in some cases co-release glutamate too, and in other cases GABA [113–116]. With the development of novel tools to target overlapping (or, if desired, nonoverlapping) neuronal populations [117], this problem will be easier to tackle.

Another problem is that most likely normal activity in brain circuits is based on the combined simultaneous contribution of several cell types, and that combinatorial manipulation of two (or even more) distinct populations will provide a deeper understanding of the natural circuit. As the optogenetic toolbox is expanding to include opsins with a wide range of spectral sensitivity, such experiments gradually become possible (e.g., [7].), as several opsins with different spectral preference are used together in different populations. Such experiments bring us closer to a more natural circuit perturbation.

3.2.3 Real-Time Manipulation, But Not Necessarily in a Physiological Pattern

The major obstacle in optogenetic excitation is that the illuminated neurons are simultaneously "locked" into a certain frequency. Optogenetic activation in vivo is currently achieved by using constant frequencies (changing between, but not within, experiments) of square light pulses, illuminating relatively large populations. Such illumination does not provide a good resemblance to real-life stimulation patterns in a specific region (and certainly not in a specific cell). This is the hardest challenge to control for, as no alternative illumination options are currently available for in vivo use. As the differences between simple tonic and phasic stimulation patterns yielded interesting insights [18], further increasing the complexity of the light stimulus by varying the frequency and intensity within the stimulation pattern, as in dynamic clamp [42], may yield interesting results. Similarly, future in vivo application of patterned illumination of specific neurons will also bring us closer to a true perturbation of complex circuits [118, 119] (*see* Chap. 10).

3.2.4 Brief Illumination—Permanent Effect?

Surprisingly little is known about the long-term effects of optogenetic stimulation. However, it is likely that a modulation that efficiently recruits a population of cells will result in plastic changes, especially as ChRs allow a direct influx of Ca^{2+} through the channel, which may directly influence long-term processes in the cell. A few studies have already shown that optogenetic activation can independently alter neuronal networks in culture [120] and support LTP in slices [121]. Recently, a causal in vivo demonstration of the effects of optogenetic LTP and LTD protocols on behavior was provided [82]: Nabavi et al. [82] inactivated auditory fear conditioning expression by optogenetic LTD stimulation to auditory projections to the amygdala, and then reactivated fear memory by delivering an LTP illumination protocol to the same projection. Interestingly, effects on synaptic plasticity were also observed following neuronal inhibition with NpHR, as a result of changes in the $GABA_A$ receptor reversal potential [122]. Thus, optogenetic manipulations may remodel the perturbed circuit during the experiment. This issue cannot be avoided, but can be monitored by comparing baseline and post-illumination neuronal activity and/or behavior whenever possible.

4   Achievements in Optogenetic Memory Research

Despite the vast repertoire of pioneering works examining learning and memory, many unsolved mysteries remained, especially concerning the real-time involvement of specific neuronal populations in normal hippocampal functioning. Some of these mysteries became clearer in recent years thanks to the incorporation of optogenetics into the field of memory research, whereas others emerged, to be tackled in the future.

While by no means intended to provide a comprehensive review of the field, this section of the chapter presents and examines a few of the major findings, revealing the roles of distinct neuronal populations, specific brain regions and defined projections in memory processes.

4.1 The Role of the Amygdala in Fear Memory

As the amygdala is the core player in fear acquisition and expression [45], extensive effort was invested in studying its role in FC using optogenetics. The first basic question to answer was - is the amygdala sufficient to independently support fear? Several studies [54, 62, 63] used optogenetics to directly stimulate, or indirectly disinhibit amygdalar nuclei to demonstrate their ability to directly induce fear [54, 62] (but see [60]). Amygdala stimulation can even serve as an independent US when paired with a tone, i.e., it induced conditioned fear of the tone at a later time point, when illumination was absent [62]. Based on previous studies using other techniques, the amygdala was thought to be necessary for FC, and indeed, optogenetic studies also supported the role of the amygdala in fear memory acquisition and retention [51, 59, 63, 81]. Of special interest are studies that pointed to specific neuronal populations that are involved in amygdalar fear learning [63, 64, 69]. For example, differential roles were shown for two different interneuron populations (PV and somatostatin) in the amygdala: These populations respond differently to the US and CS, and gate the activity of amygdalar excitatory neurons in response to these stimuli, thus supporting associative learning [69]. A recent work perturbed the specific conditions for lateral amygdala activity during conditioning and showed they abide hebbian rules: first, the lateral amygdala (LA) must be active during the time the US is presented, as if the LA is inhibited with ArchT illumination just during the few seconds the aversive stimulus is present fear memory is significantly reduced [73]. Second, if the activity in the LA is increased using ChR2 during the presentation of a weak US, the conditioning will be strengthen [73].

4.2 Diverse Roles for Hippocampal Subregion, Dorsoventral Variations, and Specific Neuronal Populations in Memory

The hippocampus is comprised of several subregions with known connections, but most past works had tested the involvement of the hippocampus as a whole in memory [123], and only few studies (e.g., [124–126]) focused on specific regions. Optogenetic experiments have shed light on differences between the roles of the primary region to receive information from the cortex (dentate gyrus, DG), CA3 region, and the final region in hippocampal processing (CA1) in memory, and demonstrated the involvement of specific cell populations in memory processes. Additionally, it showed functional differences between dorsal and ventral hippocampi.

The exclusive role of CA1 in fear memory acquisition was demonstrated in the past using CA1-specific NMDA knockout [124], and lately its real-time contribution was tested by targeting pyramidal neurons throughout the dorsal CA1 with the inhibitory opsin NpHR [51]. Another study disturbed the normal function of CA1 not by inhibiting this region, but rather by increasing its activity by NpHR inhibition of somatostatin-expressing dendrites in CA1, causing disinhibition of the pyramidal neurons [80]. This manipulation also resulted in fear acquisition impairment [80]. Together these two experiments show that any deviation from the normal function of CA1 neurons (either inhibition, or excitation via disinhibition) during FC acquisition results in impaired performance, demonstrating the importance of this region in conditioning. CA1 is also important for the process of reconsolidation, as when CA1 neurons are inhibited with ArchT right after re-exposure to the conditioning chamber, memory on the next day is impaired [74].

When light was delivered to mice expressing NpHR [51] or ArchT [17] in the CA1 during recall (rather than acquisition), the memory that had been previously present became unavailable under illumination. Recall can also be prevented using a more subtle manipulation; not by inhibiting CA1 activity in general, but by specifically inhibiting the small population of CA1 cells that were specifically involved in encoding the conditioning context [28]. Thus, CA1 excitatory neurons are involved in *both* acquisition and recall of contextual fear memory. This, however, is not the case for the DG. Kheirbek et al. [52] expressed opsins specifically in DG granular neurons using the proopiomelanocortin (POMC)-Cre mouse line [125]. When the activity of dorsal DG neurons was inhibited by NpHR during contextual FC acquisition, memory was impaired, but when the DG was inhibited during memory recall, no effect was observed [52].

Conversely, DG activation by illumination of ChR2-expressing DG granule cells during conditioning disrupted both acquisition and recall [52, 127]. When DG activation was indirectly induced, by inhibiting hilar GABAergic inputs to granular cells using NpHR [87], recall of the position of an escape platform in a water maze probe test was impaired, but no effect on the gradual acquisition of spatial memory was observed [87]. Furthermore, DG inhibition had no effect on the gradual acquisition or retrieval of active place avoidance, where mice were trained to avoid a stationary shock zone in a circular arena [52]. When the position of the shock zone was switched, requiring a rapid encoding of a new contingency and resolution of two conflicting memories, DG inhibition resulted in a marked cognitive impairment. Together, these data suggest that the DG is necessary for rapid memory acquisition and cognitive flexibility, but not for gradual memory acquisition or the recall of established memories. The latter, however, can be disturbed by nonsense hyper-activation of the DG during recall.

Interestingly, when NpHR was used to unilaterally inhibit CA3 neurons in either the left or right dorsal hippocampus, short-term memory was impaired, as demonstrated using a T maze. However, associative spatial long-term memory was only impaired by CA3 silencing on the left, whereas right CA3 silencing had no effect, showing a surprising left-right asymmetry [104].

To conclude, it seems that both DG and CA1 are necessary for the acquisition of contextual memory traces. However, during recall CA1 is necessary whereas DG silencing has no effect. What could be the reason for that difference? There are several structural and functional differences between DG and CA1 that may contribute to the different roles these regions play in contextual memory. One unique characteristic of the DG is neurogenesis; the continuous differentiation and integration of newly formed neurons into this structure. Ablation of these new neurons results in memory impairments, and their function evolves as they integrate into the existing network [128, 129]. Newly formed DG neurons can be targeted by retroviruses (which only affect dividing cells). Using this method, Arch was targeted to newly formed neurons. Four weeks later, mice were trained in a FC paradigm, and the 4-weeks old neurons were optically inhibited during recall, resulting in impaired contextual memory [58].

Based on the great variance in afferent and efferent connectivity along the dorsoventral axis of the hippocampus, it was hypothesized that the dorsal and ventral hippocampi also have different functions. Specifically, the dorsal hippocampus projects extensively to associative cortical regions, suggesting a role for this area in spatial and contextual memory, whereas the ventral hippocampus projects to the PFC, amygdala, and hypothalamus, suggesting a role in emotional processing. Indeed, most studies had shown disrupted spatial memory after dorsal hippocampus lesions, while

lesions of the ventral pole spared spatial learning but had an anxio-
lytic effect [130]. However, other studies reported opposite results
[130]. These open questions were addressed by using optogenetic
manipulation of either the dorsal or ventral DG, and testing the
effect of illumination on cognitive and emotional behaviors. As
detailed above, dorsal DG was shown to be involved in contextual
FC acquisition and in the resolution of conflicting memories [52].
Ventral-DG manipulation, on the other hand, had no effect on
contextual memory acquisition, but had a striking effect on emotional behavior: Ventral DG excitation with ChR2 exerted an anxi-
olytic effect in two different tests of anxiety, the elevated plus maze
and the open field [52]. Conversely, ventral DG inhibition with
NpHR had no effect on anxiety, suggesting this region is not
necessary for the expression of baseline anxiety levels [52]. To
conclude, this study used both inhibition and excitation of either
the dorsal or the ventral DG, to demonstrate the changing roles
performed differentially by the hippocampus along its dorsoventral
axis.

**4.3 Different
Retrieval Strategies for
Recent and Remote
Memories**

Many findings from both human studies and animal research suggest that long-term contextual fear memory consolidation requires
early involvement of the hippocampus, later replaced by the neocortex: Hippocampal lesions impair memory 1 day after training,
but the same lesions have no effect on memory several weeks after
training [131]. Conversely, nongraded retrograde amnesia was
reported in some human patients and in animal studies involving
extensive hippocampal damage. These seemingly conflicting findings have led to the "multiple trace theory", suggesting that in the
process of system-wide consolidation the memory is not merely
transferred from the hippocampus to the cortex, but rather transformed, and possibly saved in different variations in several cortical
regions, and remain available, with continuous interplay [132,
133].

The groundbreaking studies on the circuitry of remote memory involved lesion studies (physical, pharmacological and genetic),
which lack fine temporal resolution [125, 134–136]. Surprisingly,
real-time optogenetic CA1 inhibition during contextual recall also
blocks *remote* fear memory when administered 4–12 weeks after
acquisition [51], suggesting a permanent role of the hippocampus
in memory recall, as long as a memory trace exists. Furthermore,
remote fear memory expression could be interrupted even after the
context was already recalled, by optogenetic CA1 inhibition in the
midst of a recall session, which resulted in an immediate termination of the fear response [51]. Thus, CA1 is necessary not only for
the recall of remote memories, but also for their maintenance
throughout the recall session. The reason for the contradictory
results obtained using optogenetics vs. pharmacological and
genetic studies may be the temporal precision of the former

technique: pharmacological or genetic manipulations are orders of magnitudes slower than the typical neuronal activity, and thus allow compensatory mechanisms to commence before the behavioral test takes place. Indeed, when illumination was delivered in a pharmacological timescale (for 30 min before testing as well as during the test) to allow time for compensatory mechanisms to be engaged, no effect on recall was observed, whereas precise illumination disrupted recall [51]. In this case, the use of optogenetics resulted in modification, rather than validation, of a long-standing hypothesis, by showing that contrary to the prevalent belief of a gradual transition from hippocampal-mediated recall of recent memories to cortical-mediated recall of remote memories, the hippocampus is in fact the default activator of contextual memory traces at all timepoints.

4.4 Cortical Involvement in Learning and Memory

Frontal cortical regions are involved in a variety of cognitive processes (e.g., executive functions, attention, decision making) [137]; however, this short section will focus exclusively on their role in fear memory. The importance of frontal cortical areas like the ACC and PFC to systems consolidation and FC extinction was well established in the past decades [46, 131, 138]. Recent optogenetic studies support these findings and provide insight to the modulatory processes affecting them.

Fear acquisition (auditory-cued and contextual) was severely impaired by prolonged hyper-activation of pyramidal neurons using a slow ChR2 variant (stable step function opsin, SSFO) during FC [7], probably due to masking of the informative PFC trace with a noisier opsin-induced signaling.

Optogenetic inhibition of mPFC excitatory neurons with NpHR during FC acquisition also impaired associative memory formation as measured both 1 day and 30 days after conditioning [76]. The PFC was also implicated in trace FC by optogenetically inhibiting this area using ArchT during the "trace interval", a 20 s delay between the CS (auditory cue) and the US (electric shock). PFC inhibition at that specific time significantly impaired acquisition [139], providing yet another powerful demonstration of the power of temporal precision when using optogenetics.

Parvalbumin-expressing (PV) interneurons in the PFC were shown to inhibit fear expression, as their optogenetic inhibition using Arch (subsequently dis-inhibiting pyramidal PFC neurons) resulted in spontaneous fear expression, and could even induce place aversion [55]. Optogenetic activation of PFC pyramidal neurons can also serve as an independent conditioned stimulus and support delayed eye blink conditioning [111]. Furthermore, activation of PV neurons in the PFC with ChR2 (consequently inhibiting pyramidal neurons in this region), reduced conditioned fear expression [55], as did specific inhibition of the prelimbic (PL) prefrontal cortex [140]. However, directly inhibiting the infralimbic cortex (IL) with NpHR had no effect of fear expression during retrieval [72], fear acquisition was not affected by optogenetic activation of PV neurons in the PFC [7], and fear retrieval or expression was impaired in rats by activating IL during the tone presentation with ChR2 [72]. These seemingly contradicting results will surely be followed up by future experiments in an attempt to resolve them. For example, studies focusing on specific projections to or from the PFC. One such paper had shown differential effects of optogenetic silencing of different projections from PL [140]. Specifically, silencing the PL projections to the paraventricular nucleus of the thalamus (PVT) impaired retrieval days but not hours after training, whereas silencing of PL to BLA projections impaired retrieval at early, but not late, time points. These results suggest a dynamic temporal shift in retrieval circuits.

As the PFC was implicated in fear memory extinction, several optogenetic studies had addressed this role, sometimes yielding contradictory results: Two studies found that optogenetic activation of IL strengthened the expression of extinction during retrieval [72, 141]. Additionally, inhibiting IL neurons during extinction training had no effect on freezing during illumination, and impaired subsequent retrieval of extinction on the next day, with no illumination [72]. However, whereas one study reported no effect of IL inhibition during extinction retrieval [72], another study reported impaired extinction retrieval upon IL inhibition [141]. These contradictory findings are yet to be resolved, and attempts are made to do so by deconstructing the exact circuitry involved. For example, extinction memory is impaired when amygdala to IL projections were inhibited during extinction training, but increased when amygdala to PL were inhibited [24].

Another frontal region, the ACC, was shown to be involved in remote but not recent recall [138]. Indeed, NpHR-mediated optogenetic inhibition of the ACC 1 month (but not 1 day) after FC acquisition impaired contextual fear memory [51]. This deficit was not compensated for when illumination was delivered in a pharmacological timescale, suggesting that a part of the contribution of the ACC to the recall of the context is unique and cannot be compensated for by other areas, as suggested by the multiple trace theory [51].

Recently, this region was shown to exert top-down control over fear retrieval in the dorsal hippocampus [23], as activating the ACC-hippocampus projection induced contextual memory recall, and inhibiting it impaired recall [23]. This surprising potent effect of the ACC-hippocampus projection was demonstrated both in a conventional FC paradigm by measuring freezing, and in the head-fixed version by measuring licking suppression [23].

Several recent optogenetic studies described a surprising role for the primary auditory cortex in auditory-cued FC. Letzkus et al. [65] defined a circuit in which cholinergic inputs increase the activity of GABAergic cells in cortical layer 1 of the auditory cortex, which in turn inhibit the activity of PV interneurons in layers 2/3. This PV inhibition resulted in dis-inhibition of layer 2/3 pyramidal neurons, which consequently show increased firing. When PV interneurons in the auditory cortex were optogenetically activated by ChR2 following foot shock (at the time in which their activity is naturally suppressed), auditory-cued FC was dramatically impaired [65].

Surprisingly, mice can even sense optogenetic inhibition of auditory neurons, and use it as a conditioned stimulus. Thus, when neuronal inhibition is presented together with an aversive or an appetitive unconditioned stimulus, mice will exhibit fear or reward seeking, respectively [77].

To sum up, optogenetic studies to date have supported the classical involvement of frontal cortices in complex processes like systems consolidation and extinction learning, but also demonstrated their involvement in "simpler" processes like fear acquisition and expression, in which they were thought not to be involved [46]. Moreover, they shed light on the role of specific projections to and from frontal cortex, allowing a first glimpse into the detailed circuitry of memory processes.

4.5 Thinking "Outside of the Circuit"

The experiments describes above focus on brain regions that were heavily implicated in memory for decades. However, one of the strengths of optogenetics is that it allows access to probe the activity of brain regions that cannot be manipulated by lesions, for example because they have a lot of crossing fibers, or close to vital regions—e.g., in the brain stem. This section describes experiments that were performed on regions and population outside of the classical hippocampus-amygdala-cortex circuitry, and revealed many surprising results, implicating additional brain regions in the previously suspected memory circuitry [70, 71, 75, 96, 140, 142].

The thalamus, that was mostly studies in sensory and motor contexts in the past, appears to play a role in memory processes. For example, the thalamic nucleus reuniens (NR) was implicated in memory generalization, which was increased by inactivation of NR projections and decreased by constitutive NR activation [70]. Another study implicated the PVT in the recall of remote memories [140].

Hypothalamic projections were also implicated in memory using optogenetics [63, 93]. One study reported that ChR2-mediated evoked release of oxytocin from hypothalamic terminals in the central amygdala resulted in attenuated freezing [63]. Another study excited vasopressin terminals (from the hypothalamic paraventricular nucleus) in the CA2 hippocampal region, and found that this manipulation resulted in increased social memory acquisition, but had no effect on retrieval [93].

Brain stem targeting showed that inhibition of serotonergic activity by Arch in the dorsal raphe nucleus blocked the beneficial effect of stress on fear memory [71], and that activating the meso-pontine median raphe during consolidation resulted in decreased FC recall [75]. Finally, a region that was heavily investigated using optogenetic is the striatum, yielding many interesting results in addiction and motor function studies, but also in cognitive perfor-mance [33, 68, 96, 100–103, 107, 142].

4.6 Generation of Memory Traces Using Optogenetics

A basic assumption in neuroscience is that sparse discrete neuronal populations underlie specific memory traces, and indeed such populations were demonstrated in the amygdala and hippocampus [67, 143–145], and their necessity for the integrity of memory traces was established by ablating them and consequently erasing a specific memory trace [146, 147]. However, no proof that these neuronal ensembles are not only necessary but also sufficient to encode a specific memory trace was available. This section presents optogenetic approaches for the generation of false memory traces, and ask whether cracking the neuronal code is truly necessary in order to create such memories.

Liu et al. [25] expressed ChR2 specifically in the DG neurons that were activated in response to a specific context A, using the cFos-tTA mouse line. The mice were then introduced into a differ-ent context B, while the ChR2-expressing neurons were illumi-nated. Despite the fact that no aversive stimulus was ever present in context B, ChR2 expressing mice demonstrated fear, thus providing the first proof that the stimulation of a specific memory engram in the hippocampus can in fact support the memory of a specific context [25]. Such engrams, which are sufficient to support a memory trace, can be generated in the same manner in the amygdala as well [29]. Activating hippocampal engram cells can even support memory when the natural memory trace is not prop-erly created, due to the application of anisomycin after training [30].

Interestingly, not only can optogenetic activation of aversive memory engrams induce fear response, but also optogenetic acti-vation of a rewarding memory engram induces appetitive behavior response [29]. These effects, both aversive and appetitive can be mediated by optical reactivation of either hippocampal DG engrams, or BLA engrams [29]. Furthermore, activating appetitive DG engrams by light can increase activity in the nucleus accum-bens, and diminish depressive-like behaviors like struggling in the forced swim test or sucrose preference [27]. Chronic activation of an appetitive memory can even elicit a long-lasting rescue of depression-related behavior, even at times when there is no illumi-nation [27].

The engrams described above represent a "true" memory in which a context was paired with a shock in reality. Ramirez et al.

[26] then created a "false" memory trace by first tagging a specific neuronal ensemble in context A, and then activating this population with light while administering a foot shock in context B. Thus, an offline association between the engram of A and an aversive stimulus was created, which resulted in fear of the previously neutral context A upon reintroduction to this context (with no light stimulation of the ChR2 expressing cells) [26]. Interestingly, the simultaneous activation of two "true" engrams, one DG engram of a context, and one BLA engram of an aversive experience, can create an artificial offline association between the two [31].

As discussed above, CA1 and DG inhibition elicit different effects on memory recall [51, 52]. An elegant demonstration of the importance of sparse coding in the DG compared to CA1 comes from the fact that a false memory trace could be generated by DG illumination, but not by CA1 light activation [26]. The same was true for conditioned place aversion—mice showed avoidance of the falsely aversive chamber upon tagging and activation of DG, but not CA1, neurons [26].

The generation of true or false memory traces in a sophisticatedly targeted group of recently active neurons is a truly remarkable achievement. Interestingly, researchers from the Axel group managed to create a fear memory, and even more remarkably drive an appetitive behavior, in response to nonspecific neuronal targeting [112]. ChR2 was randomly expressed in neurons in the piriform cortex (independent of behavior) [112], and light stimulation of these cells served as a CS paired to a chock, resulting in a strong escape behavior in response to the light stimulation alone [112]. Importantly, randomly targeted piriform cortex neurons could also drive appetitive behavior when paired to a socially rewarding stimulus (a female mouse). When a random piriform ensemble in males is activated in a cage compartment in which a female is present, they will later prefer to be in a compartment that offers light stimulation even when no female is present [112]. Interestingly, these associations are very flexible, as the same random population can be entrained to support both appetitive and aversive behavior [112].

The studies presented in this section all used optogenetics to create memory traces that can drive behavior [25–27, 29–31, 112]. The results suggest that in order to powerfully modulate behavior in a context-specific manner one does not necessarily need to decipher the original neural code, but can simply force an alternative code onto the system, and drive behavior efficiently. Clearly, the level of permissibility is lenient in areas that provide the input on which a memory is later constructed (e.g., piriform cortex), but much less permissive in areas that are tightly related to long-term memory formation like the hippocampus.

5 Notes

5.1 Tracking System Adjustments Strategies in Animals Tethered to an Optical Fiber

A. Carefully calibrate your tracking system: Researchers often calibrate their tracking systems using naïve animals, with no fiber. It is advisable, however, to make the initial calibration with an animal attached to an optical fiber. This will allow you to better define the subject to the automated system, and to change environmental parameters like lighting modifications to decrease shadows cast by the optical system, and equipment rearrangement to prevent the optical system from hiding the subject from the camera. In some cases, it is almost impossible to use a visual tracking system. For example, freezing is very difficult to quantify, because even when the animal is completely immobile, the fiber may still be moving. Thus, it is very important to compare the automated score to a manual score of the same video, to validate the results. *B. Use a different detection technique*: Visual tracking has several automated alternatives based on breaking IR beams, which can still give information about movement and location in space. Other behaviors that can be automatically scored are lever presses, nose pokes, and touchscreen contact. In some cases, reliable manual scoring, performed in a double-blind design, is preferable to questionable automatic scoring.

6 Outlook

Every exciting new finding mentioned above is also a source of new mysteries, which will in turn stimulate further innovating studies. The combination of new research questions and hypotheses, hand in hand with rapidly evolving technological developments, is an extremely efficient driving force in memory research. Discovering the answers to these and many other challenging questions may require continuous adoption of new tools, and increased complexity in the use of current tools. Several such modulations are suggested below:

6.1 Targeting Novel Populations

An impressive body of knowledge about the function of certain populations like glutamatergic, dopaminergic, cholinergic, and GABAergic (specifically PV) cells has accumulated over an extremely short period of time, thanks to tools allowing direct and specific access to these populations. One striking example of how the development of genetic access to a specific region had rekindled interest and provided new insights can be found in the CA2 region of the hippocampus [148]. New Cre lines (DGGC-specific transgenic Cre and Amigo2-Cre) [149, 150] allowed a functional research of CA2 circuitry [149], and found it to be important in social memory [150]. Incorporation of optogenetics into CA2 function studies will surely provide important insight to its real-time involvement in memory.

The development of new Cre driver lines, as well as improved and smaller promoters that can be used for viral vectors injected to WT animals, will hopefully expand our ability to examine new populations, and their contribution to the known circuits.

6.2 Using More Challenging Cognitive Tasks

The beauty of FC is in its simplicity, but more elaborate tasks (examining spatial navigation, and the assimilation of memory schemas, for example) may provide interesting insights. This process seems more realistic thanks to the development of Cre driver rat lines [33], as these animals are capable of performing more sophisticated tasks, and carry bigger devices on their skulls. Indeed, pioneering studies are already in progress [96].

6.3 Harnessing the Power of Connectivity-Based Targeting

Optogenetic memory research to date employed mostly viruses that infect the soma, followed by illumination of cell bodies or projections. Integration of circuit-based targeting is gradually being incorporated into memory research [23, 24], using viruses that can ravel trans-synaptically or infect terminals [19, 20, 22], as explained in the technical sections above. These tools have the power to completely revolutionize memory research, allowing insight into specific circuits that were inaccessible in the past.

6.4 Combining Optogenetics and Imaging Techniques

One of the advantages of optogenetics is that it allows simultaneous optical stimulation and electrical recording in behaving animals. In the near future all-optical stimulation and imaging during the performance of cognitive tasks may become achievable. Concomitantly with the leap in tool development for neuronal manipulation, recent years had seen a tremendous progress in the development and application of neuronal activity imaging tools [118, 119, 151–153]. Future combination of calcium imaging in behaving mice [151] with new genetically encoded indicators [153] that can be optogenetics-compatible [152] can provide a better understanding of the exact spatiotemporal effect of optogenetic manipulation, and later help to refine and optimize it. Pioneering studies employing this exciting combination are already performed [23] (*see* Chaps. 9 and 10).

6.5 Summary

In an extremely short period of time an impressive variety of new findings about memory formation, consolidation, recall, and extinction was obtained using optogenetics, proving some long-standing hypotheses, modulating others, and providing insight into new and hitherto unexplored circuits. Every exciting new finding in memory research is also a source of new mysteries, which in turn stimulate further innovating studies. The combination of new research questions and hypotheses, hand in hand with rapidly evolving technological developments, is an extremely efficient driving force in memory research.

When employing new technologies, such as optogenetics, researchers must be well aware of both their power and their limitations, and be extremely careful in execution, documentation, and interpretation.

References

1. Bernstein JG, Boyden ES (2011) Optogenetic tools for analyzing the neural circuits of behavior. Trends Cogn Sci 15(12):592–600. doi:10.1016/j.tics.2011.10.003

2. Goshen I (2014) The optogenetic revolution in memory research. Trends Neurosci 37 (9):511–522. doi:10.1016/j.tins.2014.06.002

3. Deisseroth K (2010) Controlling the brain with light. Sci Am 303(5):48–55

4. Johansen JP, Wolff SB, Luthi A, LeDoux JE (2012) Controlling the elements: an optogenetic approach to understanding the neural circuits of fear. Biol Psychiatry 71 (12):1053–1060. doi:10.1016/j.biopsych.2011.10.023

5. Airan RD, Thompson KR, Fenno LE, Bernstein H, Deisseroth K (2009) Temporally precise in vivo control of intracellular signalling. Nature 458(7241):1025–1029. doi:10.1038/nature07926

6. Gunaydin LA, Yizhar O, Berndt A, Sohal VS, Deisseroth K, Hegemann P (2010) Ultrafast optogenetic control. Nat Neurosci 13 (3):387–392. doi:10.1038/nn.2495

7. Yizhar O, Fenno LE, Prigge M, Schneider F, Davidson TJ, O'Shea DJ, Sohal VS, Goshen I, Finkelstein J, Paz JT, Stehfest K, Fudim R, Ramakrishnan C, Huguenard JR, Hegemann P, Deisseroth K (2011) Neocortical excitation/inhibition balance in information processing and social dysfunction. Nature 477 (7363):171–178. doi:10.1038/nature10360

8. Mattis J, Tye KM, Ferenczi EA, Ramakrishnan C, O'Shea DJ, Prakash R, Gunaydin LA, Hyun M, Fenno LE, Gradinaru V, Yizhar O, Deisseroth K (2012) Principles for applying optogenetic tools derived from direct comparative analysis of microbial opsins. Nat Methods 9(2):159–172. doi:10.1038/nmeth.1808

9. Govorunova EG, Sineshchekov OA, Janz R, Liu X, Spudich JL (2015) NEUROSCIENCE. Natural light-gated anion channels: a family of microbial rhodopsins for advanced optogenetics. Science 349(6248):647–650. doi:10.1126/science.aaa7484

10. Berndt A, Lee SY, Ramakrishnan C, Deisseroth K (2014) Structure-guided transformation of channelrhodopsin into a light-activated chloride channel. Science 344 (6182):420–424. doi:10.1126/science.1252367

11. Wietek J, Wiegert JS, Adeishvili N, Schneider F, Watanabe H, Tsunoda SP, Vogt A, Elstner M, Oertner TG, Hegemann P (2014) Conversion of channelrhodopsin into a light-gated chloride channel. Science 344(6182): 409–412. doi:10.1126/science.1249375

12. Zhang F, Prigge M, Beyriere F, Tsunoda SP, Mattis J, Yizhar O, Hegemann P, Deisseroth K (2008) Red-shifted optogenetic excitation: a tool for fast neural control derived from Volvox Carteri. Nat Neurosci 11 (6):631–633. doi:10.1038/nn.2120

13. Szymczak AL, Workman CJ, Wang Y, Vignali KM, Dilioglou S, Vanin EF, Vignali DA (2004) Correction of multi-gene deficiency in vivo using a single 'self-cleaving' 2A peptide-based retroviral vector. Nat Biotechnol 22(5):589–594. doi:10.1038/nbt957

14. Prakash R, Yizhar O, Grewe B, Ramakrishnan C, Wang N, Goshen I, Packer AM, Peterka DS, Yuste R, Schnitzer MJ, Deisseroth K (2012) Two-photon optogenetic toolbox for fast inhibition, excitation and bistable modulation. Nat Methods 9(12):1171–1179. doi:10.1038/nmeth.2215

15. Martinez-Salas E (1999) Internal ribosome entry site biology and its use in expression vectors. Curr Opin Biotechnol 10 (5):458–464

16. Fenno L, Yizhar O, Deisseroth K (2011) The development and application of optogenetics. Annu Rev Neurosci 34:389–412. doi:10.1146/annurev-neuro-061010-113817

17. Sakaguchi M, Kim K, Yu LM, Hashikawa Y, Sekine Y, Okumura Y, Kawano M, Hayashi M, Kumar D, Boyden ES, McHugh TJ, Hayashi Y (2015) Inhibiting the activity of CA1 hippocampal neurons prevents the recall of contextual fear memory in inducible archT transgenic mice. PLoS One 10(6):e0130163. doi:10.1371/journal.pone.0130163

18. Tsai HC, Zhang F, Adamantidis A, Stuber GD, Bonci A, de Lecea L, Deisseroth K (2009) Phasic firing in dopaminergic neurons is sufficient for behavioral conditioning. Science 324(5930):1080–1084. doi:10.1126/science.1168878

19. Callaway EM (2008) Transneuronal circuit tracing with neurotropic viruses. Curr Opin Neurobiol 18(6):617–623. doi:10.1016/j.conb.2009.03.007

20. Callaway EM, Luo L (2015) Monosynaptic circuit tracing with glycoprotein-deleted rabies viruses. J Neurosci 35 (24):8979–8985. doi:10.1523/JNEUROSCI.0409-15.2015

21. Ekstrand MI, Enquist LW, Pomeranz LE (2008) The alpha-herpesviruses: molecular pathfinders in nervous system circuits. Trends Mol Med 14(3):134–140. doi:10.1016/j.molmed.2007.12.008

22. Junyent F, Kremer EJ (2015) CAV-2-why a canine virus is a neurobiologist's best friend. Curr Opin Pharmacol 24:86–93. doi:10.1016/j.coph.2015.08.004

23. Rajasethupathy P, Sankaran S, Marshel JH, Kim CK, Ferenczi E, Lee SY, Berndt A, Ramakrishnan C, Jaffe A, Lo M, Liston C, Deisseroth K (2015) Projections from neocortex mediate top-down control of memory retrieval. Nature 526(7575):653–659. doi:10.1038/nature15389

24. Senn V, Wolff SB, Herry C, Grenier F, Ehrlich I, Grundemann J, Fadok JP, Muller C, Letzkus JJ, Luthi A (2014) Long-range connectivity defines behavioral specificity of amygdala neurons. Neuron 81(2):428–437. doi:10.1016/j.neuron.2013.11.006

25. Liu X, Ramirez S, Pang PT, Puryear CB, Govindarajan A, Deisseroth K, Tonegawa S (2012) Optogenetic stimulation of a hippocampal engram activates fear memory recall. Nature 484(7394):381–385. doi:10.1038/nature11028

26. Ramirez S, Liu X, Lin PA, Suh J, Pignatelli M, Redondo RL, Ryan TJ, Tonegawa S (2013) Creating a false memory in the hippocampus. Science 341(6144):387–391. doi:10.1126/science.1239073

27. Ramirez S, Liu X, MacDonald CJ, Moffa A, Zhou J, Redondo RL, Tonegawa S (2015) Activating positive memory engrams suppresses depression-like behaviour. Nature 522(7556):335–339. doi:10.1038/nature14514

28. Tanaka KZ, Pevzner A, Hamidi AB, Nakazawa Y, Graham J, Wiltgen BJ (2014) Cortical representations are reinstated by the hippocampus during memory retrieval. Neuron 84(2):347–354. doi:10.1016/j.neuron.2014.09.037

29. Redondo RL, Kim J, Arons AL, Ramirez S, Liu X, Tonegawa S (2014) Bidirectional switch of the valence associated with a hippocampal contextual memory engram. Nature 513(7518):426–430. doi:10.1038/nature13725

30. Ryan TJ, Roy DS, Pignatelli M, Arons A, Tonegawa S (2015) Memory. Engram cells retain memory under retrograde amnesia. Science 348(6238):1007–1013. doi:10.1126/science.aaa5542

31. Ohkawa N, Saitoh Y, Suzuki A, Tsujimura S, Murayama E, Kosugi S, Nishizono H, Matsuo M, Takahashi Y, Nagase M, Sugimura YK, Watabe AM, Kato F, Inokuchi K (2015) Artificial association of pre-stored information to generate a qualitatively new memory. Cell Rep 11(2):261–269. doi:10.1016/j.celrep.2015.03.017

32. Zhang F, Gradinaru V, Adamantidis AR, Durand R, Airan RD, de Lecea L, Deisseroth K (2010) Optogenetic interrogation of neural circuits: technology for probing mammalian brain structures. Nat Protoc 5(3):439–456. doi:10.1038/nprot.2009.226

33. Witten IB, Steinberg EE, Lee SY, Davidson TJ, Zalocusky KA, Brodsky M, Yizhar O, Cho SL, Gong S, Ramakrishnan C, Stuber GD, Tye KM, Janak PH, Deisseroth K (2011) Recombinase-driver rat lines: tools, techniques, and optogenetic application to dopamine-mediated reinforcement. Neuron 72(5):721–733. doi:10.1016/j.neuron.2011.10.028

34. Yizhar O, Fenno LE, Davidson TJ, Mogri M, Deisseroth K (2011) Optogenetics in neural systems. Neuron 71(1):9–34. doi:10.1016/j.neuron.2011.06.004

35. Sidor MM, Davidson TJ, Tye KM, Warden MR, Diesseroth K, McClung CA (2015) In vivo optogenetic stimulation of the rodent central nervous system. J Vis Exp 95:51483. doi:10.3791/51483

36. Ung K, Arenkiel BR (2012) Fiber-optic implantation for chronic optogenetic stimulation of brain tissue. J Vis Exp 68:e50004. doi:10.3791/50004

37. Sparta DR, Stamatakis AM, Phillips JL, Hovelso N, van Zessen R, Stuber GD (2012) Construction of implantable optical fibers for long-term optogenetic manipulation of neural circuits. Nat Protoc 7 (1):12–23. doi:10.1038/nprot.2011.413

38. McCall JG, Kim TI, Shin G, Huang X, Jung YH, Al-Hasani R, Omenetto FG, Bruchas

MR, Rogers JA (2013) Fabrication and application of flexible, multimodal light-emitting devices for wireless optogenetics. Nat Protoc 8(12):2413–2428. doi:10.1038/nprot.2013.158

39. Montgomery KL, Yeh AJ, Ho JS, Tsao V, Mohan Iyer S, Grosenick L, Ferenczi EA, Tanabe Y, Deisseroth K, Delp SL, Poon AS (2015) Wirelessly powered, fully internal optogenetics for brain, spinal and peripheral circuits in mice. Nat Methods 12 (10):969–974. doi:10.1038/nmeth.3536

40. Geiger BM, Frank LE, Caldera-Siu AD, Pothos EN (2008) Survivable stereotaxic surgery in rodents. J Vis Exp 20:880. doi:10.3791/880

41. Adamantidis AR, Zhang F, Aravanis AM, Deisseroth K, de Lecea L (2007) Neural substrates of awakening probed with optogenetic control of hypocretin neurons. Nature 450 (7168):420–424. doi:10.1038/nature06310

42. Sohal VS, Zhang F, Yizhar O, Deisseroth K (2009) Parvalbumin neurons and gamma rhythms enhance cortical circuit performance. Nature 459(7247):698–702. doi:10.1038/nature07991

43. Warden MR, Selimbeyoglu A, Mirzabekov JJ, Lo M, Thompson KR, Kim SY, Adhikari A, Tye KM, Frank LM, Deisseroth K (2012) A prefrontal cortex-brainstem neuronal projection that controls response to behavioural challenge. Nature 492(7429):428–432. doi:10.1038/nature11617

44. Schmid F, Wachsmuth L, Albers F, Schwalm M, Stroh A, Faber C (2017) True and apparent optogenetic BOLD fMRI signals. Magn Reson Med 77(1):126–136. doi:10.1002/mrm.26095

45. LeDoux JE (2000) Emotion circuits in the brain. Annu Rev Neurosci 23:155–184. doi:10.1146/annurev.neuro.23.1.155

46. Maren S, Phan KL, Liberzon I (2013) The contextual brain: implications for fear conditioning, extinction and psychopathology. Nat Rev Neurosci 14(6):417–428. doi:10.1038/nrn3492

47. Maren S, Quirk GJ (2004) Neuronal signalling of fear memory. Nat Rev Neurosci 5 (11):844–852. doi:10.1038/nrn1535

48. Fanselow MS (2000) Contextual fear, gestalt memories, and the hippocampus. Behav Brain Res 110(1–2):73–81

49. Bouton ME, Bolles RC (1980) Conditioned fear assessed by freezing and by the suppression of three different baselines. Anim Learn Behav 8(3):429–434

50. Maren S, Holt W (2000) The hippocampus and contextual memory retrieval in Pavlovian conditioning. Behav Brain Res 110 (1–2):97–108

51. Goshen I, Brodsky M, Prakash R, Wallace J, Gradinaru V, Ramakrishnan C, Deisseroth K (2011) Dynamics of retrieval strategies for remote memories. Cell 147(3):678–689. doi:10.1016/j.cell.2011.09.033

52. Kheirbek MA, Drew LJ, Burghardt NS, Costantini DO, Tannenholz L, Ahmari SE, Zeng H, Fenton AA, Hen R (2013) Differential control of learning and anxiety along the dorsoventral axis of the dentate gyrus. Neuron 77 (5):955–968. doi:10.1016/j.neuron.2012.12.038

53. Marek R, Strobel C, Bredy TW, Sah P (2013) The amygdala and medial prefrontal cortex: partners in the fear circuit. J Physiol 591(Pt 10):2381–2391. doi:10.1113/jphysiol.2012.248575

54. Ciocchi S, Herry C, Grenier F, Wolff SB, Letzkus JJ, Vlachos I, Ehrlich I, Sprengel R, Deisseroth K, Stadler MB, Muller C, Luthi A (2010) Encoding of conditioned fear in central amygdala inhibitory circuits. Nature 468 (7321):277–282. doi:10.1038/nature09559

55. Courtin J, Chaudun F, Rozeske RR, Karalis N, Gonzalez-Campo C, Wurtz H, Abdi A, Baufreton J, Bienvenu TC, Herry C (2014) Prefrontal parvalbumin interneurons shape neuronal activity to drive fear expression. Nature 505(7481):92–96. doi:10.1038/nature12755

56. Fournier NM, Duman RS (2013) Illuminating hippocampal control of fear memory and anxiety. Neuron 77(5):803–806. doi:10.1016/j.neuron.2013.02.017

57. Garner AR, Rowland DC, Hwang SY, Baumgaertel K, Roth BL, Kentros C, Mayford M (2012) Generation of a synthetic memory trace. Science 335(6075):1513–1516. doi:10.1126/science.1214985

58. Gu Y, Arruda-Carvalho M, Wang J, Janoschka SR, Josselyn SA, Frankland PW, Ge S (2012) Optical controlling reveals time-dependent roles for adult-born dentate granule cells. Nat Neurosci 15(12):1700–1706. doi:10.1038/nn.3260

59. Huff ML, Miller RL, Deisseroth K, Moorman DE, LaLumiere RT (2013) Posttraining optogenetic manipulations of basolateral amygdala activity modulate consolidation of inhibitory avoidance memory in rats. Proc Natl Acad Sci U S A 110(9):3597–3602. doi:10.1073/pnas.1219593110

60. Jasnow AM, Ehrlich DE, Choi DC, Dabrowska J, Bowers ME, McCullough KM, Rainnie DG, Ressler KJ (2013) Thy1-expressing neurons in the basolateral amygdala may mediate fear inhibition. J Neurosci 33(25):10396–10404. doi:10.1523/JNEUROSCI.5539-12.2013

61. Jennings JH, Ung RL, Resendez SL, Stamatakis AM, Taylor JG, Huang J, Veleta K, Kantak PA, Aita M, Shilling-Scrivo K, Ramakrishnan C, Deisseroth K, Otte S, Stuber GD (2015) Visualizing hypothalamic network dynamics for appetitive and consummatory behaviors. Cell 160 (3):516–527. doi:10.1016/j.cell.2014.12.026

62. Johansen JP, Hamanaka H, Monfils MH, Behnia R, Deisseroth K, Blair HT, LeDoux JE (2010) Optical activation of lateral amygdala pyramidal cells instructs associative fear learning. Proc Natl Acad Sci U S A 107 (28):12692–12697. doi:10.1073/pnas.1002418107

63. Knobloch HS, Charlet A, Hoffmann LC, Eliava M, Khrulev S, Cetin AH, Osten P, Schwarz MK, Seeburg PH, Stoop R, Grinevich V (2012) Evoked axonal oxytocin release in the central amygdala attenuates fear response. Neuron 73(3):553–566. doi:10.1016/j.neuron.2011.11.030

64. Sears RM, Fink AE, Wigestrand MB, Farb CR, de Lecea L, Ledoux JE (2013) Orexin/hypocretin system modulates amygdala-dependent threat learning through the locus coeruleus. Proc Natl Acad Sci U S A 110 (50):20260–20265. doi:10.1073/pnas.1320325110

65. Letzkus JJ, Wolff SB, Meyer EM, Tovote P, Courtin J, Herry C, Luthi A (2011) A disinhibitory microcircuit for associative fear learning in the auditory cortex. Nature 480 (7377):331–335. doi:10.1038/nature10674

66. Sparta DR, Smithuis J, Stamatakis AM, Jennings JH, Kantak PA, Ung RL, Stuber GD (2014) Inhibition of projections from the basolateral amygdala to the entorhinal cortex disrupts the acquisition of contextual fear. Front Behav Neurosci 8:129. doi:10.3389/fnbeh.2014.00129

67. Tayler KK, Tanaka KZ, Reijmers LG, Wiltgen BJ (2013) Reactivation of neural ensembles during the retrieval of recent and remote memory. Curr Biol 23(2):99–106. doi:10.1016/j.cub.2012.11.019

68. Witten IB, Lin SC, Brodsky M, Prakash R, Diester I, Anikeeva P, Gradinaru V, Ramakrishnan C, Deisseroth K (2010) Cholinergic interneurons control local circuit activity and cocaine conditioning. Science 330 (6011):1677–1681. doi:10.1126/science.1193771

69. Wolff SB, Grundemann J, Tovote P, Krabbe S, Jacobson GA, Muller C, Herry C, Ehrlich I, Friedrich RW, Letzkus JJ, Luthi A (2014) Amygdala interneuron subtypes control fear learning through disinhibition. Nature 509 (7501):453–458. doi:10.1038/nature13258

70. Xu W, Sudhof TC (2013) A neural circuit for memory specificity and generalization. Science 339(6125):1290–1295. doi:10.1126/science.1229534

71. Baratta MV, Kodandaramaiah SB, Monahan PE, Yao J, Weber MD, Lin PA, Gisabella B, Petrossian N, Amat J, Kim K, Yang A, Forest CR, Boyden ES, Goosens KA (2015) Stress enables reinforcement-elicited serotonergic consolidation of fear memory. Biol Psychiatry 79(10):814–822. doi:10.1016/j.biopsych.2015.06.025

72. Do-Monte FH, Manzano-Nieves G, Quinones-Laracuente K, Ramos-Medina L, Quirk GJ (2015) Revisiting the role of infralimbic cortex in fear extinction with optogenetics. J Neurosci 35(8):3607–3615. doi:10.1523/JNEUROSCI.3137-14.2015

73. Johansen JP, Diaz-Mataix L, Hamanaka H, Ozawa T, Ycu E, Koivumaa J, Kumar A, Hou M, Deisseroth K, Boyden ES, LeDoux JE (2014) Hebbian and neuromodulatory mechanisms interact to trigger associative memory formation. Proc Natl Acad Sci U S A 111(51):E5584–E5592. doi:10.1073/pnas.1421304111

74. Lux V, Masseck OA, Herlitze S, Sauvage MM (2015) Optogenetic destabilization of the memory trace in CA1: insights into reconsolidation and retrieval processes. Cereb Cortex 27(1):841–851. doi:10.1093/cercor/bhv282

75. Wang DV, Yau HJ, Broker CJ, Tsou JH, Bonci A, Ikemoto S (2015) Mesopontine median raphe regulates hippocampal ripple oscillation and memory consolidation. Nat Neurosci 18(5):728–735. doi:10.1038/nn.3998

76. Bero AW, Meng J, Cho S, Shen AH, Canter RG, Ericsson M, Tsai LH (2014) Early remodeling of the neocortex upon episodic memory encoding. Proc Natl Acad Sci U S A 111 (32):11852–11857. doi:10.1073/pnas.1408378111

77. Nomura H, Hara K, Abe R, Hitora-Imamura N, Nakayama R, Sasaki T, Matsuki N, Ikegaya Y (2015) Memory formation and retrieval of neuronal silencing in the auditory cortex. Proc Natl Acad Sci U S A 112

(31):9740–9744.   doi:10.1073/pnas.
1500869112

78. Kwon JT, Nakajima R, Kim HS, Jeong Y, Augustine GJ, Han JH (2014) Optogenetic activation of presynaptic inputs in lateral amygdala forms associative fear memory. Learn Mem 21(11):627–633. doi:10.1101/lm.035816.114

79. Yiu AP, Mercaldo V, Yan C, Richards B, Rashid AJ, Hsiang HL, Pressey J, Mahadevan V, Tran MM, Kushner SA, Woodin MA, Frankland PW, Josselyn SA (2014) Neurons are recruited to a memory trace based on relative neuronal excitability immediately before training. Neuron 83(3):722–735. doi:10.1016/j.neuron.2014.07.017

80. Lovett-Barron M, Kaifosh P, Kheirbek MA, Danielson N, Zaremba JD, Reardon TR, Turi GF, Hen R, Zemelman BV, Losonczy A (2014) Dendritic inhibition in the hippocampus supports fear learning. Science 343 (6173):857–863.   doi:10.1126/science.1247485

81. Rei D, Mason X, Seo J, Graff J, Rudenko A, Wang J, Rueda R, Siegert S, Cho S, Canter RG, Mungenast AE, Deisseroth K, Tsai LH (2015) Basolateral amygdala bidirectionally modulates stress-induced hippocampal learning and memory deficits through a p25/Cdk5-dependent pathway. Proc Natl Acad Sci U S A 112(23):7291–7296. doi:10.1073/pnas.1415845112

82. Nabavi S, Fox R, Proulx CD, Lin JY, Tsien RY, Malinow R (2014) Engineering a memory with LTD and LTP. Nature 511 (7509):348–352. doi:10.1038/nature13294

83. Morris R (1984) Developments of a water-maze procedure for studying spatial learning in the rat. J Neurosci Methods 11(1):47–60

84. Morris RG, Garrud P, Rawlins JN, O'Keefe J (1982) Place navigation impaired in rats with hippocampal lesions. Nature 297 (5868):681–683

85. Sandi C, Loscertales M, Guaza C (1997) Experience-dependent facilitating effect of corticosterone on spatial memory formation in the water maze. Eur J Neurosci 9(4):637–642

86. Kolisnyk B, Guzman MS, Raulic S, Fan J, Magalhaes AC, Feng G, Gros R, Prado VF, Prado MA (2013) ChAT-ChR2-EYFP mice have enhanced motor endurance but show deficits in attention and several additional cognitive domains. J Neurosci 33 (25):10427–10438.   doi:10.1523/JNEUROSCI.0395-13.2013

87. Andrews-Zwilling Y, Gillespie AK, Kravitz AV, Nelson AB, Devidze N, Lo I, Yoon SY,

Bien-Ly N, Ring K, Zwilling D, Potter GB, Rubenstein JL, Kreitzer AC, Huang Y (2012) Hilar GABAergic interneuron activity controls spatial learning and memory retrieval. PLoS One 7(7):e40555. doi:10.1371/journal.pone.0040555

88. Clark RE, Broadbent NJ, Squire LR (2005) Hippocampus and remote spatial memory in rats. Hippocampus 15(2):260–272. doi:10.1002/hipo.20056

89. Barnes CA (1979) Memory deficits associated with senescence: a neurophysiological and behavioral study in the rat. J Comp Physiol Psychol 93(1):74–104

90. Olton DS, Samuelson RJ (1976) Remembrance of places passed: spatial memory in rats. J Exp Psychol Anim Learn Cogn 2 (2):97–116

91. Ennaceur A, Delacour J (1988) A new one-trial test for neurobiological studies of memory in rats. 1: behavioral data. Behav Brain Res 31(1):47–59

92. Gheusi G, Bluthe RM, Goodall G, Dantzer R (1994) Social and individual recognition in rodents: methodological aspects and neurobiological bases. Behav Process 33(1–2):59–87. doi:10.1016/0376-6357(94)90060-4

93. Smith AS, Williams Avram SK, Cymerblit-Sabba A, Song J, Young WS (2016) Targeted activation of the hippocampal CA2 area strongly enhances social memory. Mol Psychiatry 21(8):1137–1144.   doi:10.1038/mp.2015.189

94. Olds J (1963) Self-stimulation experiments. Science 140(3563):218–220

95. Stamatakis AM, Stuber GD (2012) Activation of lateral habenula inputs to the ventral midbrain promotes behavioral avoidance. Nat Neurosci 15(8):1105–1107. doi:10.1038/nn.3145

96. Steinberg EE, Keiflin R, Boivin JR, Witten IB, Deisseroth K, Janak PH (2013) A causal link between prediction errors, dopamine neurons and learning. Nat Neurosci 16(7):966–973. doi:10.1038/nn.3413

97. Stuber GD, Sparta DR, Stamatakis AM, van Leeuwen WA, Hardjoprajitno JE, Cho S, Tye KM, Kempadoo KA, Zhang F, Deisseroth K, Bonci A (2011) Excitatory transmission from the amygdala to nucleus accumbens facilitates reward seeking. Nature 475(7356):377–380. doi:10.1038/nature10194

98. Tai LH, Lee AM, Benavidez N, Bonci A, Wilbrecht L (2012) Transient stimulation of distinct subpopulations of striatal neurons mimics changes in action value. Nat Neurosci 15(9):1281–1289. doi:10.1038/nn.3188

99. Tan KR, Yvon C, Turiault M, Mirzabekov JJ, Doehner J, Labouebe G, Deisseroth K, Tye KM, Luscher C (2012) GABA neurons of the VTA drive conditioned place aversion. Neuron 73(6):1173–1183. doi:10.1016/j.neuron.2012.02.015

100. Canetta S, Bolkan S, Padilla-Coreano N, Song LJ, Sahn R, Harrison NL, Gordon JA, Brown A, Kellendonk C (2016) Maternal immune activation leads to selective functional deficits in offspring parvalbumin interneurons. Mol Psychiatry 21(7):956–968. doi:10.1038/mp.2015.222

101. Spellman T, Rigotti M, Ahmari SE, Fusi S, Gogos JA, Gordon JA (2015) Hippocampal-prefrontal input supports spatial encoding in working memory. Nature 522 (7556):309–314. doi:10.1038/nature14445

102. Duan AR, Varela C, Zhang Y, Shen Y, Xiong L, Wilson MA, Lisman J (2015) Delta frequency optogenetic stimulation of the thalamic nucleus reuniens is sufficient to produce working memory deficits: relevance to schizophrenia. Biol Psychiatry 77(12):1098–1107. doi:10.1016/j.biopsych.2015.01.020

103. Yamamoto J, Suh J, Takeuchi D, Tonegawa S (2014) Successful execution of working memory linked to synchronized high-frequency gamma oscillations. Cell 157 (4):845–857. doi:10.1016/j.cell.2014.04.009

104. Shipton OA, El-Gaby M, Apergis-Schoute J, Deisseroth K, Bannerman DM, Paulsen O, Kohl MM (2014) Left-right dissociation of hippocampal memory processes in mice. Proc Natl Acad Sci U S A 111 (42):15238–15243. doi:10.1073/pnas.1405648111

105. Siegle JH, Wilson MA (2014) Enhancement of encoding and retrieval functions through theta phase-specific manipulation of hippocampus. eLife 3:e03061. doi:10.7554/eLife.03061

106. Li P, Rial D, Canas PM, Yoo JH, Li W, Zhou X, Wang Y, van Westen GJ, Payen MP, Augusto E, Goncalves N, Tome AR, Li Z, Wu Z, Hou X, Zhou Y, AP IJ, Boyden ES, Cunha RA, Qu J, Chen JF (2015) Optogenetic activation of intracellular adenosine A2A receptor signaling in the hippocampus is sufficient to trigger CREB phosphorylation and impair memory. Mol Psychiatry 20 (11):1339–1349. doi:10.1038/mp.2014.182

107. McNamara CG, Tejero-Cantero A, Trouche S, Campo-Urriza N, Dupret D (2014) Dopaminergic neurons promote hippocampal reactivation and spatial memory persistence. Nat

Neurosci 17(12):1658–1660. doi:10.1038/nn.3843

108. Bussey TJ, Padain TL, Skillings EA, Winters BD, Morton AJ, Saksida LM (2008) The touchscreen cognitive testing method for rodents: how to get the best out of your rat. Learn Mem 15(7):516–523. doi:10.1101/lm.987808

109. Clelland CD, Choi M, Romberg C, Clemenson GD Jr, Fragniere A, Tyers P, Jessberger S, Saksida LM, Barker RA, Gage FH, Bussey TJ (2009) A functional role for adult hippocampal neurogenesis in spatial pattern separation. Science 325(5937):210–213. doi:10.1126/science.1173215

110. Ho JW, Poeta DL, Jacobson TK, Zolnik TA, Neske GT, Connors BW, Burwell RD (2015) Bidirectional modulation of recognition memory. J Neurosci 35(39):13323–13335. doi:10.1523/JNEUROSCI.2278-15.2015

111. Wu GY, Liu GL, Zhang HM, Chen C, Liu SL, Feng H, Sui JF (2015) Optogenetic stimulation of mPFC pyramidal neurons as a conditioned stimulus supports associative learning in rats. Sci Rep 5:10065. doi:10.1038/srep10065

112. Choi GB, Stettler DD, Kallman BR, Bhaskar ST, Fleischmann A, Axel R (2011) Driving opposing behaviors with ensembles of piriform neurons. Cell 146(6):1004–1015. doi:10.1016/j.cell.2011.07.041

113. Tritsch NX, Sabatini BL (2012) Dopaminergic modulation of synaptic transmission in cortex and striatum. Neuron 76(1):33–50. doi:10.1016/j.neuron.2012.09.023

114. Tritsch NX, Oh WJ, Gu C, Sabatini BL (2014) Midbrain dopamine neurons sustain inhibitory transmission using plasma membrane uptake of GABA, not synthesis. eLife 3:e01936. doi:10.7554/eLife.01936

115. Stuber GD, Hnasko TS, Britt JP, Edwards RH, Bonci A (2010) Dopaminergic terminals in the nucleus accumbens but not the dorsal striatum corelease glutamate. J Neurosci 30 (24):8229–8233. doi:10.1523/JNEUROSCI.1754-10.2010

116. Zhang S, Qi J, Li X, Wang HL, Britt JP, Hoffman AF, Bonci A, Lupica CR, Morales M (2015) Dopaminergic and glutamatergic microdomains in a subset of rodent mesoaccumbens axons. Nat Neurosci 18 (3):386–392. doi:10.1038/nn.3945

117. Fenno LE, Mattis J, Ramakrishnan C, Hyun M, Lee SY, He M, Tucciarone J, Selimbeyoglu A, Berndt A, Grosenick L, Zalocusky KA, Bernstein H, Swanson H, Perry C, Diester I, Boyce FM, Bass CE, Neve R, Huang ZJ,

Deisseroth K (2014) Targeting cells with single vectors using multiple-feature Boolean logic. Nat Methods 11(7):763–772. doi:10.1038/nmeth.2996

118. Deisseroth K, Schnitzer MJ (2013) Engineering approaches to illuminating brain structure and dynamics. Neuron 80(3):568–577. doi:10.1016/j.neuron.2013.10.032

119. Packer AM, Roska B, Hausser M (2013) Targeting neurons and photons for optogenetics. Nat Neurosci 16(7):805–815. doi:10.1038/nn.3427

120. Lignani G, Ferrea E, Difato F, Amaru J, Ferroni E, Lugara E, Espinoza S, Gainetdinov RR, Baldelli P, Benfenati F (2013) Long-term optical stimulation of channelrhodopsin-expressing neurons to study network plasticity. Front Mol Neurosci 6:22. doi:10.3389/fnmol.2013.00022

121. Chun S, Bayazitov IT, Blundon JA, Zakharenko SS (2013) Thalamocortical long-term potentiation becomes gated after the early critical period in the auditory cortex. J Neurosci 33(17):7345–7357. doi:10.1523/JNEUROSCI.4500-12.2013

122. Raimondo JV, Kay L, Ellender TJ, Akerman CJ (2012) Optogenetic silencing strategies differ in their effects on inhibitory synaptic transmission. Nat Neurosci 15(8):1102–1104. doi:10.1038/nn.3143

123. Wang SH, Morris RG (2010) Hippocampal-neocortical interactions in memory formation, consolidation, and reconsolidation. Annu Rev Psychol 61(49–79):C41–C44. doi:10.1146/annurev.psych.093008.100523

124. Shimizu E, Tang YP, Rampon C, Tsien JZ (2000) NMDA receptor-dependent synaptic reinforcement as a crucial process for memory consolidation. Science 290(5494):1170–1174

125. McHugh TJ, Jones MW, Quinn JJ, Balthasar N, Coppari R, Elmquist JK, Lowell BB, Fanselow MS, Wilson MA, Tonegawa S (2007) Dentate gyrus NMDA receptors mediate rapid pattern separation in the hippocampal network. Science 317(5834):94–99. doi:10.1126/science.1140263

126. Daumas S, Halley H, Frances B, Lassalle JM (2005) Encoding, consolidation, and retrieval of contextual memory: differential involvement of dorsal CA3 and CA1 hippocampal subregions. Learn Mem 12(4):375–382. doi:10.1101/lm.81905

127. Stefanelli T, Bertollini C, Luscher C, Muller D, Mendez P (2016) Hippocampal somatostatin interneurons control the size of neuronal memory ensembles. Neuron 89 (5):1074–1085. doi:10.1016/j.neuron.2016.01.024

128. Aimone JB, Deng W, Gage FH (2011) Resolving new memories: a critical look at the dentate gyrus, adult neurogenesis, and pattern separation. Neuron 70(4):589–596. doi:10.1016/j.neuron.2011.05.010

129. Arruda-Carvalho M, Sakaguchi M, Akers KG, Josselyn SA, Frankland PW (2011) Posttraining ablation of adult-generated neurons degrades previously acquired memories. J Neurosci 31(42):15113–15127. doi:10.1523/JNEUROSCI.3432-11.2011

130. Bannerman DM, Rawlins JN, McHugh SB, Deacon RM, Yee BK, Bast T, Zhang WN, Pothuizen HH, Feldon J (2004) Regional dissociations within the hippocampus–memory and anxiety. Neurosci Biobehav Rev 28 (3):273–283. doi:10.1016/j.neubiorev.2004.03.004

131. Frankland PW, Bontempi B (2005) The organization of recent and remote memories. Nat Rev Neurosci 6(2):119–130. doi:10.1038/nrn1607

132. Moscovitch M, Nadel L, Winocur G, Gilboa A, Rosenbaum RS (2006) The cognitive neuroscience of remote episodic, semantic and spatial memory. Curr Opin Neurobiol 16 (2):179–190. doi:10.1016/j.conb.2006.03.013

133. Winocur G, Moscovitch M, Bontempi B (2010) Memory formation and long-term retention in humans and animals: convergence towards a transformation account of hippocampal-neocortical interactions. Neuropsychologia 48(8):2339–2356. doi:10.1016/j.neuropsychologia.2010.04.016

134. Nakashiba T, Young JZ, McHugh TJ, Buhl DL, Tonegawa S (2008) Transgenic inhibition of synaptic transmission reveals role of CA3 output in hippocampal learning. Science 319(5867):1260–1264. doi:10.1126/science.1151120

135. Kitamura T, Saitoh Y, Takashima N, Murayama A, Niibori Y, Ageta H, Sekiguchi M, Sugiyama H, Inokuchi K (2009) Adult neurogenesis modulates the hippocampus-dependent period of associative fear memory. Cell 139(4):814–827. doi:10.1016/j.cell.2009.10.020

136. Wiltgen BJ, Zhou M, Cai Y, Balaji J, Karlsson MG, Parivash SN, Li W, Silva AJ (2010) The hippocampus plays a selective role in the retrieval of detailed contextual memories. Curr Biol 20(15):1336–1344. doi:10.1016/j.cub.2010.06.068

137. Euston DR, Gruber AJ, McNaughton BL (2012) The role of medial prefrontal cortex in memory and decision making. Neuron 76 (6):1057–1070. doi:10.1016/j.neuron. 2012.12.002

138. Frankland PW, Bontempi B, Talton LE, Kaczmarek L, Silva AJ (2004) The involvement of the anterior cingulate cortex in remote contextual fear memory. Science 304(5672): 881–883. doi:10.1126/science.1094804

139. Gilmartin MR, Miyawaki H, Helmstetter FJ, Diba K (2013) Prefrontal activity links non-overlapping events in memory. J Neurosci 33 (26):10910–10914.   doi:10.1523/ JNEUROSCI.0144-13.2013

140. Do-Monte FH, Quinones-Laracuente K, Quirk GJ (2015) A temporal shift in the circuits mediating retrieval of fear memory. Nature 519(7544):460–463. doi:10.1038/ nature14030

141. Kim HS, Cho HY, Augustine GJ, Han JH (2015) Selective control of fear expression by optogenetic manipulation of infralimbic cortex after extinction. Neuropsychopharmacology 41(5):1261–1273. doi:10.1038/npp. 2015.276

142. Brown MT, Tan KR, O'Connor EC, Nikonenko I, Muller D, Luscher C (2012) Ventral tegmental area GABA projections pause accumbal cholinergic interneurons to enhance associative learning. Nature 492 (7429):452–456. doi:10.1038/nature11657

143. Han JH, Kushner SA, Yiu AP, Cole CJ, Matynia A, Brown RA, Neve RL, Guzowski JF, Silva AJ, Josselyn SA (2007) Neuronal competition and selection during memory formation. Science 316(5823):457–460. doi:10. 1126/science.1139438

144. Reijmers LG, Perkins BL, Matsuo N, Mayford M (2007) Localization of a stable neural correlate of associative memory. Science 317 (5842):1230–1233.   doi:10.1126/science. 1143839

145. Rogerson T, Cai DJ, Frank A, Sano Y, Shobe J, Lopez-Aranda MF, Silva AJ (2014) Synaptic tagging during memory allocation. Nat Rev Neurosci 15(3):157–169. doi:10.1038/ nrn3667

146. Han JH, Kushner SA, Yiu AP, Hsiang HL, Buch T, Waisman A, Bontempi B, Neve RL, Frankland PW, Josselyn SA (2009) Selective erasure of a fear memory. Science 323 (5920):1492–1496.   doi:10.1126/science. 1164139

147. Zhou Y, Won J, Karlsson MG, Zhou M, Rogerson T, Balaji J, Neve R, Poirazi P, Silva AJ (2009) CREB regulates excitability and the allocation of memory to subsets of neurons in the amygdala. Nat Neurosci 12 (11):1438–1443. doi:10.1038/nn.2405

148. Dudek SM, Alexander GM, Farris S (2016) Rediscovering area CA2: unique properties and functions. Nat Rev Neurosci 17 (2):89–102. doi:10.1038/nrn.2015.22

149. Kohara K, Pignatelli M, Rivest AJ, Jung HY, Kitamura T, Suh J, Frank D, Kajikawa K, Mise N, Obata Y, Wickersham IR, Tonegawa S (2014) Cell type-specific genetic and optogenetic tools reveal hippocampal CA2 circuits. Nat Neurosci 17(2):269–279. doi:10.1038/ nn.3614

150. Hitti FL, Siegelbaum SA (2014) The hippocampal CA2 region is essential for social memory. Nature 508(7494):88–92. doi:10. 1038/nature13028

151. Ziv Y, Burns LD, Cocker ED, Hamel EO, Ghosh KK, Kitch LJ, El Gamal A, Schnitzer MJ (2013) Long-term dynamics of CA1 hippocampal place codes. Nat Neurosci 16 (3):264–266. doi:10.1038/nn.3329

152. Akerboom J, Carreras Calderon N, Tian L, Wabnig S, Prigge M, Tolo J, Gordus A, Orger MB, Severi KE, Macklin JJ, Patel R, Pulver SR, Wardill TJ, Fischer E, Schuler C, Chen TW, Sarkisyan KS, Marvin JS, Bargmann CI, Kim DS, Kugler S, Lagnado L, Hegemann P, Gottschalk A, Schreiter ER, Looger LL (2013) Genetically encoded calcium indicators for multi-color neural activity imaging and combination with optogenetics. Front Mol Neurosci 6:2. doi:10.3389/ fnmol.2013.00002

153. Ohkura M, Sasaki T, Sadakari J, Gengyo-Ando K, Kagawa-Nagamura Y, Kobayashi C, Ikegaya Y, Nakai J (2012) Genetically encoded green fluorescent Ca2+ indicators with improved detectability for neuronal Ca2+ signals. PLoS One 7(12):e51286. doi:10.1371/journal.pone.0051286

Chapter 13

Towards Opto-Magnetic Physiology: Concepts and Pitfalls of ofMRI

Miriam Schwalm, Eduardo Rosales Jubal, and Albrecht Stroh

Abstract

Optogenetic functional magnetic resonance imaging (ofMRI) represents the combination of optogenetic modulation of neural circuits with high-field fMRI. ofMRI enables the monitoring of neural activity across the entire brain, while precisely controlling the activation of specific neuronal elements within the neural circuit, defined by their genetic identity, location of somata, and axonal projection targets. Those exclusive features of ofMRI open a plethora of new possibilities for the in vivo characterization of neural networks, simultaneously at local and global scale. In addition, recent advances in optical readouts of intracellular Ca^{2+} within optogenetic fMRI experiments led to all-optical interrogations within the scanner. Here, we provide a guide to applying optogenetics and optical recordings in the setting of fMRI experiments. In addition, we devise a straight forward control experiment addressing the discrimination of specific network activation through the excitation of opsins from unspecific heat effects leading to an apparent BOLD effect. Lastly, we propose a method to estimate the scope of optogenetic network modulation by combining the estimations of the number of opsin-expressing neurons and light spreading in brain tissue.

Key words ofMRI, Optogenetics, Ca^{2+} recordings, BOLD fMRI, Optical recordings

1 Introduction

Since the discovery of the Blood Oxygenation Level Dependent (BOLD) effect utilizing paramagnetic deoxyhemoglobin in venous blood as a naturally occurring contrast agent [1], functional Magnetic Resonance Imaging (fMRI) became a widely used brain imaging technique. By detecting changes of cerebral blood flow associated with increased metabolism of neural tissue, BOLD fMRI provides a noninvasive and brain-wide readout of neurovascular activity. Whereas a spatial resolution of usually <1 mm^3 voxel size is commonly feasible nowadays, the intrinsic slow response properties of the hemodynamic signal impose biological constraints to the temporal resolution of BOLD fMRI. The BOLD signal relies on neurovascular coupling, a complex interaction of neurons, glia, and vascular cells, tightly coupled to neural activity, which fine

Albrecht Stroh (ed.), *Optogenetics: A Roadmap*, Neuromethods, vol. 133,
DOI 10.1007/978-1-4939-7417-7_13, © Springer Science+Business Media LLC 2018

grained mechanisms remain under investigation, but are robust enough to yield a consistent, albeit indirect, marker of neural activation [2].

Beyond its clinical application and use in human brain research, fMRI in animal models became a key tool for basic neuroscience. On the one hand, the indirect mapping of neural activity by following global neurovascular activity can be combined with local electric or optical recordings of neuronal responses to achieve a better understanding of the relationship between these signals [3–5]). On the other hand, a promising translational bridge between human and animal models of various neurological diseases and psychiatric disorders can be established by fMRI due to its applicability in rodents, nonhuman primates, and humans. Originally, rats were the model of choice in fMRI studies, because their higher cognitive abilities are evolutionarily closer to humans [6, 7] and the advantage of a larger brain, favoring more precise mapping of anatomical structures with equal spatial resolution of images. Recently, however, the use of mice has gained popularity, the extensive mapping of the mouse genome and connectome (Allen Brain Atlas, http://mouse.brain-map.org/) led to the development of transgenic mouse models for studying neurological diseases such as Alzheimer's disease [8]. Furthermore, the use of mice facilitates the combination of fMRI with optogenetics due to the variety of Cre-expressing lines allowing for rather straightforward concepts, achieving cell-type specificity of opsin expression (see Chaps. 1 and 2). Small animal fMRI became a pillar of preclinical research due to the advent of magnets with very high field strengths in the last decade (typically 7–11 Tesla), allowing for high effective resolution in anatomical and functional images.

The combination of optogenetic modulation of neural circuits with high-field fMRI led to the development of *optogenetic functional magnetic resonance imaging* (ofMRI, [9]). Simultaneous optogenetic control and fMRI readout enables the monitoring of neural activity across the entire brain, while precisely controlling the activation of specific neuronal elements within the neural circuit, defined by their genetic identity, location of somata, and axonal projection targets. Those exclusive features of ofMRI open a plethora of new possibilities for the in vivo characterization of neural networks, simultaneously at local and global scale. In addition, recent advances in optical readouts of intracellular Ca^{2+} (calcium) within optogenetic fMRI experiments lead to all-optical interrogations within the scanner [4] (Figs. 1 and 2). Here, we aim at providing a guide to applying optogenetics and optical recordings in the setting of fMRI experiments, and most importantly, to ensure the specificity of ofMRI by implementing straightforward control experiments and propose a method to estimate the scope of optogenetic network modulation by combining the estimations of the number of opsin-expressing neurons and light spreading in brain tissue.

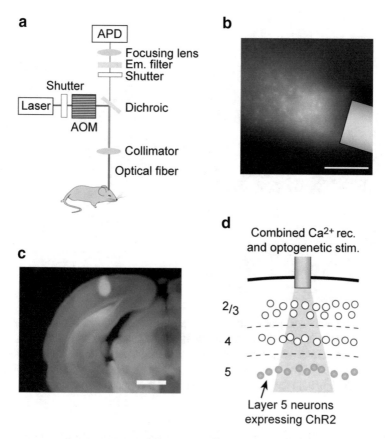

Fig. 1 (**a**) Scheme of optic fiber-based recording and stimulation setup. *AOM* acousto optic modulator, *APD* avalanche photo diode, *Em:* emission. (**b**) Confocal image of Oregon green 488 BAPTA-1 AM-(OGB-1)-stained cortical slice. Only light emitted by fiber was used for OGB-1 excitation (scale bar = 200 μm), delineating the area of optic fiber recording. (**c**) Photomicrograph of a coronal brain slice at the level of the visual cortex (scale bar = 1 mm). Overlay transmitted light and green fluorescence channel showing the OGB-1-stained region. (**d**) Schematic depicting the tip of an optic fiber implanted into a stained cortical region above neurons expressing Channelrhodopsin-2 (ChR2) (*green*). *Adapted from* [46]

1.1 Basics of fMRI and BOLD Signal

Anatomical image acquisition (MRI) is obligatory for every fMRI experiment as the anatomical image provides a template for the subsequently localized functional activity. The principle of anatomical imaging requires a permanent magnetic field combined with a temporally varying gradient field for spatial encoding of the acquired images to create axial, sagittal, or coronal images of the brain. For functional images, the commonly used BOLD signal reflects local changes of deoxyhemoglobin [1] driven by changes in blood flow, blood volume and blood oxygenation level. The BOLD signal increases with decreasing deoxyhemoglobin, resulting in functional activity maps of the brain following the assumption that the augmented metabolism is related to neuronal activity by the process of neurovascular coupling. Briefly, it is assumed that

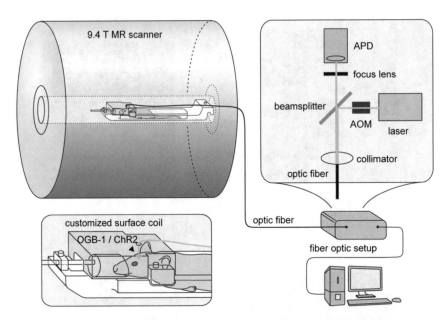

Fig. 2 Scheme of the combination of optic fiber-based recording/stimulation and fMRI. Optic fiber-based stimulation/recording setup is placed outside of the scanner cabin. Light intensities of the *blue* laser are modulated with high temporal precision by an acousto optic modulator (AOM), emitted light is recorded by an avalanche photo diode (APD). The customized surface coil has a lead-trough for the optical fiber. Electric stimulation, laser excitation, fMRI data acquisition, and optical recordings are synchronized. *Adapted from* [4].

local neural activity elevates the metabolic demand and oxygen consumption of the respective brain region, leading primarily to a local increase of deoxyhemoglobin concentration ("initial dip"), followed by an overcompensation ("overshoot") of oxygenated hemoglobin provided by increasing blood flow and vasodilation. These signal changes can be detected by the MR scanner due to the paramagnetic properties of deoxyhemoglobin. When oxygenated, the central iron atom of hemoglobin changes to a low spin state by receiving the oxygen's electrons and hemoglobin becomes diamagnetic and hence neutral to the magnetic field. Because of these different magnetic properties of oxygenated and deoxygenated hemoglobin and their effects on relaxation time, the "initial dip" leads to a decrease of the BOLD signal, followed by a signal increase originating from the subsequently provided oxygenated hemoglobin. After 5–6 s, (depending on brain region and stimulus characteristics) blood volume and vasodilation, and consequently the BOLD contrast, return to baseline. This stereotypic dynamic is called hemodynamic response (HRF), and is characterized by a delayed peak (5–6 s), dispersion in time (~4 s) and space (mm range depending on cortical size) and small amplitude (0.5–3% signal change) [10–14].

1.2 fMRI in Basic Neuroscience

Despite the widespread use of the technique, the true origin of the fMRI BOLD signal is still insufficiently understood [5, 15–18]. Hence, as a direct link between BOLD and neuronal oxygen consumption cannot be assumed [19], fMRI data interpretation remains controversial. Whereas classical models assume a linear relationship between neuronal activity and blood flow [20] over the last 10 years different neural players including astrocytes, pericytes and interneurons, as well as cellular [21, 22] and vascular [15] mechanisms have been proposed to be critically involved in the process of neurovascular coupling. Furthermore, the discovery of altered neurovascular coupling under certain conditions, as for instance in pathological states [23, 24], impacting stimulus-evoked as well as resting-state BOLD signal provided additional evidence for the limits of the quantitative interpretation of BOLD contrast [15]. Later in this chapter we will elaborate on how the use of optogenetics can lead to a better understanding of neurovascular coupling processes and the differentially involved signaling mechanisms.

Notwithstanding the incomplete understanding of the underlying signal, BOLD fMRI became the state of the art method in human neuroimaging, due to its non-invasiveness and relatively high spatial resolution. In animal research, beyond its use in comparative and preclinical studies, fMRI provides a unique technique in terms of spatial reach of the readout and the option to track functional connectivity between brain regions. Particularly, the combination of fMRI with other neuronal activity readouts, such as electrophysiological recordings [5, 9, 25, 26] has led to a considerable progress on the understanding of the underlying signal dynamics. Especially suited for the combination with fMRI, given its nonmagnetic properties, are optic fiber-based recordings of intracellular Ca^{2+} [3, 4] (Fig. 1), not disturbed by the magnetic field and mainly reflecting action potential related Ca^{2+} influx into cells, has provided additional insight on the relationship between BOLD and neuronal activity.

1.3 Advantages of the Combination of fMRI and Optogenetics

The advent of optogenetics for causally investigating network connectivity enabled a causal interrogation of aberrant functional connectivity, e.g., in Parkinson's disease [27], or depression, and schizophrenia [28]. In the framework of fMRI studies, the use of optogenetics was primarily introduced to study the distribution of cortical and subcortical activity following cell-type specific stimulation. The first proof-of-concept ofMRI study from the group of Karl Deisseroth [9] combined optogenetic stimulation of thalamo-cortical circuits with BOLD fMRI, studying spatial and temporal patterns of network recruitment, identifying long-range projections, as well as assessing whole brain responses of optogenetic stimuli.

Following the first reports linking optogenetically induced neuronal activity and BOLD signals [9, 26], ofMRI has been explored in rodents, both anesthetized and awake, as well as in primates [25, 29–37]. Aspects of the optogenetically evoked hemodynamic response, as well as the suitability of optogenetic stimulation to recruit functional networks [9, 25, 26, 34–36, 38] and to mimic sensory stimulation [4, 37] have been assessed, for example, sensory inputs to the barrel fields have been mimicked by optogenetics, under brain-wide fMRI recordings [33].

Individual aspects of the optogenetically evoked BOLD responses, such as spatial limits [29], temporal characteristics [9, 37], stimulus strength dependence [30, 37], stimulus frequency dependence [4, 37] and linearity upon repeated stimulations [25] have been investigated in rodents. In primates, the interference of optogenetic stimulation and visual pathways, affecting saccade latencies, has been demonstrated [32], also, corticothalamic networks have been explored in detail [39].

Apart from global circuit mapping, ofMRI is capable to assess the effect of the activity of transplanted stem-cell derived neurons on downstream networks [40, 41]. The precise activity control offered by optogenetic techniques not only allows to monitor whether transplanted cells have been functionally integrated into the host brain tissue, but might also increase their differentiation into mature neurons [42] (see also Chap. 5).

The simultaneous control of neuronal spiking and imaging of brain wide functional activity also enables the probing of connections and wiring of distant areas. The temporal and spatial specificity of optogenetic modulation allows the investigation of the relationship between neuronal spiking and the BOLD signal, as well as the isolation of individual cell types thought to contribute to the neurovascular response, in order to disentangle their contribution to the BOLD signal.

The complexity of the mechanism underpinning the BOLD signal led to scepticism on the physiological relevance of optogenetic elicited BOLD signal [17]. This scepticism stems from the difficulty to disentangle direct effects of optogenetic activation in local populations, from the immediate activation of adjacent interconnected circuits, given the highly recurrent network formed by cortical neurons. Nevertheless, careful observation of the latencies of the immediate optogenetic effects in the local network [4] and the combination of ofMRI and computational models [43], enables the systematic investigation of the functional projections of optogenetically stimulated sites.

1.4 Combining fMRI with Optic Fiber-Based Recordings of Neuronal Activity

Optic-fiber-based recordings of neuronal activity employing synthetic or genetically encoded indicators of intracellular Ca^{2+} offer the unique advantage of seamless and artifact-free combination with fMRI. Optical Ca^{2+} recordings, based on the monitoring of somatic Ca^{2+} concentration of a small population of neurons (typically between 30 and 1000) (Fig. 1b), permit a highly sensitive assessment of local network activity with high specificity for neuronal spiking and a high spatiotemporal resolution [3, 44–47]. While not providing single-neuron resolution, synchronous firing of as few as 30 neurons can be reliable detected [48], and spiking activity can be easily quantified, permitting the detection of even small changes in stimulus strength [4]. In addition, in contrast to 2-photon imaging methods, the temporal resolution of Ca^{2+} recordings is very high (around 1 kHz), limited only by the intrinsic dynamics of Ca^{2+} binding and release (see Chap. 9). Furthermore, unlike electric recordings, at least for synthetic indicators such as Oregon-Green BAPTA-1 (OGB-1) there is a linear correlation between amplitude of Ca^{2+} transients and the number of underlying action potentials [49]. Finally, by using the newest generation of genetically encoded Ca^{2+} indicators such as GCaMP6 [50], it is possible to selectively record from genetically defined neuronal populations, however due to the four binding sites of Ca^{2+} of GCaMP, a linearity between the number of underlying action potentials and fluorescence emission cannot be assumed.

1.5 Controlling for the Specificity of ofMRI

From the advent of ofMRI in 2010, the specificity of this novel method had been subject of debate. In particular, the following specific issues need to be considered in the design of ofMRI studies: (1) artifacts in the MR images caused by fiber implantation, and (2) a heat-induced nonspecific apparent BOLD effect.

1. Undisturbed images in MRI rely on a homogeneous magnetic field. Perturbations in the magnetic field can be caused by different materials, notably ferromagnetic metals, but also by inhomogeneities in bone or brain tissue caused by tissue-air boundaries resulting from invasive manipulations as they are necessary for simultaneous neuronal recordings or optogenetic modulation. In ofMRI, the necessity to perform a craniotomy, and the implantation of an optic fiber entails that commonly used EPI sequences—anyhow prone to susceptibility artifacts due to the sequence itself—might be disturbed. We will provide a point-by-point description on how to deal with these limitations in the protocol and notes section.

2. Using BOLD fMRI as readout of optogenetic activation, a concern may arise from the potential non specificity of the BOLD response due to heating effects [26, 51]. To resolve this concern and provide evidence for the specificity of optogenetically evoked BOLD signal, we tested the effect of constant illumination in addition to the optical stimulation pulse trains. No BOLD response was detected in ChR2 expressing animals when subjected to pulsed stimulation with a light intensity at the tip of the fiber of 80 mW/mm^2 *and additional* constant illumination of about 11 mW/mm^2 (Fig. 3c), but

BOLD signal was restored when no constant illumination was applied in subsequent experiments (Fig. 3b). This control experiment takes advantage of the intrinsic channel kinetics of $ChR2_{H134R}$ [42, 52] (Fig. 3a). Light activation leads to short high amplitude peak currents followed by low amplitude long lasting steady-state currents. Short peak currents provide far greater depolarization than steady state currents. Constant illumination results in continuous steady state currents, preventing the peak current of pulsed excitation.

1.6 Estimating the Scope of Optogenetic Network Stimulation

In order to assess the effective optogenetic modulation exerted in a given neuronal circuit, it is necessary to estimate the proportion of neurons potentially driven by optogenetic stimulation. The number of effectively optogenetically activated neurons is a function of two main factors: the number of cells expressing the opsin, and the number of cells irradiated with sufficient light power to elicit neuronal spiking (the threshold to trigger an action potential depends ultimately on the opsin, but ranges at 1 mW/mm^2).

Quantification of neurons expressing the opsins can be achieved by extrapolating empirical measures obtained from fluorescence or confocal microscopy. Total numbers of cells tagged with the fluorophore can be derived by counting all cells with a smooth, strong, and membrane-bound fluorescence in image stacks covering the entire area of opsin/fluorophore expression (see also Chap. 8). To derive densities of opsin/fluorophore-expressing neurons, brain slices can be stained with a somatic marker (e.g., DAPI or neurotrace) and histological slices in the center of the expression area can be assessed by stereological methods. Assuming homogeneous expression throughout a cortical section, a representative area in these slices can be chosen, and within this area, cells distinguished by their emission wavelength and morphological distribution of fluorescence can be counted separately. All cells exhibiting cytosolic fluorescence from the soma marker are counted to derive the number of cells in this area, and cells showing homogenous and strong membrane-bound fluorescence are counted to derive the fraction of cells presumably exhibiting functional expression of the opsin. Both sets should overlap, but only a subset of cells should exhibit both the flurophore-tagged opsin and the somatic marker. Usually, stereology software offers unbiased counting procedures allowing extrapolation of cells per mm^3 within the virus expressing area based on the total volume of the expression area per slide, defined by the slice thickness and the average number of cells in the area used for counting [4].

The number of neurons illuminated by suprathreshold light intensity to elicit an action potential requires the estimation of light diffusion through brain tissue. Early approximations to this problem used analytical models combined with ex vivo or in vitro

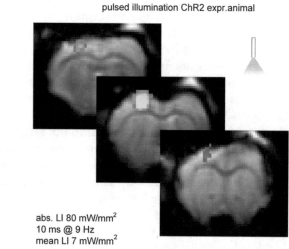

Fig. 3 (a) Kinetics of de-inactivation of the peak ChR2 current. Ten overlaid photocurrent traces obtained in voltage clamp are shown; pairs of 0.5 s light pulses (indicated by *blue bars*) were separated by increasing intervals from 1 to 10 s; traces are aligned to the initiation of the first pulse in each sweep. Note that steady-

empirical measures ([53], based on this model e.g., see [4]). How-ever, the Kubelka-Munk model is based on fundamental assumptions inconsistent with light scattering geometries relevant for optogenetics, such as isotropic light scattering and isotropic illumination [54] (Fig. 4a). Therefore, Kubelka-Munk-based approaches gave place to models based on more realistic assumptions, such as Montecarlo simulations (Fig. 4c) and analytical approaches based on the beam spread function (BSF) method (Fig. 4b) [54].

Critically for fMRI paradigms, oxygenated and deoxygenated hemoglobin absorb light differentially. Oxyhemoglobin absorbs substantially more light than deoxyhemoglobin, which suggests that a model based on ex vivo or in vitro preparation likely under-estimates blood-related light absorption, thus underlining the need to estimate light propagation in vivo. Recently, realistic conditions of energy transmission were applied to derive in vivo light absorption coefficients obtained for omnidirectional light propagation in mouse cortex [55] (Fig. 4c), such coefficients can be used in Montecarlo simulations for approximation of realistic light and heat spreading conditions.

It is known that light emitted into the brain for optogenetics is sufficient to increase local temperature and enhance neuronal firing rate [56]. Mild cortical heating can also increase slow wave rhyth-micity and thereby change intrinsic cortical states determining ongoing activity, as well as stimulus response properties of the network [57, 58]. Also it has been shown that blue light can lead to vasodilation (without excitation of neurons and astrocytes), which is important to consider for cerebral blood-flow dependent BOLD measurements [51]. To avoid heating-induced effects unre-lated to optogenetic stimulation, open source custom scripts (MATLAB, Mathworks, Natick, MA.) are available to predict light and heat spread, validated by recorded temperatures in vivo [56]. Further, a straightforward experiment has been proposed to control for the effects of heat, see above and below [59] (Fig. 3).

2   Roadmap to an ofMRI Experiment

2.1   Virus Injection

1. Perform a virus injection procedure, as described in Chaps. 8 and 9. Allocate 4 weeks to allow for functional expression of opsins.

Fig. 3 (Continued) state current is constant while the peak current shows inactivating behavior. *Adapted from* [42]. (**b**) BOLD response map showing activation upon optogenetic stimulation, pulse duration 10 ms, 9 Hz, pulse train duration 10 s in an animal virally transduced with ChR2 in S1FL. (**c**) Activation map of same animal as in (**b**) stimulated with identical light pulses but with additional continuous illumination, showing no *BOLD* response. *Adapted from* [59]

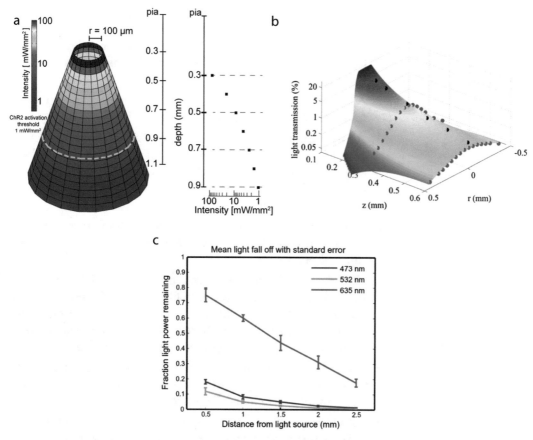

Fig. 4 (**a**) *left:* Schematic of ideal light propagation following [53] assumptions. *Right:* relationship between light intensity (mW/mm²) and cortical depth. *Adapted from* [4]. (**b**) Surface plot of simulated light distribution in slice, obtained from Beam Spread Function (BSF) model using experimentally estimated parameters. Light transmission in log scale. *Adapted from* [54]. (**c**) In-vivo measurement of light propagation, relative fraction of light power at different depths. From these empirical measures derive parameters which can inform Montecarlo simulations. *Adapted from* [55]

2. An angular injection might be useful since it avoids damaging the tissue dorsal to the expression area, where the fiber will be implanted at the day of the experiment (see Chap. 9). This is of particular importance, if simultaneous Ca^{2+} recordings are conducted which require healthy tissue without scar formation from a previous injection.

3. Regarding the choice of opsins see Chaps. 1–3, take particular notion of intrinsic temporal dynamics of the opsin, and review Chap. 9 in case you intend to co-express genetic Ca^{2+} indicators (e.g., GCaMP6).

2.2 Animal Preparation

1. Anesthetize and monitor the animal until reflexes are absent (tail pinch, eye lid) to reach surgical depth of anesthesia; shave the head of the animal (electric razor or hydroxide-based hair removal topical cream).

2. Position the animal in a stereotactic frame (e.g., from Stoelting Co, Wooddale, IL, USA; or Kopf Instruments, Tujunga, CA, USA) by fixing the animal's upper teeth on a bite bar and placing the ear bars inside the ear canal, making sure the skull is firmly stabilized.

3. Apply ointment (Bepanthen, Bayer AG, Leverkusen, Germany) to avoid dehydration of the eyes.

4. Apply local anesthetics to the skin (Xylocaine gel 2%, Astra Zeneka GmbH, Wedel, Germany) and perform a scalp incision of around 1 cm with a scalpel. Identify cranial sutures (bregma and lambda) and center the stereotactic frame according to those reference points.

5. Determine the position of the subsequent craniotomy based on stereotaxic coordinates of the brain region where the opsin was injected previously.

6. Perform a small craniotomy with a dental drill (Ultimate XL-F, NSK, Tochigi, Japan) preferentially using a 0.5 mm ceramic or diamond burr (Komet Dental, Lemgo, Germany) since metal bits spread small shrapnel which may generate distortions in the magnetic field, leading to susceptibility artifacts.

7. Attempt a craniotomy size as small as possible, in order to reduce tissue damage and to avoid air pockets, which could cause inhomogeneity artifacts later.

8. All above mentioned procedures should be performed with the aid of a dissecting microscope (e.g., Leica Microsystems, Olympus, Zeiss).

9. In case simultaneous optic fiber-based Ca^{2+} recordings with dye-loading methods are desired for your experiment, prepare e.g., Oregon Green 488 BAPTA-1 (OGB-1, Thermofisher) as described [60], pull a graduated pipette with a standard puller (e.g., Sutter Instruments, Novato, CA, USA) resulting in a long thin shaft, and a tip of 45 μm outer and 15 μm inner diameter, connect the glass pipette to the tubing and attach it to the cannula holder of the stereotactic frame. Transfer OGB-1 on a piece of parafilm and insert the tip of the glass pipette into solution applying negative pressure with the syringe until the desired amount of OGB-1 is taken up. Insert the pipette to the desired depth and inject OGB-1 as slow as possible. Wait for the OGB-1 ester to be cleaved for at least 30 min up to 1 h, trapping OGB-1 inside the cell, see [45] for review.

10. After removing the cladding, insert a multimode optic fiber (e.g., with a diameter of 200 μm from Thorlabs, Grünberg, Germany) delivering light pulses for optogenetic stimulation and continuous illumination for potential Ca^{2+} recordings [44], via the craniotomy into the targeted region and glue it

to the skull with UV glue (Polytec, PT GmbH, Waldbrunn, Germany). A nonmetal cannula (e.g., a ceramic cannula, CFMLC22L05, Thorlabs) can be used to additionally interface the bone and the fiber.

11. Mount the animal on a heated MRI cradle and supply with a mixture of 80% air and 20% oxygen. Cover the skull with a 1–2 mm thick layer of dental alginate (Weiton, Johannes Weithas dental-Kunststoffe, Lütjenburg, Germany) or 1% agar, to reduce susceptibility artifacts at the bone–air interface. Guide the fiber through an RF surface head coil which has a lead-through for the optical fiber (e.g., Bruker Biospin GmbH, Ettlingen, Germany). For animal preparation for chronic instead of acute experiments see Chap. 9.

12. For details of the optic recordings setup please see [4, 44, 46, 59]

13. Transfer the animal to the scanner.

2.3 Data Acquisition

1. For awake ofMRI measurements in rodents see [26]. For anesthetized measurements, consider that different anesthesia regimens might induce distinct types of baseline neuronal network activation [61], affecting BOLD responses upon sensory stimulation (as an example see [4]). Typical anesthesia regimens for ofMRI experiments are isoflurane [9] or medetomidine [4].

2. Start the scanner (e.g., 9.4 T small animal imaging system, Biospec 94/20, Bruker Biospin GmbH, Ettlingen, Germany) and proceed with data acquisition. As in regular fMRI experiments local shimming should be applied. For anatomical images e.g., a T_2-weighted 2D RARE sequence, TR/TE 2000/12.7 ms, RARE factor 8, 256 matrix, $110 \times 100 \ \mu m^2$ spatial resolution and slice thickness 1.2 mm with nine contiguous slices can be used. For BOLD fMRI measurements, T_2*-weighted images can be acquired with a single-shot gradient echo EPI (Echo Planar Imaging) sequence with TR = 1 s and TE = 18 ms. The spatial resolution of the MR images should range around $350 \times 325 \ \mu m^2$, with a slice thickness around 1.2 mm.

3. For optogenetic stimulation of e.g., ChR2 (for details see Chaps. 2 and 3), light pulses of blue light (470–490 nm) of 10 ms duration at 9 Hz with a light intensity of 80 mW/mm^2 at the tip of the fiber (this is below the threshold of heat-induced BOLD artifacts, see [59] and Figs. 1 and 2) can be used. For optogenetic stimulation alongside Ca^{2+} recordings, high-intensity light pulses need to be switched on during continuous low illumination with blue light (1.3 mW/mm^2; corresponding to ~0.05 mW at the tip of the fiber, not

preventing de-inactivation of the ChR2 peak current at this intensity) for excitation of the Ca^{2+} indicator (see [44], Fig. 1), e.g., by the use of an acousto-optic modulator [46] (Fig. 1a). Functional data acquisition needs to be synchronized with the stimulation paradigms.

4. For details regarding the Ca^{2+} recording setup and correspondent data sampling please refer to [4, 44].

2.4 Ensuring the Specificity of ofMRI

Here, we describe a straightforward experimental approach to control for the apparent BOLD effect caused by heating, for details on the rationale of the experiment please see also [59] and Sect. 1.5. Briefly, the experiment takes advantage of the opsin's channel kinetics. Upon light stimulation, opsins respond with a high amplitude peak current, followed by a longer lasting steady state current (Fig. 3), governed by the intrinsic photo-cycle of the opsin (see Chap. 3). The completion of the photo-cycle and recovery of the peak current requires about 10 s in the absence of stimulation light (see Chap. 3, Fig. 3 and [42, 62]). Consequently, by continuous light delivery, the de-inactivation of the peak current is prevented. Provided that the light intensity is titrated so that just the peak current suffices to evoke an AP, the spike related BOLD signal can be decoupled of any effect produced by the mean energy delivered to the tissue.

1. Outside the scanner, titrate the light intensity (single pulse stimulation) so that only the peak current evoke action potentials (see Fig. 3a), ideally using electrophysiology as readout (see Chap. 8). This will define the minimal light intensity necessary to evoke AP. In our lab, delivering 80 mW/mm^2 with a 200 μm fiber (see above), yielded specific BOLD responses (see Fig. 3), and we estimated suprathreshold power reaching all cortical layers, when the fiber is placed over the dura mater (see [4]). The Optogenetic Resource Center of Stanford University (http://web.stanford.edu/group/dlab/optogenetics/) also provides a tool for a quick check of predicted irradiance values for different fiber diameters, based on a model informed by direct measurements in mammalian brain tissue (https://web.stanford.edu/group/dlab/cgi-bin/graph/chart.php).

2. Apply this very light intensity in single pulses of 10 ms duration inside the scanner while recording fMRI.

3. This optogenetic stimulation should result in a spatially confined BOLD activation in the opsin-expressing area (Fig. 3b).

4. Apply the same pulse sequence as in the previous point, but *additionally apply constant* light with an intensity of about 11 mW/mm^2. This constant light does prevent the de-inactivation of the peak current (Fig. 3a).

5. This last stimulation paradigm (brief pulses of the minimal light intensity necessary to evoke an AP and additional illumination with 11 mW/mm^2) should lead to no BOLD response (Fig. 3c). If this is the case, then the specificity of the optogenetically evoked response is undoubtedly proven, as an apparent heat-induced BOLD effect is not sensitive towards additional constant light.

6. As an additional control experiment, apply the single-pulse optogenetic stimulation in animals injected with a fluorophore instead of an opsin.

3 Notes

1. Bio-monitoring is crucial in any in vivo experiment, but particularly in the scanner bore animal body temperature can drop severely and should be continuously controlled to maintain physiological conditions in which neuronal responses can be measured reliably. Also breathing and heart rate should be monitored closely.

2. Anesthesia needs to be carefully controlled in order to maintain a systemic state allowing for meaningful measurement. Temperature, breathing and heart rate (see above) depend closely on anesthesia levels. Also the state of the network and therefore neuronal response properties highly depend on anesthesia depth. Animals should be supplied with a mixture of 80% air and 20% oxygen.

3. The *hemodynamic response function* HRF is a cornerstone in the analysis of fMRI data, since the HRF is convolved with the raw BOLD signal in order to detect the underlying BOLD dynamics. Crucially, much analysis software relies by default on a stereotypical HRF derived from humans. Thus, in order to accurately represent BOLD dynamics, it might be necessary to use an empirical HRF, derived from BOLD dynamics from measured data of the animal model of choice. Within the animal model domain, HRF in rats have shown robust similarity between sensory and optogenetic stimulation [37]. Thus, no particular approach needs to be deployed in the analysis of BOLD signals evoked by optogenetic stimulation as compared with sensory stimulation.

4. While a combination of optogenetics and optic fiber-based Ca^{2+} recordings alongside fMRI in our view greatly expands the possibilities of ofMRI, it poses significant challenges: When "only" combining optogenetics and fMRI, fiber commutators can be used, enabling chronic experiments and repeated measures of the same animal. However, the current generation of

commutators yields to significant loss of light. This is tolerable using stimulation-only (optogenetics experiments) however, if the fiber is additionally used for light collection, as it is the case of Ca^{2+} recordings, the emission light ranges in the nanowatt intensity, so the loss of light due to the commutator can often not be afforded. Therefore, we suggest to start these experiments involving Ca^{2+} recordings in acute preparations, and only in case of very good SNR, the use of commutators can be attempted.

4   Outlook

Clearly, even though the proof-of-concept studies implementing optogenetics and fMRI were published already 7 years ago [9], the broad implementation of this technique in the preclinical imaging community is still rather slow and might require substantial more time and efforts. Similar to the first years of the advent of optogenetics in basic neuroscience (2006–2010), concerns on the specificity and selectivity of ofMRI [17, 26, 59] may be one reason for the hesitation of the vast majority of neuroimaging scientists to adopt this technique, despite the apparent advantages. With this chapter, we aimed at demonstrating, that by using the appropriate and straightforward control experiments, a specific and meaningful ofMRI experiment is a feasible endeavor. Second, the majority of neuroimaging labs are not equipped for the additional experimental techniques required for the implementation of optogenetics: injection of viral vectors, control of expression by confocal microscopy, and implantation of optic fibers. Last but not least, a specific coil with a lead-through for an optic fiber is required for conducting ofMRI; and this lead-through poses limitations on the geometry of the coil design. Currently, only surface coils with a lead-through are commercially available (Bruker, Billerica, MA; RAPID Biomedical, Rimpar, Germany), at a considerable cost. However, we strongly believe that the field of ofMRI will greatly expand in the near future. Current new concepts expand the use of ofMRI to the field of regenerative medicine (please see also Chap. 5), probing the functional integration of stem cells [40]. Furthermore, in our view, the use of the same implanted optic fiber as a *readout* of neuronal activity in addition to optogenetic manipulation has not gained appropriate attention yet [3, 4]. With the commercialization of optic fiber-based recording setups (e.g., NPI electronic, Tamm, Germany; Doric Lenses, Quebec, Canada) these approaches will in the future not be limited to labs with the capability of in-house design and construction of these setups. Along those lines, as genetically encoded Ca^{2+} indicators [50] and all optical approaches [63] are coming of age, the critical roadblocks for fMRI combined with optical techniques have been overcome.

While optogenetic modulation of neuronal circuits will never fully emulate e.g., sensory-driven activation, the initial concerns on the nonspecificity of ofMRI in terms of network activation [17] have been adequately addressed [4]. The scope of optogenetic network activation seems to be at least similar to sensory evoked activation, and by using fast ramping up of light intensities, also the artificially high degree of synchronicity can be avoided, taking advantage of differential expression strength of opsins in-between neurons: lower light intensities at the beginning of the pulse will first lead to spiking of highly expressing neurons, followed by the spiking of lower-expressing neurons at higher light intensities. Consequently, ofMRI can be an indispensable tool in our endeavor to understand how sensory afferents are being represented in neuronal network with global brain-resolution. The last critical roadblock is represented in the clinical translation of ofMRI. While this holds true for the entire field of optogenetics (see Chap. 14), this roadblock is particularly critical in the field of fMRI, in which the straightforward translation has been a major asset of the method.

Together, we might summarize that the current shortcomings and difficulties of opto-fMRI are by far excelled by the tremendous new possibilities in causally exploring neuronal networks with brain-wide resolution.

References

1. Ogawa S, Lee TM, Kay AR, Tank DW (1990) Brain magnetic resonance imaging with contrast dependent on blood oxygenation. Proc Natl Acad Sci U S A 87(24):9868–9872

2. Logothetis NK, Wandell BA (2004) Interpreting the BOLD signal. Annu Rev Physiol 66:735–769. doi:10.1146/annurev.physiol.66.082602.092845

3. Schulz K, Sydekum E, Krueppel R, Engelbrecht CJ, Schlegel F, Schroter A, Rudin M, Helmchen F (2012) Simultaneous BOLD fMRI and fiber-optic calcium recording in rat neocortex. Nat Methods 9(6):597–602. doi:10.1038/nmeth.2013. [doi] nmeth.2013 [pii]

4. Schmid F, Wachsmuth L, Schwalm M, Prouvot PH, Jubal ER, Fois C, Pramanik G, Zimmer C, Faber C, Stroh A (2016) Assessing sensory versus optogenetic network activation by combining (o)fMRI with optical Ca2+ recordings. J Cereb Blood Flow Metab 36 (11):1885–1900. doi:10.1177/0271678X15619428

5. Logothetis NK, Pauls J, Augath M, Trinath T, Oeltermann A (2001) Neurophysiological investigation of the basis of the fMRI signal. Nature 412(6843):150–157. doi:10.1038/35084005. [doi];35084005 [pii]

6. Carandini M, Churchland AK (2013) Probing perceptual decisions in rodents. Nat Neurosci 16(7):824–831. doi:10.1038/nn.3410

7. Krubitzer L (2007) The magnificent compromise: cortical field evolution in mammals. Neuron 56(2):201–208. doi:10.1016/j.neuron.2007.10.002

8. Oakley H, Cole SL, Logan S, Maus E, Shao P, Craft J, Guillozet-Bongaarts A, Ohno M, Disterhoft J, Van Eldik L, Berry R, Vassar R (2006) Intraneuronal beta-amyloid aggregates, neurodegeneration, and neuron loss in transgenic mice with five familial Alzheimer's disease mutations: potential factors in amyloid plaque formation. J Neurosci 26 (40):10129–10140. doi:10.1523/JNEUROSCI.1202-06.2006

9. Lee JH, Durand R, Gradinaru V, Zhang F, Goshen I, Kim DS, Fenno LE, Ramakrishnan C, Deisseroth K (2010) Global and local fMRI signals driven by neurons defined optogenetically by type and wiring. Nature 465:788–792. doi:10.1038/nature09108. [doi] nature09108 [pii]

10. Lindquist MA, Meng Loh J, Atlas LY, Wager TD (2009) Modeling the hemodynamic response function in fMRI: efficiency, bias and mis-modeling. Neuroimage 45(1 Suppl): S187–S198. doi:10.1016/j.neuroimage. 2008.10.065

11. Birn RM, Saad ZS, Bandettini PA (2001) Spatial heterogeneity of the nonlinear dynamics in the FMRI BOLD response. Neuroimage 14 (4):817–826. doi:10.1006/nimg.2001.0873

12. Handwerker DA, Ollinger JM, D'Esposito M (2004) Variation of BOLD hemodynamic responses across subjects and brain regions and their effects on statistical analyses. Neuroimage 21(4):1639–1651. doi:10.1016/j. neuroimage.2003.11.029

13. Marrelec G, Benali H, Ciuciu P, Pelegrini-Issac M, Poline JB (2003) Robust Bayesian estimation of the hemodynamic response function in event-related BOLD fMRI using basic physiological information. Hum Brain Mapp 19 (1):1–17. doi:10.1002/hbm.10100

14. Wager TD, Vazquez A, Hernandez L, Noll DC (2005) Accounting for nonlinear BOLD effects in fMRI: parameter estimates and a model for prediction in rapid event-related studies. Neuroimage 25(1):206–218. doi:10. 1016/j.neuroimage.2004.11.008

15. Hillman EM (2014) Coupling mechanism and significance of the BOLD signal: a status report. Annu Rev Neurosci 37:161–181. doi:10.1146/annurev-neuro-071013-014111

16. Lima B, Cardoso MM, Sirotin YB, Das A (2014) Stimulus-related neuroimaging in task-engaged subjects is best predicted by concurrent spiking. J Neurosci 34 (42):13878–13891. doi:10.1523/ JNEUROSCI.1595-14.2014

17. Logothetis NK (2010) Bold claims for optogenetics. Nature 468(7323):E3–E4. doi:10. 1038/nature09532. [doi] nature09532 [pii]

18. Sirotin YB, Das A (2009) Anticipatory haemodynamic signals in sensory cortex not predicted by local neuronal activity. Nature 457 (7228):475–479. doi:10.1038/nature07664

19. Attwell D, Iadecola C (2002) The neural basis of functional brain imaging signals. Trends Neurosci 25(12):621–625

20. Boynton GM, Engel SA, Glover GH, Heeger DJ (1996) Linear systems analysis of functional magnetic resonance imaging in human V1. J Neurosci 16(13):4207–4221

21. Fernandez-Klett F, Offenhauser N, Dirnagl U, Priller J, Lindauer U (2010) Pericytes in capillaries are contractile in vivo, but arterioles mediate functional hyperemia in the mouse brain. Proc Natl Acad Sci U S A 107 (51):22290–22295. doi:10.1073/pnas. 1011321108

22. Lindauer U, Dirnagl U, Fuchtemeier M, Bottiger C, Offenhauser N, Leithner C, Royl G (2010) Pathophysiological interference with neurovascular coupling–when imaging based on hemoglobin might go blind. Front Neuroenerg 2:25. doi:10.3389/fnene.2010.00025

23. Iadecola C (2004) Neurovascular regulation in the normal brain and in Alzheimer's disease. Nat Rev Neurosci 5(5):347–360. doi:10. 1038/nrn1387

24. Girouard H, Iadecola C (2006) Neurovascular coupling in the normal brain and in hypertension, stroke, and Alzheimer disease. J Appl Physiol 100(1):328–335. doi:10.1152/ japplphysiol.00966.2005

25. Kahn I, Desai M, Knoblich U, Bernstein J, Henninger M, Graybiel AM, Boyden ES, Buckner RL, Moore CI (2011) Characterization of the functional MRI response temporal linearity via optical control of neocortical pyramidal neurons. J Neurosci 31 (42):15086–15091. doi:10.1523/ JNEUROSCI.0007-11.2011

26. Desai M, Kahn I, Knoblich U, Bernstein J, Atallah H, Yang A, Kopell N, Buckner RL, Graybiel AM, Moore CI, Boyden ES (2011) Mapping brain networks in awake mice using combined optical neural control and fMRI. J Neurophysiol 105(3):1393–1405. doi:10. 1152/jn.00828.2010

27. Gradinaru V, Mogri M, Thompson KR, Henderson JM, Deisseroth K (2009) Optical deconstruction of parkinsonian neural circuitry. Science 324(5925):354–359. doi:10. 1126/science.1167093. [doi] 1167093 [pii]

28. Yizhar O, Fenno LE, Prigge M, Schneider F, Davidson TJ, O'Shea DJ, Sohal VS, Goshen I, Finkelstein J, Paz JT, Stehfest K, Fudim R, Ramakrishnan C, Huguenard JR, Hegemann P, Deisseroth K (2011) Neocortical excitation/inhibition balance in information processing and social dysfunction. Nature 477 (7363):171–178. doi:10.1038/nature10360. [doi] nature10360 [pii]

29. Vazquez AL, Fukuda M, Crowley JC, Kim SG (2014) Neural and hemodynamic responses elicited by forelimb- and photo-stimulation in channelrhodopsin-2 mice: insights into the hemodynamic point spread function. Cereb Cortex 24(11):2908–2919. doi:10.1093/ cercor/bht147

30. Christie IN, Wells JA, Southern P, Marina N, Kasparov S, Gourine AV, Lythgoe MF (2013) fMRI response to blue light delivery in the naive brain: implications for combined

optogenetic fMRI studies. Neuroimage 66:634–641. doi:10.1016/j.neuroimage.2012.10.074

31. Kahn I, Knoblich U, Desai M, Bernstein J, Graybiel AM, Boyden ES, Buckner RL, Moore CI (2013) Optogenetic drive of neocortical pyramidal neurons generates fMRI signals that are correlated with spiking activity. Brain Res 1511:33–45. doi:10.1016/j.brainres.2013.03.011

32. Gerits A, Farivar R, Rosen BR, Wald LL, Boyden ES, Vanduffel W (2012) Optogenetically induced behavioral and functional network changes in primates. Curr Biol 22 (18):1722–1726. doi:10.1016/j.cub.2012.07.023. [doi] S0960-9822(12)00814-7 [pii]

33. Honjoh T, Ji ZG, Yokoyama Y, Sumiyoshi A, Shibuya Y, Matsuzaka Y, Kawashima R, Mushiake H, Ishizuka T, Yawo H (2014) Optogenetic patterning of whisker-barrel cortical system in transgenic rat expressing channelrhodopsin-2. PLoS One 9(4):e93706. doi:10.1371/journal.pone.0093706

34. Weitz AJ, Fang Z, Lee HJ, Fisher RS, Smith WC, Choy M, Liu J, Lin P, Rosenberg M, Lee JH (2015) Optogenetic fMRI reveals distinct, frequency-dependent networks recruited by dorsal and intermediate hippocampus stimulations. Neuroimage 107:229–241. doi:10.1016/j.neuroimage.2014.10.039

35. Liang Z, Watson GD, Alloway KD, Lee G, Neuberger T, Zhang N (2015) Mapping the functional network of medial prefrontal cortex by combining optogenetics and fMRI in awake rats. Neuroimage 117:114–123. doi:10.1016/j.neuroimage.2015.05.036

36. Takata N, Yoshida K, Komaki Y, Xu M, Sakai Y, Hikishima K, Mimura M, Okano H, Tanaka KF (2015) Optogenetic activation of CA1 pyramidal neurons at the dorsal and ventral hippocampus evokes distinct brain-wide responses revealed by mouse fMRI. PLoS One 10(3):e0121417. doi:10.1371/journal.pone.0121417

37. Iordanova B, Vazquez AL, Poplawsky AJ, Fukuda M, Kim SG (2015) Neural and hemodynamic responses to optogenetic and sensory stimulation in the rat somatosensory cortex. J Cereb Blood Flow Metab 35(6):922–932. doi:10.1038/jcbfm.2015.10

38. Fang P, An J, Zeng LL, Shen H, Chen F, Wang W, Qiu S, Hu D (2015) Multivariate pattern analysis reveals anatomical connectivity differences between the left and right mesial temporal lobe epilepsy. NeuroImage Clin 7:555–561. doi:10.1016/j.nicl.2014.12.018

39. Galvan A, Hu X, Smith Y, Wichmann T (2016) Effects of optogenetic activation of corticothalamic terminals in the motor thalamus of awake monkeys. J Neurosci 36 (12):3519–3530. doi:10.1523/JNEUROSCI.4363-15.2016

40. Weitz AJ, Lee JH (2016) Probing neural transplant networks in vivo with optogenetics and optogenetic fMRI. Stem Cells Int 2016:8612751. doi:10.1155/2016/8612751

41. Zhuo JM, Tseng HA, Desai M, Bucklin ME, Mohammed AI, Robinson NT, Boyden ES, Rangel LM, Jasanoff AP, Gritton HJ, Han X (2016) Young adult born neurons enhance hippocampal dependent performance via influences on bilateral networks. elife 5:e22429. doi:10.7554/eLife.22429

42. Stroh A, Tsai HC, Wang LP, Zhang F, Kressel J, Aravanis A, Santhanam N, Deisseroth K, Konnerth A, Schneider MB (2011) Tracking stem cell differentiation in the setting of automated optogenetic stimulation. Stem Cells 29 (1):78–88. doi:10.1002/stem.558

43. Ryali S, Shih YY, Chen T, Kochalka J, Albaugh D, Fang Z, Supekar K, Lee JH, Menon V (2016) Combining optogenetic stimulation and fMRI to validate a multivariate dynamical systems model for estimating causal brain interactions. Neuroimage 132:398–405. doi:10.1016/j.neuroimage.2016.02.067

44. Adelsberger H, Grienberger C, Stroh A, Konnerth A (2014) In vivo calcium recordings and channelrhodopsin-2 activation through an optical fiber. Cold Spring Harb Protoc 2014 (10):prot084145. doi:10.1101/pdb.prot084145

45. Grienberger C, Konnerth A (2012) Imaging calcium in neurons. Neuron 73(5):862–885. doi:10.1016/j.neuron.2012.02.011

46. Stroh A, Adelsberger H, Groh A, Ruhlmann C, Fischer S, Schierloh A, Deisseroth K, Konnerth A (2013) Making waves: initiation and propagation of corticothalamic Ca2+ waves in vivo. Neuron 77(6):1136–1150. doi:10.1016/j.neuron.2013.01.031

47. Kerr JN, Greenberg D, Helmchen F (2005) Imaging input and output of neocortical networks in vivo. Proc Natl Acad Sci U S A 102 (39):14063–14068. doi:10.1073/pnas.0506029102. [doi] 0506029102 [pii]

48. Grienberger C, Adelsberger H, Stroh A, Milos RI, Garaschuk O, Schierloh A, Nelken I, Konnerth A (2012) Sound-evoked network calcium transients in mouse auditory cortex in vivo. J Physiol 590(Pt 4):899–918. doi:10.1113/jphysiol.2011.222513

49. Rochefort NL, Garaschuk O, Milos RI, Narushima M, Marandi N, Pichler B, Kovalchuk Y, Konnerth A (2009) Sparsification of neuronal

activity in the visual cortex at eye-opening. Proc Natl Acad Sci U S A 106 (35):15049–15054. doi:10.1073/pnas. 0907660106

50. Chen TW, Wardill TJ, Sun Y, Pulver SR, Renninger SL, Baohan A, Schreiter ER, Kerr RA, Orger MB, Jayaraman V, Looger LL, Svoboda K, Kim DS (2013) Ultrasensitive fluorescent proteins for imaging neuronal activity. Nature 499(7458):295–300. doi:10.1038/nature12354

51. Rungta RL, Osmanski BF, Boido D, Tanter M, Charpak S (2017) Light controls cerebral blood flow in naive animals. Nat Commun 8:14191. doi:10.1038/ncomms14191

52. Boyden ES, Zhang F, Bamberg E, Nagel G, Deisseroth K (2005) Millisecond-timescale, genetically targeted optical control of neural activity. Nat Neurosci 8(9):1263–1268

53. Aravanis AM, Wang LP, Zhang F, Meltzer LA, Mogri MZ, Schneider MB, Deisseroth K (2007) An optical neural interface: in vivo control of rodent motor cortex with integrated fiberoptic and optogenetic technology. J Neural Eng 4(3):S143–S156

54. Yona G, Meitav N, Kahn I, Shoham S (2016) Realistic numerical and analytical modeling of light scattering in brain tissue for optogenetic applications(1,2,3). eNeuro 3(1). doi:10. 1523/ENEURO.0059-15.2015

55. Acker L, Pino EN, Boyden ES, Desimone R (2016) FEF inactivation with improved optogenetic methods. Proc Natl Acad Sci U S A 113 (46):E7297–E7306. doi:10.1073/pnas. 1610784113

56. Stujenske JM, Spellman T, Gordon JA (2015) Modeling the spatiotemporal dynamics of light and heat propagation for in vivo optogenetics. Cell Rep 12(3):525–534. doi:10.1016/j.celrep.2015.06.036

57. Sheroziya M, Timofeev I (2015) Moderate cortical cooling eliminates thalamocortical silent states during slow oscillation. J Neurosci 35 (38):13006–13019. doi:10.1523/JNEUROSCI.1359-15.2015

58. Schwalm M, Easton C (2016) Cortical temperature change: a tool for modulating brain states? eNeuro 3(3). doi:10.1523/ENEURO.0096-16.2016

59. Schmid F, Wachsmuth L, Albers F, Schwalm M, Stroh A, Faber C (2017) True and apparent optogenetic BOLD fMRI signals. Magn Reson Med 77(1):126–136. doi:10.1002/mrm.26095

60. Stosiek C, Garaschuk O, Holthoff K, Konnerth A (2003) In vivo two-photon calcium imaging of neuronal networks. Proc Natl Acad Sci U S A 100(12):7319–7324. doi:10.1073/pnas. 1232232100

61. Constantinople CM, Bruno RM (2011) Effects and mechanisms of wakefulness on local cortical networks. Neuron 69 (6):1061–1068. doi:10.1016/j.neuron.2011. 02.040. [doi] S0896-6273(11)00162-0 [pii]

62. Fois C, Prouvot PH, Stroh A (2014) A roadmap to applying optogenetics in neuroscience. Methods Mol Biol 1148:129–147. doi:10. 1007/978-1-4939-0470-9_9

63. Deisseroth K (2015) Optogenetics: 10 years of microbial opsins in neuroscience. Nat Neurosci 18(9):1213–1225. doi:10.1038/nn.4091

Chapter 14

Optogenetics: Lighting a Path from the Laboratory to the Clinic

Hannah K. Kim, Allyson L. Alexander, and Ivan Soltesz

Abstract

The advent of optogenetics has brought about an unprecedented ability to control neuronal activity with great spatial and temporal precision. Questions about neuronal circuitry that had previously been impossible or extremely difficult to test suddenly have become viable experimental questions. Although optogenetics was originally pioneered for use in neurons, it is now clear that systems as diverse as atrial cardiomyocytes, pancreatic islet cells, and tumor cells can be manipulated with this technique. A great many of the studies that employ optogenetics propose that this new technology can be used as therapy for a variety of pathologies. For example, optogenetic strategies in animal models have restored sight to blind rodents, stopped seizures in epileptic animals, and abolished behavioral manifestations of addiction in cocaine-treated mice. It is no surprise, then, that one of the most oft-asked questions is if and when optogenetics can be used as a treatment in the clinic. Will optogenetic technology ever be safe enough to use in human therapies? In this closing chapter, we will briefly review the current state of optogenetic technology in relation to translational efforts, discuss the ethical issues surrounding the use of optogenetics in human patients, and finally examine several lines of experimental evidence that illustrate the potential clinical uses of optogenetics.

Key words Optogenetics, Clinical translation, Animal models

1 Introduction

When imagining the possibilities of optogenetic therapy in the clinic, the mind's eye might paint a futuristic picture of patients sitting at home, lights flashing in their brains to stop Parkinsonian tremors, within their eyes to allow them to read the morning paper, or perhaps within their ears to allow them to listen to their favorite song again. With the rapid progress of optogenetic tool development and its application in so many studies of human disease and disorders, will the aforementioned picture eventually become a reality, or remain in the realm of science fiction?

The advent of modern optogenetics, combined with existing genetic and viral tools for targeting specific cell types, brought about the unprecedented ability to control the activity of specific

Albrecht Stroh (ed.), *Optogenetics: A Roadmap*, Neuromethods, vol. 133,
DOI 10.1007/978-1-4939-7417-7_14, © Springer Science+Business Media LLC 2018

cell types (see Chaps. 1 and 2). Not surprisingly, the promise of translational applications of optogenetics was grasped in a short time, as evidenced by numerous new studies that applied this technology to animal models of disease, often revealing promising refinements to existing treatments or new therapeutic targets for human diseases and disorders (see Chaps. 11 and 12). However, while optogenetics has been unquestionably valuable in the laboratory, there are certainly many barriers, both technical and ethical, to translating this technology into the clinic. For example, the current optogenetic tools and experiments leading to potentially translatable results are optimized for small, genetically modifiable organisms, and not necessarily for primates (see Chap. 12). In some cases, a better understanding of human neuronal circuits and other physiological systems is needed before these discoveries in model systems can be translated for human use.

Progress towards clinical application of optogenetics is being made on numerous fronts, including improvements in optogenetic technology, and an increase in the number of translational studies that are taking advantage of the clinical potential of optogenetics. This chapter discusses recent technological advances regarding optogenetics, particularly in relation to applying the technology to the nonhuman primate brain. This section is followed by a discussion on the general practicality of optogenetic therapies, as well as ethical issues that arise from considering such a complex therapy. Finally, we will discuss clinically relevant model organisms to give a few examples of how optogenetics-based research has supported its use as a clinical therapy.

2   How Close Are We? The Current State of Optogenetic Progress Towards Clinical Use

2.1   The Tools: Opsins, Gene Expression, and Light Delivery

At this time, optogenetics is largely used as a research tool for furthering understanding of the brain and many other organ systems. From experimental and preclinical tests, it is clear that optogenetics offers a powerful method of controlling cellular activity and synaptic transmission, and much has been learned regarding neuronal circuitry and potential treatments for neurological disorders. However, just how close is the scientific and medical community to seeing this tool being applied in the clinic? There is no arguing that optogenetics offers a powerful technique for manipulating neural circuits, either by teasing apart circuits to discover previously unknown or untested components, or by testing how modulation of a circuit changes behavioral output. One can easily envision such tests leading to the discovery of new treatment targets or to the modification of current treatments to improve behavioral outcomes. However, while effective, most current optogenetic methods that are used to probe neural circuits and test treatments in animal models of disease are not directly usable in humans.

Undoubtedly, opsin technology has advanced in leaps and bounds since its conception. Opsin development and advancement have continued since the publication of the first evidence of the effectiveness of Channelrhodopsin-2 (ChR2) in murine neurons in 2005 [1]. However, while powerful in experimental settings, these opsins had significant limitations that would complicate or prevent some of the brain disorder-related applications in the clinic, such as displaying relatively slow kinetics, the need for high intensities of light stimulation to achieve physiological responses, and the fact that some opsins have peak responses to wavelengths of light not suited for penetrating deep into neural tissue [2, 3] (see Chap. 3). In order to overcome these limitations, development of newer versions of opsins is proceeding rapidly, for example the development of Chrimson, which is a channelrhodopsin that is activated by surprisingly low intensities of red light [4]. The longer wavelength of red light, which also activates another channelrhodopsin variant known as ReaChR, can penetrate deeply into tissue including through the skin and skull in mice [5]. Other newer opsins include Chronos, a fast opsin variant [4], and CheTA [6], which also achieves faster kinetics compared to the original ChR2. Chrimson and Chronos, in particular, were developed to be used simultaneously in the same organism since their absorption spectra are separated enough that light stimulation of one opsin does not affect the other, allowing for two cell populations to be controlled at a time [4]. In addition to opsins, optically activated G-protein coupled receptors (opto-GPCRs) are also being developed and they have been used effectively to control cellular processes, and the fact that there are already approximately 800 known GPCRs suggests that there may eventually be ways to control many different cellular processes as more opto-GPCRs are developed (Reviewed in [7]). With the rapid development of different tools, it is almost certain that opsins and opto-GPCRS with kinetics and responsiveness tailored to specific purposes and treatments will be feasible.

Achieving specific targeted stable expression of the opsin gene is another major hurdle. In preclinical settings, numerous methods have been used for expressing opsins in model organisms in the cells of choice, such as through genetically modified animal lines, electroporation, viral transduction, and others (see Chaps. 1 and 2). Transgenics and electroporation are not feasible in humans, and these are unlikely to be effective options even for nonhuman primates, given the difficulty in manipulating the primate genome. The use of viral vectors based on adeno-associated virus (AAV) or lentivirus, which is a well-established technique in animal models, offers the most promise for translational use. Indeed, AAV is thought to be quite safe for use in humans. Thus, it has been used in over one hundred clinical trials and an AAV-based gene therapy has even been approved for clinical use in Europe for the treatment of lipoprotein lipase deficiency (Reviewed in [8]). Therefore, AAV can potentially also be employed for future clinical trials of optogenetics. Despite this, there are many uncertainties regarding the use of viral vectors for gene expression in patients. For example, gene expression through viral infection is usually not even across all targeted cells (see Chap. 8), the virus may spread to areas that were not originally targeted, and the expression may not remain stable over a patient's lifetime. Other concerns include possibilities of mutagenesis, cytotoxicity, or immune response [9].

The combination of optogenetics and stem cell injections offers an exciting way to solve the problem of how to deliver light-sensitive opsins to the human brain without using viruses (see Chap. 5). In this model, stem cells would be transfected with a gene for an opsin and then injected into the area of interest. Three recent studies have taken major steps toward the realization of this goal. In the first, neuronal precursors were harvested from the medial ganglionic eminence of embryonic, ChR2-expressing mice. These cells were then implanted into the dentate gyrus of epileptic mice where they developed into functional light-sensitive inhibitory neurons that were functionally integrated into the circuit. Mice with these implants that received light stimulation had fewer seizures than controls [10]. In the second study, human embryonic stem cells (ESCs) were transfected with gene for the inhibitory opsin halorhodopsin, differentiated into dopamine neurons, and injected into the midbrain of hemi-parkinsonian mice. The dopamine neurons integrated into the basal ganglia circuitry and led to behavioral improvement in the mice, unless the dopamine neurons were optogenetically inactivated [11]. In the third study, ChR2-expressing human induced pluripotent stem cells were differentiated into neurons and then transplanted into rat striatum. Using a functional MRI-compatible optrode (i.e., a combination of an optical fiber and an electrode), the researchers were able to modulate neuronal activity in the rat via light pulses, and detect the changing activity both with the electrode and with the MRI signal, demonstrating that light-sensitive human stem cell-derived neurons can be effectively incorporated into functional brain networks and that this activity can be monitored noninvasively with technology that is safe for patients [12] (see Chap. 13). Although stem cell injections are not approved for human use currently, recent clinical trials of stem cell injections into the brain after stroke have demonstrated good safety profiles. Several of these trials have even shown promising neurological results [13]. Therefore, engraftment of optogenetically transformed stem cells offers a promising translational tool for clinical therapy.

Another roadblock for successful translation of optogenetics into the clinic is the effective delivery of light to the structures being targeted. This is especially true for neurological applications, considering the significant increase in scale between a rodent brain and the human brain. Current research in the laboratory often employs optical fibers coupled to lasers or LEDs to deliver light to deep brain structures. Delivery of light in this manner is invasive and, for some applications, would require multiple lights implanted into deep brain structures, such as a LED diode array that includes multiple electrodes and micro LEDs [14]. Surgical implantation of a light source deep in the brain may result in complications similar to those of deep brain stimulation (DBS), of which infection and hemorrhage around the implanted device are the most serious as they can lead to permanent neurological compromise. The risks of these complications could be even higher than in DBS if the light source has a short functional life span and requires frequent replacement. Currently used optical sources in the laboratory also require bulky power sources that would be unsuitable for human implantation. Future technology may include the development of a power source which is small and implantable, similar to the batteries for cardiac pacemakers or DBS electrodes. Even more intriguing is the possibility for an implantable mini-LED that is very light and can be charged wirelessly [15].

In some cases, it may be possible to avoid inserting optical fibers into the brain by targeting more distal areas of the nervous system such as the spinal cord or peripheral nerves. For example, preliminary data suggest that AAV8 can be used to successfully deliver opsins to sensory neurons in the spinal cord via lumbar puncture, a commonly performed procedure in humans. Using this system, light inhibition of the sensory neurons provided pain relief in a mouse model of chronic pain [16]. The spinal cord and peripheral nerves might be more approachable not only for gene delivery but for light delivery as well. For example, an intriguing soft optoelectronic system was invented by Park and colleagues that is wireless and fully implantable. This system can be implanted epidurally over the spinal cord or under the gluteus maximus to illuminate the sciatic nerve [17]. Therefore, choosing targets wisely may avoid the complications of implanting optical devices into the brain (see Chaps. 11 and 12).

There are alternative approaches to standard optogenetic options which may offer clever translational approaches for neuromodulation in human patients by using optogenetics as an inspiration. One promising technology utilizes ion channels that are activated by magnetic fields instead of light, such as Magneto2.0. This channel has been shown to be effective in changing rodent behavior in vivo. If this channel could be expressed in humans, this might obviate the need for an implanted device at all, as the magnetic field could be applied externally [18]. Another type of neuromodulation does not rely on genetic manipulation of human cells at all. Transcranial magnetic stimulation (TMS) is a noninvasive modality of neurostimulation which relies on externally applied strong magnetic fields to induce electrical currents in the cortex. This therapy offers exciting possibilities because it has already been FDA-approved to treat refractory unipolar depression, with a high degree of efficacy given the difficulty of treating these patients [19, 20]. A recent clinical trial has shown that focusing the TMS on the dorsolateral prefrontal cortex (PFC) in cocaine-addicted human patients resulted in promising increases in the number of drug-free patients in the treated versus the placebo arms [21]. The idea for this study was based on results from a prior study in mice showing that optogenetic activation of the rat medial PFC (the analogous area to human dorsolateral PFC) led to decreases in drug-seeking behavior [22, 23].

2.2 The Nonhuman Primate Bridge to Human Optogenetics

In addition to improving technical aspects of optogenetics, making the jump from model organisms such as invertebrates, zebrafish, and rodents to humans is not a simple feat, and there are many differences to consider beyond simple size scaling (see Chaps. 6 and 7). In order to address some of these, optogenetic research in nonhuman primates is a growing field that aims to provide a late preclinical foundation to develop and improve optogenetic technologies, and to overcome some of the roadblocks preventing translation to humans. As mentioned, development of genetically modified animal lines that can express opsin genes in specific cell types, while extremely useful in mice and other model organisms, is not feasible in monkeys, other nonhuman primates, and of course, humans. Currently, the remaining options are viral construct injections and stem cell grafts. The development of viral technology that targets specific cell types in monkeys and nonhuman primates was initially slow, due to limited numbers of available promoters and the challenge of developing new viral technologies for monkeys and nonhuman primates. The technical aspects of viral vector design and cell-type specific targeting strategies in primates is beyond the scope of this chapter, but has been reviewed previously [24].

Nevertheless, there has been a recent explosion of optogenetic studies in nonhuman primates. The visual system has been the target of many of these studies. Early attempts to transfer optogenetic technology to primate systems led to successful transfection of neurons with AAV-based vectors [25], and successful neuronal opsin expression resulting in local electrophysiological responses [26–28]. Later studies, using new viral constructs, were able to achieve milestones such as behavior modulation [29–31], and then pathway-specific [32] and cell-specific [33] targeting of opsins into the visual system. Success has also been gained recently in other neural systems, such as the ability to optogenetically stimulate long-range projections in the motor system, in both cortico-thalamic and thalamo-cortical directions [34]. The problem of scaling up in size from rodent to monkey brain has also been addressed by taking advantage of existing technology such as the use of convection-enhanced, MRI-guided delivery of the viral vector, a clear artificial dura to provide light access to the cortex without the need for an optrode, and micro-electrocorticographic arrays for large-scale electrical recordings [35]. With these improvements, the authors were able to achieve a large area of expression of the excitatory opsin C1V1 as measured on the surface of the sensory-motor cortex, with stable expression up to 27 months post-injection.

An exciting step in the pathway to the clinic has recently been made. A pioneering new study in human tissue resected at epilepsy surgery has shown that human neurons can be successfully transfected with ChR2 using a lentiviral vector [36]. This is the first demonstration of optogenetic manipulation of human brain tissue. This breakthrough, combined with the above-mentioned achievements will likely lead the way to the development of a library of targeting methods for specific pathways and cell-types in the primate brain. Questions that remain to be answered include the stability, specificity and coverage of virus-mediated gene transfer, methods of replacing light sources when failure occurs, and the general safety of optogenetic therapy in the primate brain.

2.3 Ethical Considerations of Optogenetics

Even as optogenetics was first unveiled as a cutting-edge tool that provides the almost unprecedented ability to finely control neuronal activity both temporally and spatially, it was touted as a potential therapeutic modality that could overcome the unknowns and deficiencies in currently available treatments for various neurological disorders. However, it is important to step back and consider not only the significant technical issues to be resolved before such a step can be taken, but also the ethical considerations to be made as well. Some ethical issues stem from the practicality of effectively employing optogenetics compared to currently available treatments, such as the dangers of needing multiple viral injections and the necessity of invasive device implants for upcoming clinical trials. Some ethical issues are philosophical, due to the potential ability of optogenetics to alter natural neurological function, and therefore change behavioral output in people. In fact, there is a novel field termed "neurogenethics" which contemplates the ethical, legal, and social implications of the modulation of the genome in the nervous system [37].

There are already general concerns about safety in using viral vectors, since they are currently the most common method of delivering opsin genes to specific cells. There have been risks found associated with viral vectors, such as immune responses, insertional mutagenesis, clonal expansion, and tumorigenesis [9]. These safety issues are particularly prominent with lentiviral vectors, as many lentiviruses, including HIV, are human pathogens. AAV, however, is a nonpathogenic virus. In addition to this, there is the concern that the virus-delivered genes are introducing nonhuman DNA into patients, with unknown long-term consequences.

In addition to gene therapy, successful application of optogenetics in the clinic will require invasive modifications to the brain, such as the implantation of a light source and perhaps an optical fiber in cases where the brain areas are not readily accessible, and possibly multiple virus injections in case the original treatment is not effective or does not cover the necessary volume or surface area needed [38]. To justify these complex series of surgeries, the therapeutic potential would have to far outstrip the benefits offered by currently available, less invasive treatments. In addition, such a complex treatment has the potential to be exceedingly expensive, and there could be concerns that it will ultimately become a treatment only for the wealthy.

Most of the ethical questions raised by optogenetics are not new ones, but ones that have been raised previously whenever a therapy can alter human DNA and behavior [39]. One can imagine the possibilities considering the behavioral manipulations that have already been accomplished: implanting false fear memories [40], artificially changing aggression levels [41, 42], enhancing recognition memory [43], among others. As with all behavior-modifying treatments, there is as much the potential for abuse as there is the potential for treatment and healing. For example, there are rodent studies that optogenetically manipulate hypothalamic circuits and cause mice to attack benign objects due to increased aggression [41]. Perhaps such a technology could be used to decrease violence and aggression in patients who have difficulty controlling aggressive tendencies, but on the other hand it is not difficult to imagine such a technology being used to increase aggressiveness when needed, such as in the military [44]. Another possibility is that optogenetics could be used to treat neuronal disorders such as Parkinson's and schizophrenia. However, once the infrastructure of optogenetics has been implemented in a person, whether for aggression or disease treatment, this essentially creates a platform by which a person's thoughts or behavior can be altered [39]. Is it ethical to cause this altered state of mind? Who will be responsible for the settings of such mind-altering tools [39]? Can this be taken further into the realm of cognitive enhancement [45]? Will the alteration of neural activity lead to unintended or "off-target" behavioral effects [37]? Though it remains to be seen in the upcoming years if optogenetics will be a viable clinical treatment, due to the ethical questions raised, there is still great value in the discussion and any answers that are reached.

3   Why Is Optogenetics So Promising as a Therapy? Evidence from Many Animal Models

As groups from a diverse number of laboratories and fields adopted optogenetics in a rapid fire manner over the past years, a stunning number of studies in animal models has emerged that all point to the effectiveness of optogenetics as a potential treatment for various diseases and disorders. This section addresses several such potential areas of optogenetic therapy: (1) optogenetics as a method of overriding and controlling uncontrolled activity, utilizing temporal and spatial specificity to bypass side effects of many drugs and direct stimulation methods; (2) optogenetics as a tool to modify and refine currently existing and approved therapies (e.g., deep brain stimulation); (3) optogenetics as a method to restore sensory systems; (4) restoration of motor function with optogenetics. Whether or not these studies ultimately lead to clinical applications, they provide strong evidence of the translational potential of optogenetics.

3.1   Taking Control of the Uncontrolled

Optogenetics derives its power from the ability to control cellular activity. Therefore, an obvious application of optogenetics is to exert control in cases where cellular activity is aberrant. Experimental studies have successfully used optogenetics to modulate such misbehaving cells in models of epilepsy and cardiac arrhythmias.

Normal brain function requires excitation and inhibition of neurons that is tightly controlled at the millisecond time scale. However, when this control is lost due to one of a number of different causes, such as injury, fever, or genetic mutations, neuronal activity can become excessive or hypersynchronous and cause seizures. In the epileptic brain, such loss of control is usually spontaneous and recurrent, leading to loss of motor control, awareness, and even memory. One of the touted strengths of optogenetics is the ability to exert control over specific subtypes of neurons, and as such, was quickly considered as a potential method of stopping seizures in their tracks. Indeed, in a model of temporal lobe epilepsy, spontaneous focal seizures were significantly decreased in duration and generalized seizures decreased in occurrence when hippocampal principal cells were inhibited or subsets of interneurons were excited through optogenetics [46]. In another example, cortical seizures induced by stroke were optogenetically suppressed utilizing long-range thalamocortical neurons [47], and tetanus toxin-induced neocortical seizures could also be suppressed by optogenetically inhibiting principal cells [48]. These findings demonstrate that optogenetics is undoubtedly a powerful tool that has great potential for translational application in seizure control.

Ultimately, the cell-type specificity and spatial precision of optogenetic stimulation should be able to eliminate unwanted side effects that plague many currently available treatments. Most patients are treated with anti-convulsant medications, which can affect the entire body, and cause side effects ranging from blurry vision and unsteadiness, to liver problems and low blood cell counts. In addition, many of these medications used to treat epilepsy suppress neural activity and therefore can cause problems with cognitive functions, such as thinking, learning, and attention. Surgical treatments are options for a selected group of epilepsy patients, but may also lead to complications such as infection, hemorrhage, and cognitive decline. Though the actual application of optogenetics to treat epilepsy patients is not feasible at this time, the results from the above preclinical studies already have profound implications for future epilepsy treatments or modifications of current treatments. For example, probing neural circuits in epileptic brains can lead to the discovery of new targets, which may be remote from the seizure focus that can act as a chokepoint for halting seizures. Such chokepoints could be used as targets for currently-approved treatments such as DBS or responsive neurostimulation. For example, optogenetic intervention during seizures has shown that there are regions such as the dentate gyrus that can serve as a gate to stop the spread of seizure activity [49]. Other studies of epileptic rodents revealed that optogenetic manipulation of neurons in areas physically distant from the ictal focus—such as the cerebellum, thalamus, or superior colliculus–may act as chokepoints to suppress seizure propagation [46, 47, 50–53]. Computer modeling of the human cortex has provided additional insights into how optogenetics might be able to control seizures in patients [54, 55]. The knowledge from these studies will give insights into how different kinds of epilepsies can be targeted for treatment. Even in cases where the epileptic focus is unknown or diffuse, optogenetic targeting of these gates or chokepoints may be able to provide treatment.

Loss of control of electrical propagation between cells is not limited to the nervous system, but is the problem underlying cardiac arrhythmias such as atrial fibrillation. Cardiovascular biologists have begun to look towards optogenetics as a method to bypass some of the negative effects of current treatments. For example, typical treatment for arrhythmias may include drugs, such as warfarin or other anti-coagulants, with dangerous side-effects, or may consist of electrotherapy to cause defibrillation. However, this process causes pain due to the stimulation of skeletal muscle near the heart, and long electrical stimuli can lead to faradaic charge transfer due to redox reactions at the electrode, which can cause corrosion of electrodes and production of toxic chemical species, limiting the length of stimulus that can be used by current methods.

Thus far, a number of different approaches in applying optogenetics to animal models of cardiac arrhythmias have been progressing. In an initial study by Bruegmann et al., ChR2 was expressed in embryonic stem cells under an actin promoter. Light activation of the resulting cell line and mouse strain was able to successfully modulate electrical behavior in cardiomyocytes both in vitro and in vivo [56]. Further work using transgenic expression of ChR2 and halorhodopsin showed successful control of zebrafish cardiomyocytes in vivo [57], and following this, viral transfer of ChR2 into human ESCs resulted in cardiomyocytes whose activity could be controlled by light stimulation in vitro [58]. Exciting new work has shown that heart rate can be directly modulated in vivo by optogenetic stimulation in Drosophila larvae [59]. Researchers have now been able to optogenetically stop pathological rhythms in an in vitro model of atrial fibrillation [60]. Simulations of a biophysically realistic human cardiac ventricle demonstrated that optical defibrillation is possible if the cardiomyocytes express a red light-sensitive opsin [61].

In addition to transgenics and viral transfer of opsins, another method of opsin delivery is to engraft opsin-expressing cells onto native tissue, as it has been shown that grafted cardiac tissues can electrically couple to existing heart tissue [62]. Grafting offers an obvious translational advantage as this technique does not require germ-line mutation or in vivo viral transfection. Other advances include melanopsin-expressing ESCs differentiated into cardiomyocytes, which are of note in that these require extremely low levels of light stimulation in comparison to traditional ChR2 methods, which could allow for light stimulation of cardiomyocytes using much more compact light and power sources than currently would be required [63]. Considering these options, optogenetics-based tools could potentially be applied not only to better understand cardiac dysfunction, but to also be used as a therapeutic in the future, perhaps correcting the timing of action potentials found in heart rhythm disorders, ending arrhythmias in a pain-free manner, and developing optical pacemakers that are more specific and powerful than current versions.

3.2 Teaming Up Optogenetics and Deep Brain Stimulation: Fast Lane to the Clinic?

There is no question that it will take time and much refinement of optogenetic tools before safe translation into the clinic is possible, for most types of neurological disorders, especially as many of these are caused by abnormalities deep in the brain, as discussed earlier. It has been proposed that, while efforts are being made to translate optogenetics to human treatments, other efforts should be directed towards using optogenetics to "inspire" modifications and improvements to currently existing and approved clinical treatments [64]. In particular, this current movement targets DBS, which has been FDA approved for treatment in disorders such as Parkinson's disease, obsessive compulsive disorder, essential tremor, and dystonia (Reviewed in [65]). DBS can provide alleviation from symptoms in these disorders, but the mechanism that underlies the therapeutic nature of DBS is essentially unknown, and the treatment also comes with a number of side effects due to the nonspecific nature of electrical stimulation. Perhaps it is possible, then, to simulate DBS in animal models using optogenetics to achieve the desired treatment effects but without undesirable side effects, and then use this new knowledge to produce a new, modified DBS therapy. In this section, we discuss the first studies that clearly demonstrated the potential of modifying DBS based on optogenetic manipulation of neural circuits in the study of addiction. Then we will briefly discuss whether a similar strategy may be feasible in Parkinson's Disease and clinical depression.

Drug addiction, in general, is difficult to treat because of pathological changes in neuronal plasticity that lead to patients often suffering from relapses. With cocaine addiction, past work has shown that addiction-related behavior can be suppressed by the application of optogenetic suppression to the medial prefrontal cortex, which projects to the nucleus accumbens [66, 67]. However, a classic high-frequency DBS protocol to the shell of the nucleus accumbens only showed a transient effect on addiction-related behavior, perhaps because the DBS paradigm did not affect synaptic plasticity [68]. Optogenetic manipulation has the advantage of being limited to one cell type—in this study, ChR2 is expressed in the medial prefrontal cortex, which projects excitatory afferents to the nucleus accumbens. By comparing the differences between optogenetic stimulation and DBS on the nucleus accumbens, the authors of this study realized that low frequency optogenetic stimulation caused long term depression (LTD) in the medium spiny neurons (MSNs), though this did not work for a similar low frequency DBS protocol [68]. Previous work had shown that a blockade of dopamine receptor D1 (D1Rs) was necessary for this LTD to occur [69]. Therefore, to replicate the effect of optogenetically induced LTD on the MSNs, chemical antagonists of D1Rs were added systemically, finally allowing DBS stimulation to eliminate the behavioral sensitization to cocaine after a single stimulus treatment. These types of discoveries, made using insights gained from optogenetic studies, have great potential for clinical application through modification of current technologies, even while waiting for actual optogenetic tools to move into clinical trials.

While other studies have not yet actively tested modified DBS paradigms based on new findings from optogenetic studies, there are many other neurological disorders that rely on DBS therapy that may benefit from modifications discovered through the use of optogenetics. For example, Parkinson's disease patients can receive DBS to the subthalamic nucleus to treat the persistent tremors that are characteristic of this disease. As with most other DBS treatments, the mechanism that underlies this therapeutic effect is in the process of being researched, and chronic DBS stimulation produces unwanted side effects, such as pathological motor function, since the subthalamic nucleus modulates motor activity. Perhaps optogenetics can provide the means to this knowledge, such as with addiction research. For example, the optogenetic excitation of the D1R-mediated direct pathway in the basal ganglia completely rescued the motor symptoms of 6-hydroxydopamine lesioned parkinsonian mice [70]. Understanding exact mechanisms and pathways by which DBS is useful in Parkinson's may help to tailor this technique in the future.

Clinical depression is another neurological disorder in which various brain regions have been targeted with DBS in patients in an attempt to provide antidepressant effects [71]. The treatment is still experimental, and the first randomized controlled trial did not show a significant response, but pilot studies using different targets have shown promising results [72, 73]. Research is underway, using optogenetic experiments to better understand the therapeutic mechanisms of DBS that lead to antianxiolytic and antidepressant effects (Reviewed in [74]). Two such experiments revealed that specific optogenetic manipulation of dopamine neurons in the ventral tegmental area could exert antidepressant effects. Strikingly, different models of depression required opposing types of ventral tegmental area (VTA) neuron modulation. In a mild but chronic model of depression, optogenetic excitation of dopaminergic neurons in the VTA improved the depressive symptoms of treated rodents [75]. In contrast, optogenetic inhibition of the same subtype of neurons exerted antidepressant effects on an acute model of stress [76]. While optogenetics is primarily being used as a method to better understand changes in neural circuitry underlying depression, it is not difficult to hypothesize that it may be possible to use new targets found through these experiments or to combine a modified DBS stimulation protocols with pharmacological agents to emulate optogenetics and isolate a particular pathway for DBS effects, in order to specifically receive benefits without the accompanying side effects.

4 Making Blind Eyes See and Deaf Ears Hear

Sensory systems such as the visual and auditory systems present excellent potential targets for translation of optogenetic technology. The eye and ear are more accessible for implantation or injection than deep structures of the brain, and they will benefit from the neuronal specificity conferred by optogenetics.

4.1 Channel-rhodopsins to do the Duty of Rhodopsins

Opsins used in optogenetics are channels or pumps that are sensitive to light. Similarly, the rods and cones of the human retina also contain light-sensitive opsins, essential for light detection in the visual system. In retinitis pigmentosa, there is an initial phase of rod photoreceptor degeneration, followed by a phase of degeneration of the cone photoreceptors and the retinal pigment epithelium. However, even after degeneration of rods and most of the cones, the macula at the center of the retina still retains a layer of cone photoreceptors, though they are missing or have shortened outer segments (Reviewed in [77]). Therefore, it is possible that in cases when the visual system fails due to damage to photoreceptors so they are no longer able to photoconduct, but still have surviving cell bodies, optogenetics could be a tool to restore the lost light sensitivity by inserting opsins into these photoreceptors.

The visual system is perhaps an almost perfect starting point for testing optogenetics as a clinical application. For one, retinal gene therapies are already in use: In the case of Leber's congential amaurosis, which is a severe form of early onset retinitis pigmentosa, two genes are known to be involved in up to 17% of patients (Reviewed in [78, 79]). A clinical trial using recombinant adeno associated virus 2/2 (rAAV2/2) to express one of these genes, retinal pigment epithelium-specific protein 65 kDa (RPE65), in the retina, was performed on a number of test subjects, and it achieved some temporary improvements in vision, though it was compounded by decline in vision due to the progressive degeneration of photoreceptors [80, 81]. In another clinical trial to attempt gene therapy treatment of choroideremia, in which the degeneration of choriocapillaries and the retinal pigment epithelium leads to blindness, expression of the Rab-escort protein 1 in the retina using AAV2, led to improvements in vision in a subset of subjects [82]. In addition to the accepted use of viral gene transfer in these clinical trials, the retina is particularly suited for optogenetics as a treatment because it overcomes the need for an invasive light delivery method, and the outcomes can be observed noninvasively.

Optogenetic experiments for vision restoration in animal models showed great promise even in the earliest studies. In one study, ChR2 expressed in the retinal ganglion cells in a model of retinitis pigmentosa, the retinal degeneration (rd1) mouse, allowed the retina to obtain light sensitivity [83]. However, it is important to note that nonspecific or ectopic expression of ChR2 does not restore the function of complex circuitry mediated by retinal interneurons, so improved targeting is required if this is to be a feasible treatment. Later studies attempted to target opsin expression to other cells within the retinal circuitry, but were often confounded by limitations of various gene transfer methods. Electroporation was used successfully to target ChR2 to a specific type of cell, the ON-bipolar cells in rd1 mice, in an effort to develop potential treatments for cases where all photoreceptors have degenerated. Visually-evoked activity was restored in the cortices of rd1 mice and allowed animals to respond behaviorally to visual stimuli [84]. However, electroporation is not a method that can be translated to human gene therapy. Therefore, subsequent studies focused on using viral gene transfer to express opsins in ON-bipolar cells. One major limitation to this approach is that current viral technology does not result in uniform expression of rhodopsin throughout the retina. The most concerning aspect of this is that when AAV-mediated gene transfer is applied to the retinas of nonhuman primates, most of the expression is around and not within the fovea [85], which has the highest density of cone receptors and is crucial for visual acuity. Another major concern is that AAV does not effectively transduce the inner nuclear layer of the retina, therefore groups having been developing modified viral variants of rAAV which have a much greater transduction efficiency for ON-bipolar cells [86, 87]. For a more comprehensive review, see [78, 88].

While improvements in viral transfection methods are essential, there must be a corresponding improvement in opsin technology. An ideal optogenetic treatment for the retina would allow a person to see in normal levels of light, to restore vision using constructs that are physiologically compatible with surviving cells of the retinal inner layer, display normal trigger features of retinal ganglion cells, and do not produce toxins or induce immune responses [89]. While there have been advances in producing variants of ChR2 that have much greater sensitivity to light, as described earlier [90, 91] (Chap. 3), there may be more optimal ways to construct a light sensitive protein that is more in line with the ideal treatment. To attempt to develop a retinal therapy that could achieve as many of these ideals as possible, one group developed a chimeric protein, called Opto-mGluR which contains the mGluR specific to ON-bipolar cells as well as the light sensing domains of the photopigment melanopsin [89]. This protein was successfully able to partially restore vision in a rodent model using only natural levels of daylight. Perhaps tailoring optical components to specific applications such as in this case for treating blindness, should be a major consideration for future clinical applications for optogenetics.

The most exciting development in the clinical translation of optogenetics is that the very first clinical trial of optogenetic technology is currently underway with an attempt to restore vision to patients with retinitis pigmentosa. The trials will be performed by expressing ChR2 in retinal ganglion cells through viral gene transfer, in an effort to allow the patient to regain some vision. Though at very early stages and with unknown outcomes, this trial will likely have broad impacts both on vision gene therapy, the neuroscience community, and the field of optogenetics and its potential as a therapy in the clinic [92–94]. Now that a clinical trial has been approved for optogenetic therapy in the eye, this will pave the way for future trials of optogenetics in other areas such as the brain, the heart, or the ear as well.

4.2 Hearing the Light   While the application of optogenetics to the visual system seems almost too obvious, other sensory systems also stand to gain much from the temporal and spatial specificity provided by optogenetics.

Another sensory system in which optogenetics is gaining traction is the auditory system, where there has been significant progress in restoring hearing in animal models of deafness. Electrical cochlear implants are widely used and are able to restore hearing and speech comprehension to many children and adults who are deaf or hard of hearing. However, the quality of hearing provided by these implants has limitations, such as in sound intensity encoding, and in pitch discrimination (Reviewed in [95]). Optogenetics can offer potential solutions to these limitations due to a more specific area of excitation, which can lead to better resolution of pitch and intensity, leading to an improved quality of life.

The first step in creating an optogenetic cochlear implant implant is to develop a cochlea that expresses an appropriate opsin. The first proof of principle experiments showed that the spiral ganglion neurons (SGNs), which are responsible for transmitting sound representation information to the brain, could be made to express ChR2 via either transgenic techniques or viral transfection [96], the latter of which would be more clinically relevant. In addition, expression of ChR2 has been achieved postnatally through AAV gene transfer, and expression of a different modified opsin was achieved prenatally through electroporation [97]. However, these proof of principle experiments were not able to show ChR2 expression evenly across the cochlea, and instead most expression was limited to the base of the cochlea [96]. The expression of these opsins was clearly effective since deaf rodents were able to perform behaviors that depended on sound, such as being trained to cross an arena when a sound was played [96]. However, one of the current limitations to optogenetic hearing therapy is that adult transfection has not yet been achieved. This is essential, as many that could potentially benefit from this treatment are adults who have lost most or all their hearing later in life. Also, accurate light delivery to the cochlea is necessary as well. A flexible microLED has been developed which is suitable for use in cochlear implants [98]. While there is more development of tools and refinement of opsin expression pattern to achieve, these results indicate that as in the case for vision, optogenetic therapy has great potential in other sensory systems as well.

5   Restoring Movement Using Optogenetics

The ability to repair or restore motor circuits after injury to the spinal cord or spinal nerves through nerve regeneration is an extensively researched yet still an unachieved goal. In the meantime, other alternate approaches and potential solutions are emerging, from grafting stem cells or Schwann cells to transplanting a biological scaffold to allow axonal growth to using brain-powered prosthetic frames [99–102]. Optogenetics also presents a compelling solution to controlling muscle function, especially considering that functional electrical stimulation has already been in use for contracting muscle fibers. However, lack of specificity of electrical stimulation causes problems similar to those that arise from DBS. These include the activation of sensory afferents, which can cause pain, and unlike in normal muscle contraction, fibers are recruited in a random manner than in the proper order, which can cause rapid fatigue [99].

Early demonstrations of motor control using optogenetics are very promising indeed. The activation of ChR2-containing neurons in the vibrissal motor cortex of rats was able to elicit whisking activity expressed in motor cortex [103]. Transgenic mice expressing ChR2 in peripheral nerves under the Thy1 promoter, when given light stimulus to motor axons, showed muscle contractions tightly controlled by the stimulus timing. In addition, muscle fiber recruitment occurred in the correct order which is essential for avoiding rapid muscle fatigue [104]. Importantly, the ability to express opsins in specific cell types would be able to allow optogenetic stimulation of motor neurons and avoid sensory afferents altogether, eliminating another drawback of functional electrical stimulation. Another study used a combination of transgenic and optogenetic techniques to demonstrate that axons, newly sprouted after spinal cord injury, were successfully activated with optogenetic stimulation of the corticospinal tract, leading to limb movement [105]. Finally, in an elegant study, ChR2-expressing, stem-cell derived motor neurons were implanted into damaged sciatic nerve in the mouse. After engraftment, these motor neurons re-innervated hindlimb muscles and, when activated by blue light, were able to cause forceful muscle contractions [106].

6 Outlook

Considering the rapid advances in optogenetic technology and the push for translation towards clinical use, perhaps the image of flashing lights in the brain to improve symptoms of neurological disorders will soon not simply be science fiction. Optogenetic technology is well on its way to becoming a household term in the neurosciences, and many other fields as well, and there may be even more experimental evidence for therapeutic possibilities in many of these fields. Even from current works, there are many directions that can potentially be taken, for example, optogenetics could be used to override abnormal and uncontrolled cellular activity, such as in epilepsy and cardiac arrhythmias. It can also be used to learn more about neural circuits in order to modify or combine currently existing therapeutic technologies, such as DBS and drugs. And it can perhaps be used to restore activity to damaged sensory systems. The possibilities are numerous, and in fact there are studies looking at using optogenetics in a vast array of diseases that are beyond the scope of this review: enhancing memory in Alzheimer's disease by activating engram cells [107], promoting neurogenesis after stroke [108], control of pancreatic islet cells in diabetes [109, 110], and even altering of resting membrane potential of tumor cells to treat cancer [111]. However, these are all experimental results, and while exciting, there are still barriers even at the preclinical stages in attempting to achieve functional optogenetics in humans and nonhuman primates, and many ethical considerations to take as the first clinical trials are taking place. These are exciting times for the future of optogenetics, and we look forward to new discoveries and its first steps into the realm of human therapies.

Acknowledgments

Supported by the US National Institutes of Health grants NS94668 to I.S. and R25NS065741-04S1 to A.A.

References

1. Boyden ES, Zhang F, Bamberg E, Nagel G, Deisseroth K (2005) Millisecond-timescale, genetically targeted optical control of neural activity. Nat Neurosci 8:1263–1268. doi:10.1038/nn1525

2. Lin JY (2011) A user's guide to channelrhodopsin variants: features, limitations and future developments. Exp Physiol 96:19–25. doi:10.1113/expphysiol.2009.051961

3. Tromberg BJ, Shah N, Lanning R, Cerussi A, Espinoza J, Pham T, Svaasand L, Butler J (2000) Non-invasive in vivo characterization of breast tumors using photon migration spectroscopy. Neoplasia 2:26–40. http://www.pubmedcentral.nih.gov/articlerender.fcgi?artid=1531865&tool=pmcentrez&rendertype=abstract (Accessed 22 May 2016)

4. Klapoetke NC, Murata Y, Kim SS, Pulver SR, Birdsey-Benson A, Cho YK, Morimoto TK, Chuong AS, Carpenter EJ, Tian Z, Wang J, Xie Y, Yan Z, Zhang Y, Chow BY, Surek B, Melkonian M, Jayaraman V, Constantine-Paton M, Wong GK-S, Boyden ES (2014) Independent optical excitation of distinct neural populations. Nat Methods 11:338–346. doi:10.1038/nmeth.2836

5. Lin JY, Knutsen PM, Muller A, Kleinfeld D, Tsien RY (2013) ReaChR: a red-shifted variant of channelrhodopsin enables deep transcranial optogenetic excitation. Nat Neurosci 16:1499–1508. doi:10.1038/nn.3502

6. Gunaydin LA, Yizhar O, Berndt A, Sohal VS, Deisseroth K, Hegemann P (2010) Ultrafast optogenetic control. Nat Neurosci 13:387–392. doi:10.1038/nn.2495

7. Kleinlogel S (2016) Optogenetic user's guide to Opto-GPCRs. Front Biosci Landmark Ed 21:794–805. http://www.ncbi.nlm.nih.gov/pubmed/26709806 (Accessed 25 Feb 2016

8. Salganik M, Hirsch ML, Samulski RJ (2015) Adeno-associated virus as a mammalian DNA vector. Microbiol Spectr 3(4). doi:10.1128/microbiolspec.MDNA3-0052-2014

9. Kotterman MA, Chalberg TW, Schaffer DV (2015) Viral vectors for gene therapy: translational and clinical outlook. Annu Rev Biomed Eng 17:63–89. doi:10.1146/annurev-bioeng-071813-104938

10. Henderson KW, Gupta J, Tagliatela S, Litvina E, Zheng X, Van Zandt MA, Woods N, Grund E, Lin D, Royston S, Yanagawa Y, Aaron GB, Naegele JR (2014) Long-term seizure suppression and optogenetic analyses of synaptic connectivity in epileptic mice with hippocampal grafts of GABAergic interneurons. J Neurosci 34:13492–13504. doi:10.1523/JNEUROSCI.0005-14.2014

11. Steinbeck JA, Choi SJ, Mrejeru A, Ganat Y, Deisseroth K, Sulzer D, Mosharov EV, Studer L (2015) Optogenetics enables functional analysis of human embryonic stem cell–derived grafts in a Parkinson's disease model. Nat Biotechnol 33:204–209. doi:10.1038/nbt.3124

12. Byers B, Lee HJ, Liu J, Weitz AJ, Lin P, Zhang P, Shcheglovitov A, Dolmetsch R, Pera RR, Lee JH (2015) Direct in vivo assessment of human stem cell graft-host neural circuits. Neuroimage 114:328–337. doi:10.1016/j.neuroimage.2015.03.079

13. Azad TD, Veeravagu A, Steinberg GK (2016) Neurorestoration after stroke. Neurosurg Focus 40:E2. doi:10.3171/2016.2.FOCUS15637

14. Wu F, Stark E, Ku P-C, Wise KD, Buzsáki G, Yoon E (2015) Monolithically integrated μLEDs on silicon neural probes for high-resolution optogenetic studies in behaving animals. Neuron 88:1136–1148. doi:10.1016/j.neuron.2015.10.032

15. Montgomery KL, Yeh AJ, Ho JS, Tsao V, Mohan Iyer S, Grosenick L, Ferenczi EA, Tanabe Y, Deisseroth K, Delp SL, Poon ASY (2015) Wirelessly powered, fully internal optogenetics for brain, spinal and peripheral circuits in mice. Nat Methods 12:969–974. doi:10.1038/nmeth.3536

16. Towne C, Aguado J, Arguello A, Discenza C, Gal T, Gehrke S, Khan S, Kaplitt M (2016) Translating an optogenetic gene therapy approach for treatment of neuropathic pain in humans. Am Soc Gene Cell Ther 186:2016

17. Il Park S, Brenner DS, Shin G, Morgan CD, Copits BA, Chung HU, Pullen MY, Noh KN, Davidson S, Oh SJ, Yoon J, Jang K-I, Samineni VK, Norman M, Grajales-Reyes JG, Vogt SK, Sundaram SS, Wilson KM, Ha JS, Xu R, Pan T, Kim T-I, Huang Y, Montana MC, Golden JP, Bruchas MR, Gereau RW, Rogers JA (2015) Soft, stretchable, fully implantable miniaturized optoelectronic systems for wireless optogenetics. Nat Biotechnol 33:1280–1286. doi:10.1038/nbt.3415

18. Wheeler MA, Smith CJ, Ottolini M, Barker BS, Purohit AM, Grippo RM, Gaykema RP, Spano AJ, Beenhakker MP, Kucenas S, Patel MK, Deppmann CD, Güler AD (2016) Genetically targeted magnetic control of the nervous system. Nat Neurosci 19:756–761. doi:10.1038/nn.4265

19. Connolly KR, Helmer A, Cristancho MA, Cristancho P, O'Reardon JP (2012) Effectiveness of transcranial magnetic stimulation in clinical practice post-FDA approval in the United States: results observed with the first 100 consecutive cases of depression at an academic medical center. J Clin Psychiatry 73:e567–e573. doi:10.4088/JCP.11m07413

20. Janicak PG, Nahas Z, Lisanby SH, Solvason HB, Sampson SM, McDonald WM, Marangell LB, Rosenquist P, McCall WV, Kimball J, O'Reardon JP, Loo C, Husain MH, Krystal A, Gilmer W, Dowd SM, Demitrack MA, Schatzberg AF (2010) Durability of clinical benefit with transcranial magnetic stimulation (TMS) in the treatment of pharmacoresistant major depression: assessment of relapse during a 6-month, multisite, open-label study. Brain Stimul 3:187–199. doi:10.1016/j.brs.2010.07.003

21. Terraneo A, Leggio L, Saladini M, Ermani M, Bonci A, Gallimberti L (2015) Transcranial magnetic stimulation of dorsolateral prefrontal cortex reduces cocaine use: a pilot study. Eur Neuropsychopharmacol 26:37–44. doi:10.1016/j.euroneuro.2015.11.011

22. Ferenczi E, Deisseroth K (2016) Illuminating next-generation brain therapies. Nat Neurosci 19:414–416. doi:10.1038/nn.4232

23. Chen BT, Yau HJ, Hatch C, Kusumoto-Yoshida I, Cho SL, Hopf FW, Bonci A (2013) Rescuing cocaine-induced prefrontal cortex hypoactivity prevents compulsive cocaine seeking. Nature 496:359–362. doi:10.1038/nature12024

24. El-Shamayleh Y, Ni AM, Horwitz GD (2016) Strategies for targeting primate neural circuits with viral vectors. J Neurophysiol 116 (1):122–134. doi:10.1152/jn.00087.2016

25. Masamizu Y, Okada T, Kawasaki K, Ishibashi H, Yuasa S, Takeda S, Hasegawa I, Nakahara K (2011) Local and retrograde gene transfer into primate neuronal pathways via adeno-associated virus serotype 8 and 9. Neuroscience 193:249–258. doi:10.1016/j.neuroscience.2011.06.080

26. Han X, Qian X, Bernstein JG, Zhou H-H, Franzesi GT, Stern P, Bronson RT, Graybiel AM, Desimone R, Boyden ES (2009) Millisecond-timescale optical control of neural dynamics in the nonhuman primate brain. Neuron 62:191–198. doi:10.1016/j.neuron.2009.03.011

27. Diester I, Kaufman MT, Mogri M, Pashaie R, Goo W, Yizhar O, Ramakrishnan C, Deisseroth K, Shenoy KV (2011) An optogenetic toolbox designed for primates. Nat Neurosci 14:387–397. doi:10.1038/nn.2749

28. Han X, Chow BY, Zhou H, Klapoetke NC, Chuong A, Rajimehr R, Yang A, Baratta MV, Winkle J, Desimone R, Boyden ES (2011) A high-light sensitivity optical neural silencer: development and application to optogenetic

control of non-human primate cortex. Front Syst Neurosci 5:18. doi:10.3389/fnsys.2011. 00018

29. Cavanaugh J, Monosov IE, McAlonan K, Berman R, Smith MK, Cao V, Wang KH, Boyden ES, Wurtz RH (2012) Optogenetic inactivation modifies monkey visuomotor behavior. Neuron 76:901–907. doi:10.1016/j.neuron. 2012.10.016

30. Gerits A, Farivar R, Rosen BR, Wald LL, Boyden ES, Vanduffel W (2012) Optogenetically induced behavioral and functional network changes in primates. Curr Biol 22:1722–1726. doi:10.1016/j.cub.2012.07. 023

31. Jazayeri M, Lindbloom-Brown Z, Horwitz GD (2012) Saccadic eye movements evoked by optogenetic activation of primate V1. Nat Neurosci 15:1368–1370. doi:10.1038/nn. 3210

32. Inoue K-I, Takada M, Matsumoto M (2015) Neuronal and behavioural modulations by pathway-selective optogenetic stimulation of the primate oculomotor system. Nat Commun 6:8378. doi:10.1038/ncomms9378

33. Klein C, Evrard HC, Shapcott KA, Haverkamp S, Logothetis NK, Schmid MC (2016) Cell-targeted optogenetics and electrical microstimulation reveal the primate koniocellular projection to supra-granular visual cortex. Neuron 90:143–151. doi:10.1016/j. neuron.2016.02.036

34. Galvan A, Hu X, Smith Y, Wichmann T (2016) Effects of optogenetic activation of corticothalamic terminals in the motor thalamus of awake monkeys. J Neurosci 36:3519–3530. doi:10.1523/JNEUROSCI. 4363-15.2016

35. Yazdan-shahmorad A, Diaz-botia C, Hanson TL, Kharazia V, Ledochowitsch P, Maharbiz MM, Sabes PN, Yazdan-shahmorad A, Diaz-botia C, Hanson TL, Kharazia V, Ledochowitsch P (2016) A large-scale interface for optogenetic stimulation and recording in nonhuman primates neuroresource a large-scale interface for optogenetic stimulation and recording in nonhuman primates. Neuron 89:927–939. doi:10.1016/j.neuron. 2016.01.013

36. Andersson M, Avaliani N, Svensson A, Wickham J, Pinborg LH, Jespersen B, Christiansen SH, Bengzon J, Woldbye DPD, Kokaia M (2016) Optogenetic control of human neurons in organotypic brain cultures. Sci Rep 6:24818. doi:10.1038/srep24818

37. Canli T (2015) Neurogenethics: an emerging discipline at the intersection of ethics,

neuroscience, and genomics. Appl Transl Genomics 5:18–22. doi:10.1016/j.atg. 2015.05.002

38. Gilbert F, Harris AR, Kapsa RMI (2014) Controlling brain cells with light: ethical considerations for optogenetic clinical trials. AJOB Neurosci 5:3–11. doi:10.1080/ 21507740.2014.911213

39. R Starkman (2012) Optogenetics: a novel technology with questions old and new. Huffington Post. http://www.huffingtonpost. com/ruth-starkman/optogenetics-a-new-techno_b_1700219.html (Accessed 17 May 2016)

40. Ramirez S, Liu X, Lin P-A, Suh J, Pignatelli M, Redondo RL, Ryan TJ, Tonegawa S (2013) Creating a false memory in the hippocampus. Science 341:387–391. doi:10.1126/ science.1239073

41. Lin D, Boyle MP, Dollar P, Lee H, Lein ES, Perona P, Anderson DJ (2011) Functional identification of an aggression locus in the mouse hypothalamus. Nature 470:221–226. doi:10.1038/nature09736

42. Yu Q, Teixeira CM, Mahadevia D, Huang Y, Balsam D, Mann JJ, Gingrich JA, Ansorge MS (2014) Dopamine and serotonin signaling during two sensitive developmental periods differentially impact adult aggressive and affective behaviors in mice. Mol Psychiatry 19:688–698. doi:10.1038/mp.2014.10

43. Benn A, Barker GRI, Stuart SA, Roloff EVL, Teschemacher AG, Warburton EC, Robinson ESJ (2016) Optogenetic stimulation of prefrontal glutamatergic neurons enhances recognition memory. J Neurosci 36:4930–4939. doi:10.1523/JNEUROSCI. 2933-15.2016

44. Berryessa CM, Cho MK (2013) Ethical, legal, social, and policy implications of behavioral genetics. Annu Rev Genomics Hum Genet 14:515–534. doi:10.1146/annurev-genom-090711-163743

45. Jarvis S, Schultz SR (2015) Prospects for optogenetic augmentation of brain function. Front Syst Neurosci 9:157. doi:10.3389/ fnsys.2015.00157

46. Krook-Magnuson E, Armstrong C, Oijala M, Soltesz I (2013) On-demand optogenetic control of spontaneous seizures in temporal lobe epilepsy. Nat Commun 4:1376. doi:10. 1038/ncomms2376

47. Paz JT, Davidson TJ, Frechette ES, Delord B, Parada I, Peng K, Deisseroth K, Huguenard JR (2012) Closed-loop optogenetic control of thalamus as a tool for interrupting seizures

after cortical injury. Nat Neurosci 16:64–70. doi:10.1038/nn.3269

48. Wykes RC, Heeroma JH, Mantoan L, Zheng K, Macdonald DC, Deisseroth K, Hashemi KS, Walker MC, Schorge S, Kullmann DM (2012) Optogenetic and potassium channel gene therapy in a rodent model of focal neocortical epilepsy. Sci Transl Med 4:161ra152. doi:10.1126/scitranslmed.3004190

49. Krook-Magnuson E, Armstrong C, Bui A, Lew S, Oijala M, Soltesz I (2015) In vivo evaluation of the dentate gate theory in epilepsy. J Physiol 593:2379–2388. doi:10.1113/JP270056

50. Krook-Magnuson E, Szabo GG, Armstrong C, Oijala M, Soltesz I (2014) Cerebellar directed optogenetic intervention inhibits spontaneous hippocampal seizures in a mouse model of temporal lobe epilepsy. Eneuro 1:ENEURO.0005–ENEURO.0014. doi:10.1523/ENEURO.0005-14.2014

51. Kros L, Eelkman Rooda OHJ, Spanke JK, Alva P, van Dongen MN, Karapatis A, Tolner EA, Strydis C, Davey N, Winkelman BHJ, Negrello M, Serdijn WA, Steuber V, van den Maagdenberg AMJM, De Zeeuw CI, Hoebeek FE (2015) Cerebellar output controls generalized spike-and-wave discharge occurrence. Ann Neurol 77:1027–1049. doi:10.1002/ana.24399

52. Paz JT, Huguenard JR (2015) Microcircuits and their interactions in epilepsy: is the focus out of focus? Nat Neurosci 18:351–359. doi:10.1038/nn.3950

53. Soper C, Wicker E, Kulick CV, N'Gouemo P, Forcelli PA (2016) Optogenetic activation of superior colliculus neurons suppresses seizures originating in diverse brain networks. Neurobiol Dis 87:102–115. doi:10.1016/j.nbd.2015.12.012

54. Selvaraj P, Sleigh JW, Kirsch HE, Szeri AJ (2015) Optogenetic induced epileptiform activity in a model human cortex. Springerplus 4:155. doi:10.1186/s40064-015-0836-7

55. Selvaraj P, Sleigh JW, Kirsch HE, Szeri AJ (2016) Closed-loop feedback control and bifurcation analysis of epileptiform activity via optogenetic stimulation in a mathematical model of human cortex. Phys Rev E 93:12416. doi:10.1103/PhysRevE.93.012416

56. Bruegmann T, Malan D, Hesse M, Beiert T, Fuegemann CJ, Fleischmann BK, Sasse P (2010) Optogenetic control of heart muscle in vitro and in vivo. Nat Methods 7:897–900. doi:10.1038/nmeth.1512

57. Arrenberg AB, Stainier DYR, Baier H, Huisken J (2010) Optogenetic control of cardiac function. Science 330:971–974. doi:10.1126/science.1195929

58. Abilez OJ, Wong J, Prakash R, Deisseroth K, Zarins CK, Kuhl E (2011) Multiscale computational models for optogenetic control of cardiac function. Biophys J 101:1326–1334. doi:10.1016/j.bpj.2011.08.004

59. Zhu YC, Uradu H, Majeed ZR, Cooper RL (2016) Optogenetic stimulation of drosophila heart rate at different temperatures and Ca2+ concentrations. Physiol Rep 4:e12695. doi:10.14814/phy2.12695

60. Bingen BO, Engels MC, Schalij MJ, Jangsangthong W, Neshati Z, Feola I, Ypey DL, Askar SFA, Panfilov AV, Pijnappels DA, De Vries AAF (2014) Light-induced termination of spiral wave arrhythmias by optogenetic engineering of atrial cardiomyocytes. Cardiovasc Res 104:194–205. doi:10.1093/cvr/cvu179

61. Boyle PM, Karathanos TV, Trayanova NA (2015) "beauty is a light in the heart": the transformative potential of optogenetics for clinical applications in cardiovascular medicine. Trends Cardiovasc Med 25:73–81. doi:10.1016/j.tcm.2014.10.004

62. Jia Z, Valiunas V, Lu Z, Bien H, Liu H, Wang H-Z, Rosati B, Brink PR, Cohen IS, Entcheva E (2011) Stimulating cardiac muscle by light: cardiac optogenetics by cell delivery. Circ Arrhythm Electrophysiol 4:753–760. doi:10.1161/CIRCEP.111.964247

63. Beiert T, Bruegmann T, Sasse P (2014) Optogenetic activation of Gq signalling modulates pacemaker activity of cardiomyocytes. Cardiovasc Res 102:507–516. doi:10.1093/cvr/cvu046

64. Lüscher C, Pollak P (2016) Optogenetically inspired deep brain stimulation: linking basic with clinical research. Swiss Med Wkly 146:w14278. doi:10.4414/smw.2016.14278

65. Udupa K, Chen R (2015) The mechanisms of action of deep brain stimulation and ideas for the future development. Prog Neurobiol 133:27–49. doi:10.1016/j.pneurobio.2015.08.001

66. Pascoli V, Terrier J, Espallergues J, Valjent E, O'Connor EC, Lüscher C (2014) Contrasting forms of cocaine-evoked plasticity control components of relapse. Nature 509:459–464. doi:10.1038/nature13257

67. Pascoli V, Turiault M, Lüscher C (2012) Reversal of cocaine-evoked synaptic potentiation resets drug-induced adaptive behaviour.

Nature 481:71–75. doi:10.1038/nature10709

68. Creed M, Pascoli VJ, Luscher C (2015) Refining deep brain stimulation to emulate optogenetic treatment of synaptic pathology. Science 347:659–664. doi:10.1126/science.1260776

69. Shen W, Flajolet M, Greengard P, Surmeier DJ (2008) Dichotomous dopaminergic control of striatal synaptic plasticity. Science 321:848–851. doi:10.1126/science.1160575

70. Kravitz AV, Freeze BS, Parker PRL, Kay K, Thwin MT, Deisseroth K, Kreitzer AC (2010) Regulation of parkinsonian motor behaviours by optogenetic control of basal ganglia circuitry. Nature 466:622–626. doi:10.1038/nature09159

71. Wrobel S (2015) Targeting depression with deep brain stimulation, Emory University, Atlanta, GA, Emory News Cent. http://news.emory.edu/stories/2015/04/hspub_brain_hacking_depression/campus.html (Accessed 21 May 2016)

72. Dougherty DD, Rezai AR, Carpenter LL, Howland RH, Bhati MT, O'Reardon JP, Eskandar EN, Baltuch GH, Machado AD, Kondziolka D, Cusin C, Evans KC, Price LH, Jacobs K, Pandya M, Denko T, Tyrka AR, Brelje T, Deckersbach T, Kubu C, Malone DA (2015) A randomized sham-controlled trial of deep brain stimulation of the ventral capsule/ventral striatum for chronic treatment-resistant depression. Biol Psychiatry 78:240–248. doi:10.1016/j.biopsych.2014.11.023

73. Naesström M, Blomstedt P, Bodlund O (2016) A systematic review of psychiatric indications for deep brain stimulation, with focus on major depressive and obsessive-compulsive disorder. Nord J Psychiatry 70(7):483–491. doi:10.3109/08039488.2016.1162846

74. Lobo MK, Nestler EJ, Covington HE (2012) Potential utility of optogenetics in the study of depression. Biol Psychiatry 71:1068–1074. doi:10.1016/j.biopsych.2011.12.026

75. Tye KM, Mirzabekov JJ, Warden MR, Ferenczi EA, Tsai H-C, Finkelstein J, Kim S-Y, Adhikari A, Thompson KR, Andalman AS, Gunaydin LA, Witten IB, Deisseroth K (2013) Dopamine neurons modulate neural encoding and expression of depression-related behaviour. Nature 493:537–541. doi:10.1038/nature11740

76. Chaudhury D, Walsh JJ, Friedman AK, Juarez B, Ku SM, Koo JW, Ferguson D, Tsai H-C, Pomeranz L, Christoffel DJ, Nectow AR, Ekstrand M, Domingos A, Mazei-Robison MS, Mouzon E, Lobo MK, Neve RL, Friedman JM, Russo SJ, Deisseroth K, Nestler EJ, Han M-H (2013) Rapid regulation of depression-related behaviours by control of midbrain dopamine neurons. Nature 493:532–536. doi:10.1038/nature11713

77. Milam AH, Li ZY, Fariss RN (1998) Histopathology of the human retina in retinitis pigmentosa. Prog Retin Eye Res 17:175–205. http://www.ncbi.nlm.nih.gov/pubmed/9695792 (Accessed 16 May 2016

78. Dalkara D, Duebel J, Sahel J-A (2015) Gene therapy for the eye focus on mutation-independent approaches. Curr Opin Neurol 28:51–60. doi:10.1097/WCO.0000000000000168

79. den Hollander AI, Roepman R, Koenekoop RK, Cremers FPM (2008) Leber congenital amaurosis: genes, proteins and disease mechanisms. Prog Retin Eye Res 27:391–419. doi:10.1016/j.preteyeres.2008.05.003

80. Jacobson SG, Cideciyan AV, Roman AJ, Sumaroka A, Schwartz SB, Heon E, Hauswirth WW (2015) Improvement and decline in vision with gene therapy in childhood blindness. N Engl J Med 372:1920–1926. doi:10.1056/NEJMoa1412965

81. Bainbridge JWB, Mehat MS, Sundaram V, Robbie SJ, Barker SE, Ripamonti C, Georgiadis A, Mowat FM, Beattie SG, Gardner PJ, Feathers KL, Luong VA, Yzer S, Balaggan K, Viswanathan A, de Ravel TJL, Casteels I, Holder GE, Tyler N, Fitzke FW, Weleber RG, Nardini M, Moore AT, Thompson DA, Petersen-Jones SM, Michaelides M, van den Born LI, Stockman A, Smith AJ, Rubin G, Ali RR et al (2015) N Engl J Med 372:1887–1897. doi:10.1056/NEJMoa1414221

82. Edwards T, Jolly J, Groppe M, Barnard A, Cottrial C, Tolmachova T, Black G, Webster A, Lotery A, Holder G, Xue K, Downes S, Simunovic M, Seabra M, MacLaren R (2016) Visual acuity after retinal gene therapy for choroideremia. N Engl J Med 374(20):1996–1998. doi:10.1056/NEJMc1509501

83. Bi A, Cui J, Ma Y-P, Olshevskaya E, Pu M, Dizhoor AM, Pan Z-H (2006) Ectopic expression of a microbial-type rhodopsin restores visual responses in mice with photoreceptor degeneration. Neuron 50:23–33. doi:10.1016/j.neuron.2006.02.026

84. Lagali PS, Balya D, Awatramani GB, Münch TA, Kim DS, Busskamp V, Cepko CL, Roska B (2008) Light-activated channels targeted to ON bipolar cells restore visual function in

retinal degeneration. Nat Neurosci 11:667–675. doi:10.1038/nn.2117

85. Yin L, Greenberg K, Hunter JJ, Dalkara D, Kolstad KD, Masella BD, Wolfe R, Visel M, Stone D, Libby RT, Diloreto D, Schaffer D, Flannery J, Williams DR, Merigan WH (2011) Intravitreal injection of AAV2 transduces macaque inner retina. Invest Ophthalmol Vis Sci 52:2775–2783. doi:10.1167/iovs.10-6250

86. Cronin T, Vandenberghe LH, Hantz P, Juttner J, Reimann A, Kacsó A-E, Huckfeldt RM, Busskamp V, Kohler H, Lagali PS, Roska B, Bennett J (2014) Efficient transduction and optogenetic stimulation of retinal bipolar cells by a synthetic adeno-associated virus capsid and promoter. EMBO Mol Med 6:1175–1190. doi:10.15252/emmm.201404077

87. Macé E, Caplette R, Marre O, Sengupta A, Chaffiol A, Barbe P, Desrosiers M, Bamberg E, Sahel J-A, Picaud S, Duebel J, Dalkara D (2015) Targeting channelrhodopsin-2 to ON-bipolar cells with vitreally administered AAV restores ON and OFF visual responses in blind mice. Mol Ther 23:7–16. doi:10.1038/mt.2014.154

88. Duebel J, Marazova K, Sahel J-A (2015) Optogenetics. Curr Opin Ophthalmol 26:226–232. doi:10.1097/ICU.0000000000000140

89. van Wyk M, Pielecka-Fortuna J, Löwel S, Kleinlogel S (2015) Restoring the ON switch in blind retinas: opto-mGluR6, a next-generation, cell-tailored optogenetic tool. PLoS Biol 13:e1002143. doi:10.1371/journal.pbio.1002143

90. Kleinlogel S, Terpitz U, Legrum B, Gökbuget D, Boyden ES, Bamann C, Wood PG, Bamberg E (2011) A gene-fusion strategy for stoichiometric and co-localized expression of light-gated membrane proteins. Nat Methods 8:1083–1088. doi:10.1038/nmeth.1766

91. Pan Z-H, Ganjawala TH, Lu Q, Ivanova E, Zhang Z (2014) ChR2 mutants at L132 and T159 with improved operational light sensitivity for vision restoration. PLoS One 9:e98924. doi:10.1371/journal.pone.0098924

92. K Bourzac (2016) In first human test of optogenetics, doctors aim to restore sight to the blind. MIT Technol. Rev. https://www.technologyreview.com/s/600696/in-first-human-test-of-optogenetics-doctors-aim-to-restore-sight-to-the-blind/ (Accessed 20 May 2016)

93. Reardon S (2016) Light-controlled genes and neurons move into the clinic. Nature. doi:10.1038/nature.2016.19886

94. RetroSense Therapeutics (2015) RST-001 Phase I/II Trial for Retinitis Pigmentosa–Full Text View–ClinicalTrials.gov, ClinicalTrials.gov. https://clinicaltrials.gov/ct2/show/NCT02556736. (Accessed 5 Mar 2017)

95. Jeschke M, Moser T (2015) Considering optogenetic stimulation for cochlear implants. Hear Res 322:224–234. doi:10.1016/j.heares.2015.01.005

96. Hernandez VH, Gehrt A, Reuter K, Jing Z, Jeschke M, Mendoza Schulz A, Hoch G, Bartels M, Vogt G, Garnham CW, Yawo H, Fukazawa Y, Augustine GJ, Bamberg E, Kügler S, Salditt T, de Hoz L, Strenzke N, Moser T (2014) Optogenetic stimulation of the auditory pathway. J Clin Invest 124:1114–1129. doi:10.1172/JCI69050

97. Brigande JV, Gubbels SP, Woessner DW, Jungwirth JJ, Bresee CS (2009) Electroporation-mediated gene transfer to the developing mouse inner ear. Methods Mol Biol 493:125–139. doi:10.1007/978-1-59745-523-7_8

98. Goßler C, Bierbrauer C, Moser R, Kunzer M, Holc K, Pletschen W, Köhler K, Wagner J, Schwaerzle M, Ruther P, Paul O, Neef J, Keppeler D, Hoch G, Moser T, Schwarz UT (2014) GaN-based micro-LED arrays on flexible substrates for optical cochlear implants. J Phys D Appl Phys 47:205401. doi:10.1088/0022-3727/47/20/205401

99. Bryson JB, Machado CB, Lieberam I, Greensmith L (2016) Restoring motor function using optogenetics and neural engraftment. Curr Opin Biotechnol 40:75–81. doi:10.1016/j.copbio.2016.02.016

100. King CE, Wang PT, McCrimmon CM, Chou CC, Do AH, Nenadic Z (2015) The feasibility of a brain-computer interface functional electrical stimulation system for the restoration of overground walking after paraplegia. J Neuroeng Rehabil 12:80. doi:10.1186/s12984-015-0068-7

101. Lin X, Lai B, Zeng X, Che M, Ling E, Wu W, Zeng Y-S (2016) Cell transplantation and neuroengineering approach for spinal cord injury treatment: a summary of current laboratory findings and review of literature. Cell Transplant 25(8):1425–1438. doi:10.3727/096368916X690836

102. Yousefifard M, Rahimi-Movaghar V, Nasirinezhad F, Baikpour M, Safari S, Saadat S, Jafari AM, Asady H, Razavi Tousi SM, Asady H, Tousi SMTR, Hosseini M, Hosseini M (2016) Neural stem/progenitor cell transplantation for spinal cord injury treatment; a systematic review and meta-analysis. Neuroscience 322:377–397. doi:10.1016/j.neuroscience.2016.02.034

103. Aravanis AM, Wang L-P, Zhang F, Meltzer LA, Mogri MZ, Schneider MB, Deisseroth K (2007) An optical neural interface: in vivo control of rodent motor cortex with integrated fiberoptic and optogenetic technology. J Neural Eng 4:S143–S156. doi:10.1088/1741-2560/4/3/S02

104. Llewellyn ME, Thompson KR, Deisseroth K, Delp SL (2010) Orderly recruitment of motor units under optical control in vivo. Nat Med 16:1161–1165. doi:10.1038/nm.2228

105. Jin D, Liu Y, Sun F, Wang X, Liu X, He Z (2015) Restoration of skilled locomotion by sprouting corticospinal axons induced by co-deletion of PTEN and SOCS3. Nat Commun 6:8074. doi:10.1038/ncomms9074

106. Bryson JB, Machado CB, Crossley M, Stevenson D, Bros-Facer V, Burrone J, Greensmith L, Lieberam I (2014) Optical control of muscle function by transplantation of stem cell-derived motor neurons in mice. Science 344:94–97. doi:10.1126/science.1248523

107. Roy DS, Arons A, Mitchell TI, Pignatelli M, Ryan TJ, Tonegawa S (2016) Memory retrieval by activating engram cells in mouse models of early Alzheimer's disease. Nature 531:508–512. doi:10.1038/nature17172

108. He X, Lu Y, Lin X, Jiang L, Tang Y, Tang G, Chen X, Zhang Z, Wang Y, Yang G-Y (2016) Optical inhibition of striatal neurons promotes focal neurogenesis and neurobehavioral recovery in mice after middle cerebral artery occlusion. J Cereb Blood Flow Metab 37(3):837–847. doi:10.1177/0271678X16642242

109. Kushibiki T, Okawa S, Hirasawa T, Ishihara M (2015) Optogenetic control of insulin secretion by pancreatic β-cells in vitro and in vivo. Gene Ther 22:553–559. doi:10.1038/gt.2015.23

110. Reinbothe TM, Safi F, Axelsson AS, Mollet IG, Rosengren AH (2014) Optogenetic control of insulin secretion in intact pancreatic islets with β-cell-specific expression of Channelrhodopsin-2. Islets 6:e28095. doi:10.4161/isl.28095

111. Chernet BT, Adams DS, Lobikin M, Levin M (2016) Use of genetically encoded, light-gated ion translocators to control tumorigenesis. Oncotarget 7:19575–19588. doi:10.18632/oncotarget.8036

INDEX

Albrecht Stroh (ed.), *Optogenetics: A Roadmap*, Neuromethods, vol. 133, DOI 10.1007/978-1-4939-7417-7, © Springer Science+Business Media LLC 2018

Printed in the United States
By Bookmasters